The Mechanization of Natural Philosophy

BOSTON STUDIES IN THE PHILOSOPHY OF SCIENCE

VOLUME 300

For further volumes:
http://www.springer.com/series/5710

Daniel Garber • Sophie Roux
Editors

The Mechanization of Natural Philosophy

Edited by Daniel Garber and Sophie Roux

Springer

EUROPEAN SCIENCE FOUNDATION

Editors
Daniel Garber
Princeton University
Princeton, New Jersey, USA

Sophie Roux
Université Grenoble II/
Institut universitaire de
Grenoble, France

ISSN 0068-0346
ISBN 978-94-007-4344-1 ISBN 978-94-007-4345-8 (eBook)
DOI 10.1007/978-94-007-4345-8
Springer Dordrecht Heidelberg New York London

Library of Congress Control Number: 2012947414

Printed on acid-free paper

Springer is part of Springer Science+Business Media (www.springer.com)

Preface

This book is the quite distant result of a workshop entitled "The Mechanization of Natural Philosophy," held at the Maison des Sciences de l'Homme-Alpes (Grenoble), November 17–19, 2005. The workshop was part of a program on the general theme "From Natural Philosophy to Science," generously sponsored by the European Science Foundation; additional subsidies were given by the Université Grenoble II, Ville de Grenoble, Métro, and Ministère des Affaires Étrangères. With the exception of the essay by Gideon Manning, who kindly agreed to contribute to the volume without having been present in Grenoble, all the papers here were read at the workshop in preliminary form and then thoroughly revised for publication.

In addition to the scholars whose essays follow in this volume, participants at the workshop also included Sylvia Berryman, Antonio Clericuzio, Egidio Festa, Alan Gabbey, Ofer Gal, Elzbieta Jung, Walter Roy Laird, Cees Leijenhorst, and Christiane Vilain. Their contributions to the workshop in Grenoble, whether through the presentation of a paper or through the participation to discussions, whether oral or written, whether mentioned in the footnotes or not, have left their mark throughout the arguments of this book. We hope that they all will be happy with the final result.

For her efficiency in helping to organize the Grenoble workshop, we thank Loredana Truong, administrator of the group "Philosophie, Langage et Cognition," Université Grenoble II. For their constant support from one workshop to the other, but also from one book to the next, we thank Hans Thijssen, Chairman of the European Science Foundation program "From Natural Philosophy to Science," and Cees Leijenhorst, the coordinator of the program.

Finally, this book would not have been possible without Mark Naimark, who, through the generosity of the Région Rhône-Alpes (contrat de plan État-Région, Sciences Humaines et Sociales, appel d'offres 2003), translated Jacques Lambert's paper originally written in French. We also thank Birgit Kolboske, Dorian Rolston and Jeremy Wolos, and an anonymous referee who helped us at different stages of quite a long editing process.

Contents

Contributors

Victor Navarro-Brotons Consejo Superior de Investigaciones Científi cas, Universidad de Valencia, 9 Mestral, 46110 Godella, Valencia, Spain

Frédéric de Buzon Faculté de Philosophie, Université de Strasbourg, Strasbourg, France

Rémi Franckowiak Université Lille I, Villeneuve d'Ascq, France

Daniel Garber Department of Philosophy, Princeton University, Princeton, USA

Guido Giglioni University of London, Warburg Institute, London, UK

Susana Gómez Faculdad de Filosofia, Universidad Complutense de Madrid, Madrid, Spain

Jacques Lambert Département de philosophie, Université Grenoble II, Grenoble, France

Antoni Malet Department d'Humanitats, Universitat Pompeu Fabra, Barcelona, Spain

Gideon Manning California Institute of Technology, Pasadena, CA, USA

Craig Martin History Department, Oakland University, Rochester, MI, USA

Carla Rita Palmerino Center for the History of Philosophy and Science, Radboud University Nijmegen, The Netherlands

Sophie Roux Université Grenoble II/Institut universitaire de France, Grenoble, France

Introduction

Daniel Garber and Sophie Roux

Mechanical philosophy has been well-established as a historiographical category for some years now. However, recent historical research has revealed considerable complexity behind the often uncritical use of the term.

For a generation of historians, the mechanical philosophy was regarded as the principal alternative to Aristotelian orthodoxy in the period of the so-called Scientific Revolution. This has been shown to be a significant oversimplification of the situation, as historians of science have come to appreciate more and more the diversity of non- and anti-Aristotelian views available in the period. In addition to mechanical philosophies, in the period various alchemical, magical, and Platonistic philosophies were available, as well as alternatives to the dominant Aristotelianism that do not fit into any neat categories. Recent scholarship has also emphasized that even though the mechanical philosophy is closely linked to corpuscularianism, there existed alchemical and even Aristotelian corpuscularianisms that were not mechanical by any reasonable criterion. Though our focus in this volume is specifically on the mechanical philosophy, we certainly do not want to claim that it was the only alternative view on the table. But that said, many (though perhaps not all) important figures in the period did endorse some version of a mechanical philosophy. It is therefore important to investigate what exactly the mechanical philosophy may have been as well as the roles that it may have played in the intellectual life of the early-modern period.

The expression "mechanical philosophy" is burdened with multiple ambiguities. Because of its frequent use, it has sometimes slipped into becoming some kind of broad umbrella that may refer to at least four different enterprises: (1) the general program of substituting for the "common philosophy," i.e. the scholastic philosophy, a new philosophy, still to be identified; (2) the more specific rejection of Aristotelian hylemorphism and the correlated adoption of an ontology according to which all natural phenomena can be understood in terms of the matter and motion of the small corpuscles that make up the gross bodies of everyday experience alone; (3) the comparison of natural phenomena, most specifically the world and animals, to existing or imaginary machines; (4) lastly, the ontology associated with mechanics as a new mathematical science of motion, the laws of which are described as the laws of nature in general. Of course, these four enterprises were *sometimes*, in *some* places and in *some* respects associated, but it should be clear that a natural philosopher engaged in the first enterprise does not necessarily have the same profile as a natural philosopher engaged in the fourth one.

However mechanical philosophy is defined, its ambition was greater than its real successes. Even if commentators still tell the triumphal story of the victory of mechanical philosophers over their Aristotelian counterparts, it is now admitted that the Scientific Revolution, if there was one, cannot be reduced to the rejection of Aristotelianism, the valorization of machines or the emergence of mechanics. In most of the natural sciences, explanations in terms of matter and motion alone failed to provide satisfactory accounts of phenomena, and turned out to be as circular or vacuous as their Aristotelian counterparts. The machines the mechanical philosophers proposed were not only imaginary, but most of the time had little to do with the machines actually constructed and used by mechanics or practitioners. Lastly, phenomena that could be subject to mathematical treatment seem to have been small islands scattered in a nature that appeared at this time fundamentally resistant to mathematization.

In the face of these complexities, one might consider simply abandoning the historiographical category of the mechanical philosophy. But to dispense with historiographical categories is not that easy; most of the time when we pretend to do so they keep lingering in the background. Moreover, it cannot be denied that by the end of the seventeenth century many natural philosophers did not speak any longer of substantial forms or occult qualities, that mechanisms and machines were systematically used by many as metaphors or models in the apprehension of natural phenomena, and that a new science of motion had emerged. Thus, our goal in this volume is not to argue for rejecting the historiographical category, but to problematize it and to explore its subtleties.

The inquiry is organized into three sections. First, how were our historiographical categories constructed? Second, how were the fundamental notions of mechanical philosophy, matter and motion, articulated in physics and in mathematics? Third, what kind of mechanization took place in domains usually considered as peripheral, such as meteorology, anatomy, medicine or chemistry?

Key questions are addressed in the first part of this volume, "The Construction of Historiographical Categories." In what respect is it useful for historians of scientific and philosophical ideas to qualify an author as a "mechanical philosopher," as a "Baconian" or as a "new philosopher"? How, when and why were such categories elaborated? Note moreover that over four centuries Aristotelianism not only endured many modifications, but it successfully confronted new problems. Thus, the question is not only the question of mechanical philosophers, but of their enemies, the Aristotelians. How did they react in the seventeenth century, when they were challenged by natural philosophers who explicitly claimed to be against the old philosophy of the School? Were there doctrines, arguments or intellectual practices that neither could accept? Through these questions, we hope to clarify the historical negotiations through which the opposition between "Aristotelian philosophy" and "mechanical philosophy," or between the "old philosophy" and the "new philosophy" were constructed.

At the very beginning of the book, Garber offers a general survey that may be seen as a preliminary warning. Recalling that the official "birth certificate" of mechanical philosophy is a celebrated passage of Boyle's *The Origin of Forms and Qualities According to the Corpuscular Philosophy* (1666), Garber wonders if the expression "mechanical philosophy" referred to anything at all before Boyle. Given that some of those usually considered as mechanical philosophers should not be treated as such (Francis Bacon, Galileo Galilei, Marin Mersenne), and the differences that exist among some others (René Descartes, Thomas Hobbes, Pierre Gassendi), Garber's thesis is that we should think of these pre-Boylean authors, and some others, as belonging to the pre-history of mechanical philosophy, rather than to its history proper: they did not see themselves as belonging to the same paradigm. Boyle's achievement would have been precisely to create, out of pre-existing elements, a new paradigm able to bring together different thinkers and to formulate a consistent program under which they could be united.

Giglioni backs up Garber's conclusion by examining a figure that is sometimes enlisted among the tutelary deities of mechanical philosophy and who is certainly presented as the spiritual father of the experimental philosophy practiced at the Royal Society, namely Francis Bacon. Though he does use words like *mechanicus* or "mechanical," when Bacon spoke of mechanical motions, it had little to do with the new science of mechanics. And indeed, he defended a metaphysics founded on material struggling and blind appetites that, notwithstanding current interpretations, underlies, for example, his description of the nature of heat. At this point, the question of the early reception of Bacon's works necessarily arises, in particular among the English natural philosophers who made his fame. Giglioni shows that there were actually different receptions, and that, except for Francis Glisson, they systematically set aside the metaphysics of appetites of the Lord Chancellor. In Samuel Hartlib's hands, Bacon was transformed into a religious utopian, but in John Webster's hands, into a Fluddian cabalist; Ralph Cudworth suspected him of atheism while Robert Boyle appreciated him for his methodological suggestions, and naturally for his praise of experiments. As for Boyle's reception of Bacon, while not the most faithful to his actual texts, it turned out to be the most productive at the Royal Society and beyond.

If Giglioni's essay is not sufficient to persuade the reader that *Wirkungsgeschichte* matters, she should turn to Roux's paper, which is devoted to the reception of Descartes' physics in the late seventeenth-century France. Roux starts with the fact that during the seventeenth century there were numerous different ways of opposing the new mechanical philosophy and the old Aristotelian philosophy. She argues, however, that, remarkably enough, Descartes eventually emerged as the benchmark by which the works of other natural philosophers of the seventeenth century fall either on the side of the old or the new. She consequently examines the French debate where this historiographical representation emerges, a debate that took place along with the development of a Cartesian propaganda in the 1660s and the ensuing official condemnations of the philosophy of Descartes. She shows quite systematically that the criticisms of Cartesian philosophers by the Oratorian Jean-Baptiste de La Grange, the bishop Pierre-Daniel Huet, and various Jesuits, Ignace Pardies,

Antoine Rochon, Louis Le Valois, Gabriel Daniel, René Rapin, and Honoré Fabri respond to the mockeries of Gérauld de Cordemoy, Jacques Rohault, Louis de La Forge, Bernard Lamy, Nicolas Malebranche or Antoine Arnauld concerning the scholastic entities. Not only does she contrast their philosophical arguments concerning entities and the norms to be respected in physics, but also their ways of defining the philosophical enterprise and its public.

Taken as a whole, the three essays demonstrate the importance of the late decades of the seventeenth century for understanding the emergence of the category of mechanical philosophy, the idea of Baconianism, the opposition of the ancients and the moderns. The late seventeenth century was not so much a period of invention as a period of stabilization, where the works of Descartes, Bacon, Galileo were received, commented on, interpreted. It is no small conquest of recent historiography to have shown the relevance, for the history of ideas, of periods of this kind and, in them, of *minores*. As we will now see, *minores* are part of the story in the second part of the book as well.

Mechanical philosophy is commonly described as having replaced the hylemorphic theory of bodies, grounded in the notions of matter, form and privation, with a corpuscular theory of matter, in which material corpuscles obey laws of motion. But how this substitution took place, the various research programs associated to it, and the tensions that might exist between the mathematical description of phenomena, their corpuscular reduction and their empirical investigation, have yet to be explored. This is what is at stake in the papers of the second part, "Matter, Motion, Physics and Mathematics."

Navarro focuses on the little-known Valencian physician Bernat d'Olesa Rovera. His *Summa totius philosophiae et medicinae* (1536) belongs to the as yet only partly explored category of books that were undoubtedly conceived in an Aristotelian atmosphere, but that open the way for corpuscular explanations, in particular with respect to what was known among Aristotelians as the problem of *mixtio*. After exploring some aspects of Olesa's theories, Navarro demonstrates that no literary continuity exists between Olesa and two later Spanish alleged corpuscularists, Gomez Pereira and Francisco Valles, but that none of them are to be classified among corpuscularists. Thus, he concludes, d'Olesa remains an intriguing exception with no intellectual offspring.

In a sense, Palmerino deals with the same problem as Navarro, namely the problem of continuity between the Aristotelians and the new mechanical philosophers, but with a completely different method. Instead of focusing on what turns out to be a singularity, she emphasizes that certain structures of thought concerning both the theory of matter and the science of motion endure throughout the seventeenth century, no matter which camp. She first shows that authors as different as Galileo, Roderigo de Arriaga, Sébastien Basson, Pierre Gassendi, Jean Chrysostome Magnen, Fabri or Libertus Fromondus assumed what she calls the isomorphism theory, namely the theory according to which what holds for the structure of space, time and motion, holds for the structure of matter as well, "structure" referring here to the alleged continuity or discontinuity of these magnitudes. This is why these authors treated acceleration and deceleration of motion on the one hand, rarefaction and condensation of matter on the other hand, as if they were similar phenomena. Second, through a careful reconstruction, she explains when, how and why, towards the end of the seventeenth century, the isomorphism theory was called into question by Newton. As she argues finally, this was not a definitive victory, however, since Newton's pupil John Keill reasserted the isomorphism of space, time and matter in a course on natural philosophy held in Oxford during the first decade of the eighteenth century.

While Palmerino emphasizes the existence of structures of thought common to both Aristotelians and mechanical philosophers, de Buzon points out differences that exist between two putative spokesmen for mechanical philosophy by confronting Isaac Beeckman's and Descartes' principles of conservation of motion. Inasmuch as Beeckman had touched on the three laws of nature proposed by Descartes in *Le monde*, namely the persistence of motion in a vacuum, the persistence of direction, and the global conservation of motion in impact, it is a difficult question to determine what Descartes exactly borrowed from Beeckman. De Buzon's angle of attack is however somewhat different: he wants to point out that Beeckman's laws and Descartes' laws were not only conceptually distinct, but determined by different conceptions of the relationship between physics and mathematics. Notwithstanding the unsystematic state of Beeckman's *Journal*, the main characteristic of his way of dealing with the problem of motion is his reliance on geometric considerations, well illustrated by his principle of isoperimetric figures. On the contrary, Descartes insists that his laws of motion are laws of nature, motion having become the first object of Cartesian science. By way of conclusion, de Buzon can thus explain why the expression *physico-mathematicus*, well-attested by Beeckman, has disappeared by the time of Descartes: it is because physics and mathematics have for him fundamentally the same object.

With Malet's paper, we are not confronted with the continuity issue, but rather with the vexing question of national traditions. Traditionally, the English and French traditions in natural philosophy are contrasted, the first being supposedly more experimental, the second more mathematical. Malet begins with a discussion of Blaise Pascal's hydrostatics that ends up qualifying his alleged commitment to mixed mathematics. In contrast, the manuscript *Hydrostatica* by James Gregorie (1638–1675) is written in a clearly mathematical style with respect to its concepts, to its deductive organization, and to the subordinate place it devotes to experiments. This raises the question of the interplay of mathematics and experiment that Malet analyses in the case of the doll experiment that was used in different ways in three different hydrostatical treatises, Gregorie's *Hydrostatica* of course, Robert Boyle's *Hydrostatical Paradoxes* (1666) and Willem Jacob Gravesande's *Mathematical Elements of Natural Philosophy, Confirmed by Experiments* (1715). Finally, the question of the discussions concerning the status of hydrostatics, between mathematics and experimental philosophy, is contextualized both with respect to the Royal Society and to the Scottish Enlightenment.

The third and last part of this book, "Mechanical Philosophy Applied," is devoted to the mechanization of specific domains. It is now established that the successes of the mechanical philosophy were actually few in number and that in some domains, most notably chemistry and biology, it failed to provide satisfactory accounts of phenomena.

One can consequently wonder what were the strategies and attitudes in these domains: did the old ways of thinking last, or were new alternative models looked for? As in the second part of this book, the nature of mechanization and the question of the relationship between mechanical philosophers and their Aristotelian predecessors (or contemporaries) is at issue, but this time viewed through the constraints imposed by the domains of application.

The relevance of the object under consideration is central to Gómez's paper. Gómez concentrates on Galileo's mechanization of light and relates it to the well-known Galilean shift from physical (extended) atoms to mathematical (non-extended) atoms. She first discusses the intellectual context and theoretical implications of Galileo's letter to Piero Dini from 23 March 1615, where one finds a description of a very spiritual substance that is diffused through the whole universe and animates it. *Il saggiatore* presents a mixed and transitory stage, since bodies are composed of extended particles, while light is composed of non-extended atoms; this theory allows Galileo to account for what are, according to him, the peculiarities of light, in particular its capacity to travel instantaneously. Finally, in the *Discorsi*, Galileo generalized his idea of non-extended atoms to all kinds of bodies. The paradox at this point is that he uses such entities, normally confined to mathematics, to compose physical bodies, with one of the consequences being that, now, a set velocity is ascribed to light.

By focusing on meteorology, Martin succeeds in giving a reappraisal of Descartes' enterprise in this field. Since Étienne Gilson, it has been known that the structure and topics of the Cartesian *Météores* are quite similar to their Aristotelian counterparts; however, it was assumed that some differences exist as well, inasmuch as *Météores* illustrate some general characteristics of Descartes' physics, like the dispensability of substantial forms and final causes. A closer examination of the major Aristotelian treatises on meteorology shows that the situation is, once again, somewhat more complicated. Craig begins by an assessment of the specificity of meteorology according to Aristotle himself: because it is a field dedicated to natural phenomena that lack clear order, it is best understood by material and efficient causes rather than formal and final ones. In the Renaissance, the intractability of meteorological phenomena even leads Agostino Nifo and Pietro Pomponazzi to insist that meteorology cannot provide true causes, but only conjectures. In the early seventeenth century, Aristotelian commentators like John Poinsot, Eustachius a Sancto Paulo, Libertus Fromondus, Francesco Resta or Daniel Sennert were divided on the question of knowing if final and formal causes should be used in meteorology, the most interesting case being the one of the Jesuit Niccolò Cabeo, who explicitly wrote a comment on Aristotelian meteorology (1646) because he considered that neither form and privation nor final and formal causes were needed in this field. In that sense, Craig concludes, Descartes' meteorology should not be read as a revolutionary work, but rather as a contribution to an on-going debate on the nature of meteors and on the methods to be applied in their study.

Manning's paper is devoted to a key question for another applied part of Cartesian natural philosophy, namely medicine: the human body is described as a machine, but can one say that a machine is just a corporeal substance? And how would it be possible to speak of sickness and health if human bodies were only machines in this sense? After having recalled that the interests of medicine and of natural philosophy were closely linked in the seventeenth century in general and in the Cartesian corpus in particular, Manning comments in a detailed way on the two texts that are, according to him, pertinent for this question. First, he examines what the *Traité de l'homme* tells us about machines and about the human bodies. Second, commenting the famous text of the *Meditatio VI* on a body suffering from dropsy, Manning suggests that the expression *denominatio extrinseca* that it uses should be traced back to its scholastic origins. A close analysis of the couple *denominatio extrinsica/denominatio intrinseca* finally leads him to the conclusion that the human body cannot be said to be healthy except in relation to the human being, that is the union of a body with a mind.

With Lambert, we continue with the human body, but this time not reflected in the mind of a philosopher, but rather put into the hands of the *démonstrateur* in anatomy and surgery at the *Jardin du Roi* in a period of conflict between the medical and the surgical corporations, Pierre Dionis (1643–1718). According to Dionis's *Anatomie de l'homme suivant la circulation du sang et les nouvelles découvertes* (1690), the art of surgery is founded in the science of anatomy, and the science of anatomy is governed by mechanical principles: dissection is legitimate for understanding the living bodies; the structure should explain the function of the organs, whether this structure is apparent or to be revealed by micrography; lastly, effect is proportionate to its cause. Hence, it is not surprising that Dionis defends a truly mechanical program consisting in analyzing the human body without appealing to faculties and specific substances. But, as many others, as Lambert thoroughly explains, Dionis sometimes walks into the common traps of finality when he tries to connect structures and functions of the human body.

Finally, Franckowiak examines the reaction of Du Clos to Boyle's attempt to find in mechanical philosophy some foundations for chemistry as well as the principles of his own chemical philosophy. An early member of the class of physics of the *Académie royale des sciences*, Samuel Cottereau Du Clos (1598–1685) was charged with examining Boyle's *Tentamina chimica.* For him this was the opportunity to give what could be seen as the answer to the "vulgar chemist" Boyle: he reproached Boyle for not having taken experiments seriously and for not having found the proper causes of the disintegration of saltpeter. Du Clos actually formulated a natural philosophy that combines the actions of a mechanical principle (the passive and corporeal "body"), a chemical principle (the active and incorporeal "nature") and a mediating principle (the igneous "spirit"). In that sense, he wished paradoxically to defend through his visible and sensory experiments a natural philosophy relying on invisible and spiritual principles that, according to Franckowiak, could be interpreted as complementary to mechanical philosophy.

It is usual to conclude introductions to books that, like this one, gather a collection of essays, with an ode to the inexhaustible diversity of historical case studies. We think however that, in the present case, it is possible to say something more substantial. Whichever definition you admit, the mechanical philosophy was clearly a polemical category. Many of the essays in this collection explore exactly that, and the role that that polemic played in the debate over Aristotelian natural philosophy.

In doing so, they deal not only with the analysis of central texts of the central figures in the tradition, but with many of the lesser figures, the foot soldiers, as it were, of the Scientific Revolution. But to appreciate fully the opposition between the mechanical philosophy and the Aristotelian orthodoxy, one must see the variety of domains in which the opposition was played out. For that reason a number of the essays deal with the extension of mechanist ideas to domains where their application may not be quite so obvious and unproblematic. While we have certainly not exhausted the infinite diversity of historical case studies, this collection defends and illustrates a certain way of writing the history of the mechanization of natural philosophy.

Part I
The Construction of Historical Categories

Chapter 1
Remarks on the Pre-history of the Mechanical Philosophy

Daniel Garber

The mechanical (or corpuscular) philosophy has been well established as a historiographical category for some years now.[1] While it certainly began as an actor's category, it has slipped into being something else, a kind of broad catch-all category that is taken to include most of those who opposed the Aristotelian philosophy of the schools throughout the entire seventeenth century, part of a broad master narrative about the demise of the scholastic Aristotelian philosophy of the schools and the rise of modern mathematical and experimental science, the titanic intellectual clash that gave birth to modernity.

I am deeply indebted to the many audiences on whom I tried out various versions of these ideas, some more than once, but especially to Sophie Roux, and to Dana Jalobeanu, Vlad Alexandrescu, and the participants of various versions of the Bucharest-Princeton Seminar in Early Modern Philosophy, our annual Transylvanian seminar.

[1] The two names "mechanical philosophy" and "corpuscular philosophy" are interchangeable, according to Boyle, who introduced the terms, as I shall later show. Among contemporaries, the two names are also virtually synonymous. The *Oxford English Dictionary* (q.v. mechanical) cites John Harris's *Lexicon Technicum* (1704) on this question: "*Mechanical Philosophy*, is the same with the Corpuscular, which endeavours to explicate the Phenomena of Nature from Mechanical Principles." Robert Boyle seems to identify the two in his *Of the excellency and grounds of the corpuscular or mechanical philosophy* (1674). Calling it "corpuscular" emphasizes that the manifest properties of bodies are to be explained in terms of their smaller parts, and calling it "mechanical" emphasizes that the principles used in explanation are broadly mechanical. However, I will give precedence to the mechanical aspect rather than the corpuscular. In the figures who have been emphasized in the literature, corpuscles enter into explanations largely as a way of working out mechanical explanations. There are other traditions of corpuscularianism, though, in which corpuscles have inherent alchemical and non-mechanical properties. On this tradition, see especially Newman, *Atoms and Alchemy*, who focuses there on Daniel Sennert. In this essay, I shall not be dealing with such views.

D. Garber (✉)
Department of Philosophy, Princeton University, Princeton, USA
e-mail: dgarber@princeton.edu

D. Garber and S. Roux (eds.), *The Mechanization of Natural Philosophy*, Boston Studies in the Philosophy of Science 300, DOI 10.1007/978-94-007-4345-8_1,

As a historiographical category, the mechanical or corpuscular philosophy seems to have come into fashion in the middle of the twentieth century. There is, of course, Marie Boas Hall's seminal article, "The Establishment of the Mechanical Philosophy," taken largely from her 1949 Cornell University dissertation, and published in *Osiris* in 1952. Also important is Eduard Jan Dijksterhuis's *De Mechanisering van het Wereldbeeld*, published in the Netherlands in 1950, though it was certainly more widely read in its English translation, *The Mechanization of the World Picture*, published in 1961. Dijksterhuis begins his fat tome as follows:

> Among the numerous modifications that scientific thought about nature has undergone in the course of the centuries, it would be difficult to point to one that has had a more profound and far-reaching effect than the emergence of the conception of the world usually called mechanical or mechanistic.[2]

Both Boas Hall and Dijksterhuis push the mechanical conception of the world back to the ancients, though Boas Hall's focus is particularly on its development in the seventeenth century. The mechanical philosophy, in turn, became one of the central tools for understanding the development of science in the early-modern period in Richard Westfall's classic monograph on the Scientific Revolution, *The Construction of Modern Science: Mechanisms and Mechanics*, published first in 1971.

In these classic works, the mechanical philosophy is generally conceived of as the main alternative to the Aristotelian natural philosophy taught in the schools in the early seventeenth century. We are much more historiographically savvy now. More recent accounts of the period realize the complexity of the intellectual world, and the importance of other non- and anti-Aristotelian currents of thought throughout the seventeenth century. But even so, the mechanical philosophy still has a large role to play in current histories of science in the period. Few still subscribe to the master narrative that sees Aristotelianism and the mechanical philosophy as the only major players in the field.[3] But it still has a major role to play in such recent books as Steven Shapin's *The Scientific Revolution* (1996) and Peter Dear's *Revolutionizing the Sciences* (2001) and *The Intelligibility of Nature* (2006).[4] Even if it does not have the historiographical centrality that it once had, it is generally acknowledged as an important trend in seventeenth-century scientific thought.

In this essay, I would like to take a closer look at the concept of the mechanical philosophy. My question is this: what was the mechanical philosophy? And, most importantly, what role should it play in our understanding of the history of science? With respect to the second question, my interest will be focused on the first half of

---

[2] Dijksterhuis, *Mechanization of the World Picture*, p. 3.

[3] A recent author who still seems to hold something close to this view is Stephen Gaukroger. See his *Emergence of a Scientific Culture*, chaps. 8 and 9.

[4] "Of all the mechanical constructions whose characteristics might serve as a model for the natural world, it was the clock more than any other that appealed to many early modern natural philosophers. Indeed, to follow the clock metaphor for nature through the culture of early modern Europe is to trace the main contours of the mechanical philosophy, and therefore of much of what has been traditionally construed as central to the Scientific Revolution" (Shapin, *Scientific Revolution*, p. 32). See also Dear, *Revolutionizing the Sciences*, chap. 5, and *Intelligibility of Nature*, chap. 1.

the seventeenth century. While there are questions as to what exactly mechanism or the mechanical philosophy meant to thinkers in the second half of the century, there can be little doubt that it is important for understanding that period. But, I shall argue, it is not so clear that it is an appropriate way of thinking about figures earlier in the century.

1.1 What Was the Mechanical Philosophy?

It will be helpful to begin with an account of what is generally meant by the mechanical philosophy.[5] This question turns out to be rather delicate: depending on what we take the mechanical philosophy to be, we may well come up with rather different answers as to when it begins and who belongs to it. And rather different answers as to how useful it is as an historiographical category.

Let me begin, though, in a somewhat provisional way with Robert Boyle. Boyle is often considered to have been the first to introduce the term 'mechanical philosophy' into the seventeenth-century scientific vocabulary, and his version of the program is often assumed to be the canonical version in the period and the most influential among his contemporaries. These claims are not altogether well-founded. Henry More actually introduced the very similar term 'mechanick philosophy' in 1659, shortly before Boyle introduced his term, though it is not entirely certain what he meant by the term.[6] Furthermore, I know of no study which even purports to establish that it was specifically Boyle's particular conception of the mechanical philosophy was the one that was generally known and accepted in the period.[7] But even so, it is Boyle's conception of the mechanical philosophy that has dominated the recent historiography, and it is generally that idea of the mechanical philosophy

[5] For an excellent account of the history of the mechanical philosophy, see Sophie Roux, *La philosophie mécanique*. My own thought on this question has been much influenced by Roux's work.

[6] See More, *The Immortality of the Soul*, Preface, [b6r], [b8r], et passim. See Gabbey, "Philosophia Cartesiana Triumphata," p. 221 and "What was 'Mechanical' about 'The Mechanical Philosophy'?" p. 14. Gabbey also notes an earlier passage in Descartes, which might mistakenly be interpreted as attributing the use of the term 'mechanical philosophy' to Descartes.

[7] This is a point that Ursula Goldenbaum has repeatedly emphasized to me in conversation. See Hunter and Davis's remarks on the impact of the *Origin of Forms and Qualities*, Boyle's manifesto for the mechanical philosophy, in their introduction to the text, in Boyle, *Works*, vol. V, pp. xxx–xxxi. In the literature, it is Locke who is usually cited as having been influenced by Boyle's conception of the mechanical philosophy. See, e.g., Mandelbaum, *Philosophy, Science and Sense Perception*, chaps. 1 and 2; Alexander, *Ideas, Qualities and Corpuscles*; and Sargent, *The Diffident Naturalist*, pp. 103–108. Occasionally Leibniz as well, though it is difficult to relate Leibniz directly to Boyle on the question of the mechanical philosophy. See Clericuzio, *Elements, Principles and Corpuscles*, pp. 104f. Roux, *La philosophie mécanique*, chap. 5 shows how Leibniz adopted at least Boyle's language in talking about the mechanical and corpuscular philosophy. But outside of that, I know of no study that traces Boyle's influence on later conceptions of the mechanical philosophy.

that historians of science have in mind when they talk about it in connection with early-modern science. So it is that, with which we shall begin.

Leaving aside More's 'mechanick philosophy,' the first explicit uses of the term 'mechanical philosophy' seem to be in Boyle's writings, in the preface to *Some specimens of an attempt to make chemical experiments useful to illustrate the notions of the corpuscular philosophy*, part of *Certain Physiological Essays*, published in 1661, though probably written earlier.[8] But the canonical statement of the view is found in a widely influential work that Boyle published in 1666, *The Origin of Forms and Qualities according to the Corpuscular Philosophy*, a kind of manifesto for the new corpuscular or mechanist program that was beginning to dominate certain learned circles.[9] Before entering into a long series of experiments that, for Boyle, show the superiority, indeed the inevitability of the mechanical philosophy, he offers a careful characterization of the position that he means to support. He writes:

> That then which I chiefly aime at, is to make it Probable to you by Experiments .... That allmost all sorts of Qualities, most of which have been by the Schooles either left Unexplicated, or Generally referr'd, to I know not what Incomprehensible Substantiall Formes, *may* be produced Mechanically, I mean by such Corporeall Agents, as do not appear, either to Work otherwise, then by vertue of the Motion, Size, Figure, and Contrivance of their own Parts (which Attributes I call the Mechanicall Affections of Matter, because to Them men willingly Referre the various Operations of Mechanical Engines:) or to Produce the new Qualities exhibited by those Bodies, their Action changes, by any other way, then by changing the *Texture*, or *Motion*, or some other *Mechanical Affection* of the Body wrought upon.[10]

Boyle explicated this view in a number of basic theses. The first three (and probably the most important) of the eight theses he gives are the following:

I. I agree with the generality of Philosophers so far, as to allow, that there is one Catholick or Universal Matter common to all Bodies, by which I mean a Substance extended, divisible, and impenetrable.
II. But because this Matter being in its own Nature but one, the diversity we see in Bodies must necessarily arise from somewhat else, then the Matter they consist of. ... [T]o discriminate the Catholick Matter into variety of Natural Bodies, it must have Motion in some or all its designable Parts.
III. These two grand and most Catholick Principles of Bodies, Matter and Motion, being thus established, it will follow, both that Matter must be actually divided into Parts, that being the genuine Effect of variously determin'd Motion, and that each of the primitive Fragments ... must have two Attributes, its own Magnitude, or rather *Size*, and its own *Figure* or *Shape*.[11]

In this way, the mechanical or corpuscular philosophy rejected the explanation of physical phenomena in terms of Aristotelian forms and qualities. It also sought

---

[8] See Boyle, *Works*, vol. II, pp. 87f. On the origin and early uses of the term, see Roux, *La philosophie mécanique*, pp. 19–26. She cites Boyle as having coined the term.

[9] The text can be found in Boyle, *Works*, vol. V, pp. 281–491.

[10] Boyle, *Works*, vol. V, p. 302.

[11] Ibid., pp. 305–307.

to eliminate all sensible qualities from objects themselves; the Aristotelian's hot and cold, wet and dry are eliminated as real qualities in things, as are sensible qualities such as color and taste.[12] For the mechanical philosopher everything, be it terrestrial or celestial, natural motion or constrained, must be explained in terms of the size, shape, and motion of the parts that make it up, just as we explain the behavior of a machine.

This is what we might call the *explanatory program* of Boyle's mechanical philosophy, the explanation of everything in terms of size, shape and motion. But Boyle's mechanical philosophy was characterized not only by the doctrines it included, but also by the doctrines that it excluded. An important feature of the mechanical philosophy as Boyle articulated it was the fact that certain doctrines and debates were considered off limits. When Boyle introduced the general principles of the mechanical philosophy, he quite explicitly put aside differences among different sects, claiming to write "rather for the Corpuscularians in general, than any party of them."[13] In particular, he set aside differences about atoms and the void. For Boyle's new mechanical philosophy, it did not matter whether the smallest parts of matter were in principle unsplittable, as Gassendi held, or just not split, as Descartes held. For Boyle's new mechanical philosophy it did not matter whether there *really* is a vacuum, a genuinely empty space, or whether the interior of his vacuum pump is filled with some kind of ether.[14] He set aside these issues and concentrated on the issues that *he* thought were important: that there is one matter in the physical world, that it is divided into parts through motion, and that all the phenomena of the natural world could be explained in terms of size, shape and motion. Writing in the preface to his *Some Specimens* Boyle notes:

> I esteem'd that notwithstanding these things wherein the Atomists and the Cartesians differ'd, they might be thought to agree in the main, and their Hypotheses might by a Person of a reconciling Disposition be look'd on as, upon the matter, one Philosophy, which because it explicates things by Corpuscles, or minute Bodies, may (not very unfitly) be call'd Corpuscular; though I sometimes stile it the Phoenician Philosophy, because some antient Writers inform us, not only before Epicurus and Democritus, be ev'n before Leucippus taught in Greece, a Phoenician Naturalist was wont to give an account of the Phaenomena of Nature by Motion and other Affections of the minute Particles of Matter, which because they are obvious and very powerfull in Mechanical Engines, I sometimes also term it the Mechanical Hypothesis or Philosophy.[15]

Boyle is precisely someone "of a reconciling Disposition": he wants to convince adherents of the competing positions to put aside their disagreements, and recognize one another as adhering to a single philosophy. In the experimental part of the essay,

---

[12] Ibid., pp. 309–315.

[13] Ibid., p. 292.

[14] Boyle, *New Experiments Physico-Mechanical, Touching the Spring of the Air and its Effects*, *Works*, vol. I, pp. 197f.

[15] Boyle, *Works*, vol. II, p. 87. Cf. the Advertisements to the *History of Fluidity and Firmness*, *Works*, vol. II, p. 117.

and in much of Boyle's other chemical writings, his focus is precisely on the elements of the mechanical philosophy that he thinks should be uncontroversial, making plausible through experiment that everything can be explained in terms of size, shape and motion, and setting aside questions relating to infinite divisibility or the real existence of empty space.[16] Boyle's project is at least in part to undermine disagreement between various camps, particularly between those who advocated atoms and the void, and those who preferred infinite divisibility and the plenum, by focusing on fundamentals. And his point is to direct actual empirical research on those parts of the project that fall into this common area. Indeed, Boyle goes farther still and suggests that the points of disagreement go beyond the possibility of settling through empirical means, and are thus not appropriate subjects for discussion and debate. In the preface to *Some Specimens* he notes that the difference between "the Cartesians and the Atomists" over "the Notion of Body in general, and consequently about the Possibility of a true Vacuum, as also about the Origine of Motion, the indefinite Divisibleness of Matter … seem to be rather Metaphysical than Physiological Notions…."[17] In his 1660 *New Experiments Physico-Mechanical, Touching the Spring of the Air and its Effects*, where Boyle presents the results of the experiments with his air-pump, he is even more explicit about excluding debates about the nature of the vacuum from the realm of what is empirically meaningful, and thus from serious debate. In that work he refuses to come down on one side or the other as to whether or not the chamber of his air-pump "be truly empty, that is, devoid of all Corporeal Substance." After going through the arguments on both sides, he concludes that in the end, the controversy is not over anything that can be determined experimentally, but over the notion of a body. He concludes:

> This Reason, I say, being thus desum'd seems to make the Controversie about a *Vacuum*, rather a Metaphysical, then a Physiological Question; which therefore we shall here no longer debate, finding it very difficult either to satisfie Naturalists with this Cartesian Notion of a Body, or to manifest wherein it is erroneous, and substitute a better in its stead.[18]

This move, by which Boyle eliminates certain questions from the domain of inquiry in order to promote agreement among different factions is what might be called the *irenic program* of the mechanical philosophy.

Such is the mechanical philosophy, at least as conceived by Robert Boyle in the 1660s. It is important here to distinguish the mechanical philosophy from the science of mechanics as such. It is a traditional trope in the history of science that a central feature of the transition between the old philosophy and the new

---

[16] On this, see Newman, *Atoms and Alchemy*, pp. 79–81. Newman is here addressing arguments by Clericuzio and Chalmers that there is a divide between the theoretical writings, which support the mechanical philosophy, and his chemical practice, which does not. On this see Clericuzio, "A Redefinition of Boyle's Chemistry and Corpuscular Philosophy," and *Elements, Principles and Corpuscles*, pp. 103–148; and Chalmers, "The Lack of Excellency of Boyle's Mechanical Philosophy."

[17] Boyle, *Works*, vol. II, p. 87.

[18] Boyle, *Works*, vol. I, pp. 197f. On Boyle and the question of the divisibility of matter, see Anstey, *The Philosophy of Robert Boyle*, pp. 43f. On the question of the existence of a true vacuum, see Shapin and Shaffer, *Leviathan and the Air-Pump*, pp. 45f.

concerned the transition between a qualitative Aristotelian approach to nature, and the incompatible mathematical approach due primarily to Archimedes, though also associated with Plato and Platonism. But matters are somewhat more complicated.

Mechanics, the study of machines, was practiced from the ancients on down as one of the so-called middle sciences or mixed mathematics, along with other mathematical disciplines such as optics, astronomy, and music. The revival of Archimedes, particularly in the sixteenth century was, indeed, important, particularly for the science of mechanics. But we must not forget that there was an Aristotelian mechanics as well, the *Mechanica* or *Mechanica Problemata*. Though now known to be a somewhat later text, it was considered to be a genuine text of Aristotle throughout our period. Aristotle and Archimedes offer somewhat different treatments of machines. Aristotelian mechanics is, in general, very, very concrete, and deals with machines as actually embodied in real material things that bend, break, and wiggle. Archimedean mechanics, on the other hand, is much more abstract and mathematical, reducing machines to their geometrical bare bones.

Later figures such as Benedetti, Tartaglia, and Guidobaldo del Monte in the sixteenth century, and Galileo in the seventeenth, combined elements of these two traditions, along with some pieces of others and some original speculations, to produce a genuine renaissance in the science of mechanics in the sixteenth century. While some of these figures certainly set themselves against the Aristotelian philosophy, it is a mistake to see their work in mechanics as forging an alternative to the Aristotelian natural philosophy. The science of mechanics itself was completely neutral with respect to Aristotelian natural philosophy. Natural philosophy, physics, treats *natural things*, things that have internal principles of motion and rest. Mechanics, on the other hand, treats *machines*, *artificial things*, things that *lack* nature, properly speaking. The two studies are complementary and in no way in conflict with one another. A sixteenth-century mathematician would have had no problem at all with the idea of an Aristotelian mechanics: indeed, it was generally believed that there *was* an Aristotelian mechanics, that written by the author now known as pseudo-Aristotle. The science of mechanics was a branch of mixed mathematics that had no pretensions to be a complete physics. Boyle's mechanical philosophy, though, did: it was intended to be a program in natural philosophy that drew on mechanical models for explaining things in the physical world. Though the *science of mechanics* was not, in and of itself, anti-Aristotelian, Boyle's *mechanical philosophy* will later set itself against Aristotelian natural philosophy.[19]

It is also worth pointing out that Boyle's version of the mechanical philosophy might not be the only conception in play in the seventeenth century. Influential as he may have been among his contemporaries, and influential as he definitely was on the more recent historiography of the period, one must at least consider the possibility

[19] On early-modern mechanics and its relations with physics and the mechanical philosophy, see, e.g., Garber, "Descartes, Mechanics, and the Mechanical Philosophy"; Gabbey, "Between *ars* and *philosophia naturalis*"; Laird, "The Scope of Renaissance Mechanics." This question is one of the central themes of Laird and Roux, *Mechanics and Natural Philosophy*.

that there are other, somewhat different conceptions of the mechanical philosophy in the period. This is a consideration to which we will return later.

By the late seventeenth century it is fair to say that the mechanical philosophy, however it was understood, was the going program in many scientific circles. Its emergence provided, perhaps for the first time, an attractive alternative around which a variety of figures could gather, and gather they did. But was there a mechanical philosophy before Boyle's manifesto? More particularly, I would like to explore two questions: (1) To what extent did thinkers in the first part of the seventeenth century subscribe to the mechanical philosophy as Boyle understood it, before he articulated it? To what extent is the mechanical philosophy as Boyle understood it a useful historiographical category for the first half of the seventeenth century? And (2): even if it turns out that Boyle's particular conception of the mechanical philosophy is not a useful historiographical category for the early part of the century, to what extent is it useful to categorize the moderns, those who agreed in rejecting Aristotelian natural philosophy, as mechanists in a perhaps broader sense? To what extent do they form a club, even if maybe not in the precise sense that Boyle understood the mechanist program?

1.2 The Mechanical Philosophy Before Boyle

Were there Boylean mechanists before Boyle? Boyle himself suggests a number. In the preface to *Some specimens*, the main figures that he cites are the Cartesians and the "atomists," presumably Gassendi and his followers. In the *Origin of Forms and Qualities*, the list is fuller. Though he does not exactly call them mechanical philosophers, the names he lists with approval are: Lucretius, Verulam (i.e. Bacon), Basso, Des Cartes "and his Followers," Gassendus, the two Boots, Magnenus, Pemble, and [von] Helmont.[20] Some of these names are rather strange to us now, but many of them are not. At least the better-known names link up tolerably well with the list that Westfall gives in his account:

> No one man created the mechanical philosophy. Throughout the scientific circles of western Europe during the first half of the 17th century we can observe what appears to be a spontaneous movement toward a mechanical conception of nature in reaction against Renaissance Naturalism. Suggested in Galileo and Kepler, it assumed full proportions in

[20] See Boyle, *Works*, vol. V, p. 295. Lucretius, Bacon, Gassendi, and "Des Cartes" need no explanation, of course. The van Helmont in question is the alchemist and physician Johann Baptista van Helmont, the elder van Helmont. Sebastian Basso (or Basson) is the author of *Philosophiae naturalis adversus Aristotelem libri XII* (1621). "The two Boots" are Arnold and Gerard Boate, who jointly authored *Philosophia naturalis reformata, id est, Philosophae aristotelicae accurate examination ac solida confutatio et novae ac verioris introductio* (1641). Johann Chrysostom Magnenus was the author of *Democritus reviviscens: Sive de vita et philosophia Democriti* (1646b). William Pemble is the author of *De formarum origine ... tractatus* (1629).

the writing of such men as Mersenne, Gassendi, and Hobbes, not to mention less well known philosophers. Nevertheless, René Descartes (1596–1650) exerted a greater influence toward a mechanical philosophy of nature than any other man ....[21]

Marie Boas Hall gives a similar list. For her the main mechanical philosophers before Boyle include Gassendi, Galileo, Bacon, and Descartes, to which she adds Isaac Beeckman.[22]

Can we see the origins of the mechanical philosophy in such figures as Bacon and Galileo, Mersenne and Hobbes, Gassendi and Descartes? (I will leave out the others that Boyle lists, i.e. the two Boots, Magnenus, Pemble, and von Helmont, who are more rarely treated in the contemporary literature, if at all. It would be interesting and relevant to examine these figures as well, but I do not think that they would change the conclusion in any significant way.) Well, yes and no. There is no doubt that elements of what comes to be the later mechanical philosophy can be found in all of them, to some degree or another. But when we look more carefully at these figures, often identified as the founding fathers of the mechanical philosophy, we see something rather more complex. Bacon, Galileo and Mersenne, as it turns out, are not really mechanical philosophers at all, under the most generous construal insofar as they violate the principle that everything in nature must be explained in terms of size, shape, and motion, what I called Boyle's explanatory program. Hobbes, Gassendi, and Descartes come closer to fitting Boyle's conception of a mechanical philosophy, insofar as they hold that everything can be explained in terms of size, shape and motion. However, it is important to understand that these earlier figures did not see matters in that way. They saw themselves as competitors in the effort to replace the Aristotelian philosophy with something new, and not as practitioners of a common program. In this respect they violate what I called the irenic program of the mechanical philosophy.

Let me go briefly through each of these figures to make my case, and show that at least on Boyle's conception of the program, it is inappropriate to consider these earlier thinkers to be practitioners of the mechanical philosophy.

---

[21] Westfall, *The Construction of Modern Science*, pp. 30f. It is interesting here that Westfall includes Hobbes, who is definitely not on Boyle's list. Sophie Roux has suggested to me that Boyle excluded Hobbes, because he saw the objective of the mechanical philosophy as a defense of the Christian religion, and saw Hobbes as radically heterodox, if not atheist. This is certainly consistent with the general view of Hobbes by his contemporaries. Even if the mechanical philosophy was not as closely linked with the defense of Christianity as Roux suggests, many of his contemporaries actively distanced themselves from Hobbes on account of his suspect religious views. Furthermore, we have to remember that the moment when Boyle is articulating his mechanical philosophy is the exact moment when he is in open conflict with Hobbes over the air-pump experiments. Under the circumstances, it is not surprising that he would not want to endorse publicly Hobbes's philosophical views. On this see Shapin and Schaffer, *Leviathan and the Air-Pump* (though I have some serious reservations about their treatment of Hobbes). That said, though, it is not inappropriate for Westfall to include him: Hobbes fits the conception of the mechanical philosopher as well as do Descartes and Gassendi.

[22] See Boas Hall, "The Establishment of the Mechanical Philosophy," § IV, "Early Mechanical Philosophies."

1.3 Bacon

Bacon was the first modern on Boyle's list. He is on Boas Hall's list, although, conspicuously enough, not on Westfall's. Bacon deserves a long and detailed discussion; his natural philosophy is too complex and embedded in late Renaissance thought about nature to admit of a short exposition.[23] But it is worth saying a few words about him.

Boyle's inclusion of Bacon on his list of "good guys" for the mechanical philosophy attests to the fact that Bacon's thought and his intellectual prestige were freely appropriated by those thinkers in the circle of founding members of the Royal Society. Boyle does not cite any texts, though.

To be sure, there are some passages of Bacon that strongly suggest a mechanical/corpuscular account of nature. For example, the one example of the method that Bacon works out in his *Novum organum* is that of the form of heat. After carefully constructing a model natural history, his first conclusion (the "first vintage") is that heat is a kind of motion.[24] It is quite natural to read this as suggesting the view of heat as the local motion (agitation) of the corpuscles that make up a body, a view very much in line with idea with mechanistic conceptions of fire, and more generally, consistent with mechanist conceptions of the natural world.[25] But Bacon's view is rather more complex than that. His remarks on how to understand the notion of motion considerably clarify how he conceives of heat. Earlier in the *Novum organum* he had written:

> Nor is there any value in those vulgar distinctions of motion which are observed in the received system of natural philosophy, as generation, corruption, augmentation, diminution, alteration, and local motion. … But if, leaving all this, any one shall observe (for instance) that there is in bodies a desire of mutual contact, so as not to suffer the unity of nature to be quite separated or broken and a vacuum thus made; or if any one say that there is in bodies a desire of resuming their natural dimensions or tension, so that if compressed within or extended beyond them, they immediately strive to recover themselves, and fall back to their old volume and extent; or if any one say that there is in bodies a desire of congregating towards masses of kindred nature,—of dense bodies, for instance, towards the globe of the earth, of thin and rare bodies towards the compass of the sky; all these and the like are truly physical kinds of motion;—but those others are entirely logical and scholastic, as is abundantly manifest from this comparison.[26]

---

[23] In recent years, Bacon's matter theory in particular has gotten considerable attention. Graham Rees has been especially important in unearthing Bacon's views on the material world. See especially his introductions and commentaries in vols. VI and XIII of the *Oxford Francis Bacon*, and "Matter Theory: a Unifying Factor in Bacon's Natural Philosophy." See also Manzo, "Francis Bacon and Atomism"; Weeks, "Francis Bacon and the Art-Nature Distinction"; and the essay (Chap. 2) by Guido Giglioni in this volume.

[24] Bacon, *Novum organum* II 20.

[25] This, for example, is why Boas Hall puts Bacon among the early mechanists. See "The Establishment of the Mechanical Philosophy," pp. 440f.

[26] Bacon, *Novum organum* I 66.

This passage suggests that Bacon's view is a natural world governed by appetites and desires. If this is what he has in mind by motion in his account of heat as a kind of motion, then it is quite clear that it is not in any way a mechanist account.

Recent studies have suggested exactly such a view of nature in Bacon, an animistic conception of nature deeply influenced by Italian thinkers such as Telesius.[27] What has emerged from these studies is something quite distant from anything that Boyle would recognize as a mechanical philosopher: Bacon clearly rejects what I called the explanatory program that is at the heart of Boyle's project. This, of course, raises the question as to how someone like Boyle could have been so mistaken about Bacon's real views.[28] But however it may have happened, Bacon was not a mechanical philosopher in Boyle's sense.

1.4 Galileo

Let me now turn to Galileo. Interestingly enough, Galileo was not on Boyle's list of mechanical philosophers, though he appears on virtually all modern lists of early mechanical philosophers. Galileo was, of course, notably hostile toward the philosophy of the schools. However, Galileo's debt to the Aristotelian tradition in mechanics would also have been evident to any contemporary reader. The Aristotelian influence is particularly clear in days 1 and 2 of his *Discorsi* (*Two New Sciences*), where Galileo is discussing the strength of materials. Indeed, the material embodiment of machines is precisely what interests Galileo in the first 2 days of the dialogue. The task he sets before himself is very similar to the question first broached in the Aristotelian *Mechanica*, that is, why larger and smaller machines, built on the same principles, behave differently.[29] Furthermore, the centerpiece of Day 1 of the *Discorsi* is "Aristotle's Wheel," a problem first posed in the *Mechanica*.[30]

The Aristotelian quest for the causes of the behavior of real machines leads Galileo in the *Discorsi* from the theory of machines to a genuine question in physics: what is the make-up of matter? The account of the make-up of the world that Galileo suggests is, in a very broad sense, what might be called atomistic or

[27] See the references given above in note 24.

[28] This is the subject of Giglioni's essay (Chap. 2) in this volume.

[29] See Aristotle, *Mechanica* 848b1 f with Galileo, *Discorsi*, in *Opere*, vol. VIII, p. 50f., translated in *Two New Sciences*, p. 12.

[30] See Aristotle, *Mechanica* 855a29 f with Galileo, *Discorsi*, in *Opere*, vol. VIII, pp. 68–72, translated in *Two New Sciences*, pp. 29–33. On the first 2 days of Galileo's *Discorsi* as mixed mathematics, see Biener, "Galileo's First New Science."

corpuscular. In a celebrated passage from the *Il Saggiatore* (*The Assayer*, 1623), Galileo asserted:

> To excite in us tastes, odors, and sounds I believe that nothing is required in external bodies except shapes, numbers, and slow or rapid movements. I think that if ears, tongues, and noses were removed, shapes and numbers and motions would remain, but not odors or tastes or sounds.[31]

However, it is important to note that his ultimate particles seem not to have been the small but finite corpuscles Boyle had in mind, but "infinitely many unquantifiable atoms," suggesting an infinitesimal conception, though this idea was not worked out in great detail.[32] Coordinate with the infinitesimal particles, there were infinitesimal voids. The consistency of bodies, Galileo argued, is due to these tiny voids, interspersed in bodies, together with "the repugnance nature has against allowing a void to exist."[33] Corpuscular, yes, in a rather broad sense. But is this really mechanical? It is difficult to see the repugnance of the void as a real mechanical explanation for the coherence of bodies.

But this excursus into natural philosophy proper is rather unusual for Galileo. In a famous passage in the letters on sunspots, Galileo writes in opposition to Scheiner, who claimed to know the nature of the sunspots in question:

> For we either want to try to penetrate the true and intrinsic essence of natural substances by speculating, or we want to satisfy ourselves by finding out a few of their properties. I consider investigating the essence of the nearest elementary substances an undertaking no less impossible and a labor no less vain than that of the most remote and celestial ones. … And likewise I do not understand any more of the true essence of Earth or of fire than I do that of the Moon or of the Sun. Such knowledge is available to our comprehension when we enter the state of beatitude, and not before then.[34]

More generally, though Galileo offered mathematical accounts of some specific phenomena, there is nothing that can be called a Galilean *physics* properly speaking, a general account of the way the world is, of the ultimate natures and causes.

Galileo was interested in motion and mechanics from a mathematical point of view, and he was certainly an anti-Aristotelian. But was he a mechanical philosopher? I would hesitate to call him that. While he is certainly interested in mechanics, he nowhere says that everything in nature can be explained in terms of machines. Indeed, he does not make *any claims at all* about how everything is to be explained. In that sense I would hesitate to attribute to Galileo a natural philosophy at all, strictly speaking. And while he does seem to hold, at least on occasion, that matter can be resolved into smaller parts, those smaller parts do not look very much at all like the corpuscles of Boyle's mechanical philosophy. Their coherence is explained

---

[31] Galileo, *Il Saggiatore*, in *Opere*, vol. VI, p. 350, translated in Drake, *Discoveries and Opinions of Galileo*, pp. 276f. On Galileo's atomism, see Shea, "Galileo's Atomic Hypothesis"; Smith, "Galileo's Theory of Indivisibles"; Nonnoi, "Galileo Galilei: quale atomismo?"; and most recently, Galluzzi, *Tra atomi e indivisibili*.

[32] Galileo, *Discorsi*, in *Opere*, vol. VIII, pp. 71f., translated in *Two New Sciences*, p. 33.

[33] Galileo, *Discorsi*, in *Opere*, vol. VIII, p. 59, translated in *Two New Sciences*, p. 19.

[34] Galilei, *Opere*, vol. V, pp. 187f., translated in *On Sunspots*, p. 254.

not through mechanical interaction but through the power of the vacuum. Though Galileo was a major figure in the history of mechanics and motion, it seems anachronistic to include Galileo among the mechanical philosophers, on Boyle's conception. Though for very different reasons, like Bacon, Galileo does not seem to subscribe to Boyle's explanatory program for the mechanical philosophy.

1.5 Mersenne

Marin Mersenne, a rough contemporary of Galileo's, has often been included among the fathers of the mechanical philosophy.[35] But was Mersenne really a mechanical philosopher? Not, I think, on Boyle's conception.

Mersenne was certainly a great supporter of Galileo's program, as he understood it and interpreted it. It was through Mersenne's published paraphrases of Galileo that many learned of the Italian's work on motion and mechanics. In his own work Mersenne made use of Galileo's science of motion to try to explain the motion of the air that constitutes sound.[36] Mersenne was a great enthusiast for many thinkers, such as Descartes, Hobbes and Gassendi who rejected Aristotle and Aristotelianism and adopted what have been later characterized as mechanist approaches to nature.

But even though Mersenne wrote extensively about motion in general, sound in particular, and acoustics, he never formulated anything like a natural *philosophy* of his own. Indeed, he seemed quite skeptical that one could ever come to have real knowledge, in the strict sense, of the physical world. (In this respect he seems rather faithful to Galileo's own attitudes.) This makes it somewhat misleading to include Mersenne among the mechanical philosophers: in the strict sense, he is not a philosopher at all. In the context of a discussion of the Galilean law of free-fall, Mersenne notes:

> Since we cannot know the true reasons, or the true knowledge [*science*] of that which happens in nature, since there are always some circumstances or instances which make us doubt if the causes which we imagine are true ... I don't see that one should require anything more from scientists than their observations and the remarks that they have made on the different effects or phenomena of nature.[37]

More generally, Mersenne argues that "there is nothing certain in physics."[38] Indeed, he argues that it is only mathematics, a science of the pure imagination, and metaphysics, a science of the pure intellect, that can give us true certainty: "...one can say that men can only possess and enjoy [*jouir*] these two sciences while they

[35] I have in mind here the title of Lenoble's seminal, and still useful book, *Mersenne ou la naissance du mécanisme*.

[36] For the details concerning Mersenne's relations with Galileo, see Garber, "On the Frontlines of the Scientific Revolution."

[37] Mersenne, *Questions Inouyes*, p. 224.

[38] Ibid., p. 54.

are alive."[39] To take a concrete example, Mersenne expresses complete agnosticism about the ultimate make-up of bodies. Whether the world is made up of matter and form, atoms, the general soul, ideas, hot and cold, dense and rare, is all impossible to say. "That which one can hope from those [i.e. philosophers] (in the case where they desire to aid in the establishment of the true philosophy) consists only in setting out faithful reports of their observations and experiment."[40] It is interesting the way Mersenne characterizes his academy in a letter to Peiresc on 23 May 1635: "It will be the most noble academy in the world … which it will undoubtedly be, since it is entirely mathematical."[41] That is to say, he thinks of what he is doing not as physics, or as philosophy, but as mathematics. While Mersenne certainly favored experiment and the application of mathematics to the world, I do not see how he could be considered as a mechanical philosopher.

1.6 Descartes/Gassendi/Hobbes: Mechanical Philosophers?

At this point I would like to turn to the three remaining figures: Descartes, Gassendi, and Hobbes. With the earlier figures under discussion, Bacon, Galileo and Mersenne, I showed how each failed to fit Boyle's conception of a mechanical philosopher insofar as each, in his own way, held views that violate Boyle's explanatory program. With Descartes, Gassendi and Hobbes, this is not true. For these three, everything in the world is explicable in terms of size, shape, and motion.

Descartes saw the physical world and its contents as a collection of machines. At the end of his *Principia Philosophiae*, Descartes tells the reader that "I have described this earth and indeed the whole visible universe as if it were a machine: I have considered only the various shapes and movements of its parts."[42] Later in the *Principia* he writes:

> I do not recognize any difference between artifacts and natural bodies except that the operations of artifacts are for the most part performed by mechanisms which are large enough to be easily perceivable by the senses—as indeed must be the case if they are to be capable of being manufactured by human beings. The effects produced in nature, by contrast, almost always depend on structures which are so minute that they completely elude our senses.[43]

Similarly, Descartes suggests to an unknown correspondent, seeking to clarify his position that "all the causes of motion in material things are the same as in artificial machines."[44] In his *Traité de l'homme* of 1633, Descartes even assimilated

---

[39] Ibid. See pp. 51–54, 357.

[40] Ibid., p. 78.

[41] "Il verra la plus noble academie du monde qui se fait depuis peu en ceste ville, dont il sera sans doute, car elle est toute mathematique" (Mersenne, *Correspondance*, vol. V, p. 209).

[42] *Principia* IV 188.

[43] *Principia* IV 203.

[44] Descartes to?, March 1642, in *Œuvres*, vol. V, p. 546.

the *human* body to a machine. On Descartes' philosophy, mechanics subsumes physics, in a way: *everything* in physics now receives a mechanical explanation, that is to say, everything is explained as if it were a machine.

Gassendi's large intellectual project was the revival of Epicurean atomism. For Gassendi, as for Epicurus and his disciple Lucretius, the ultimate constituents of the world are atoms, naturally unsplittable, and separated from one another by empty space, that is, vacuum.[45] Following Epicurus, the only properties atoms really have are size, shape, and heaviness (*gravitas* or *pondus*).[46] But heaviness is "nothing but the natural and internal faculty or force by which an atom moves itself through itself, and can move."[47] And so, insofar as everything in the natural world is made up of atoms, everything is explicable in terms of size, shape, and motion for Gassendi as well.

Hobbes' natural philosophy begins with body and motion. Body is defined as "that, which having no dependence upon our thought, is coincident or coextended with some part of space."[48] Motion, in turn, is defined as "a continual relinquishing of one place, and acquiring of another."[49] But the central principle of Hobbes' philosophy is that "there can be no cause of motion, except in a body contiguous and moved."[50] As a consequence, "it is necessary that mutation can be nothing else but motion of the parts of that body which is changed."[51] Behind these deceptively bland statements, though, there is a radical doctrine. If motion is the only cause, then a body at rest can offer no resistance to motion: "nothing but motion gives motion to such things as be at rest, and takes it from things moved."[52] In this way, there is no force in the world except for motion itself.[53] As a consequence, the laws of the impact of bodies reduce simply to the law of the composition (combination) of velocities. What this means, in essence, is that there is no difference between the behavior of a body in physics, and the way in which geometrical objects (points, lines, surfaces) trace out curves, surfaces and solids in pure geometry. In this way, for Hobbes *there is no substantive distinction between physics and geometry*. Though he differs in interesting ways from Descartes and Gassendi, for Hobbes as well, everything in the natural world is explicable in terms of size, shape, and motion.

In this way, all three, Descartes, Gassendi, and Hobbes satisfy the doctrines that I have identified as the explanatory program of Boyle's mechanical philosophy. It is on account of this, no doubt, that all three are often considered mechanical philosophers

---

[45] For Gassendi's Epicurean project, see Rochot, *Les travaux de Gassendi*, and Joy, *Gassendi the Atomist*.

[46] Gassendi, *Opera*, vol. I, pp. 266f.; see p. 366.

[47] Ibid., p. 273.

[48] Hobbes, *De corpore*, 8.1. Translations are taken from the 1656 English translation.

[49] Ibid., 8.10.

[50] Ibid., 9.7.

[51] Ibid., 9.9.

[52] Ibid., 15.3.

[53] Ibid., 8.19, 15.2, 15.8.

by modern historians. But even so, there are important differences among them: all three considered themselves to be advancing distinct and competing intellectual programs. Indeed, many of the issues over which they clashed were the very issues that Boyle explicitly excluded from the mechanical philosophy.

It is no surprise that when Boyle notes disputes his mechanical philosophy wants to avoid, it is Descartes and Gassendi whom he mentions most prominently. Descartes and Gassendi were famously at odds with one another. Gassendi's *Fifth Objections* were only the beginning of an acrimonious exchange, filled with ironic phrases and sarcasm. Gassendi referred to his antagonist as "O Mind,"[54] to which Descartes responded in the *Fifth Replies* by referring to Gassendi as "O Flesh."[55] Gassendi followed with his reply to Descartes' reply in his *Disquisitio* (1644). Descartes was so incensed by this that he pulled Gassendi's *Fifth Objections* from the French translation of the *Meditations* in 1647, adding a letter to Clerselier in its place, in which he answered some of the objections raised in the *Disquisitio*.

There were many areas of dispute between the two over the metaphysics and account of knowledge in the *Meditations*. But their differences in natural philosophy were no less serious. At the end of the *Principia*, Descartes made it clear to his readers that his project was quite distinct from Gassendi's revival of ancient atomism: "The Philosophy of Democritus differs no less from our philosophy than does the common [i.e. scholastic Aristotelian] philosophy."[56] The issues, of course, were atoms and the void. On atoms, Descartes came out explicitly against them, arguing that unsplittable particles of matter are inconsistent with divine omnipotence.[57] Matter for him was continuous and infinitely divisible. And his rejection of the void is no less firm. In addition to arguments directed against the void,[58] he argued that even if God were to annihilate all of the matter within a vase, there would be no vacuum, since necessarily, the sides would have to touch, since there would be no thing between them.[59] Gassendi, for his part, rejected both of these views. In addition to his defense of atomism, he explicitly attacked Descartes' conception of body in the *Disquisitio*.[60] In his *Syntagma*, Gassendi presented an extended defense of the existence of a void, which includes a direct attack on Descartes' "vase" argument.[61] Indeed, the polemics

---

[54] Descartes, *Œuvres*, vol. VII, pp. 265, 275, 321, etc.

[55] Ibid., pp. 352, 354, 357, etc.

[56] *Principia* IV 202.

[57] *Principia* II 20.

[58] *Principia* II 16–17.

[59] *Principia* II 18.

[60] Gassendi, *Works*, vol. III, pp. 384b, 305b; see LoLordo, *Pierre Gassendi and the Birth of Early Modern Philosophy*, p. 52. An anonymous referee of this chapter pointed out that Gassendi's account of body also recognizes material souls and *semina*. While these are ultimately made up of atoms, they do seem to have some emergent properties that look resistant to a simple reduction to size, shape and motion. See LoLordo, *Pierre Gassendi*, chap. 8.

[61] The vacuum is discussed in Gassendi, *Works*, vol. I, pp. 179ff. The vase argument is discussed in I 184a.

about the vacuum in the period are not limited to the disputes between Descartes and Gassendi. The existence of the void was one of the most contentious issues in natural philosophy in the period. Inspired by the famous Torricelli experiments, Pascal performed a series of experiments intending to show the existence of the vacuum. This sparked a vigorous debate between advocates of the void and advocates of plenist views, including Descartes himself in his last years. The debate continued into the 1660s with Boyle's famous air-pump experiments and the disputes between Boyle, Hobbes, and others over just what they showed.[62]

Hobbes never disputed with Gassendi, with whom he remained personally close, despite their philosophical differences. However, his relations with Descartes were rather more hostile. Like Gassendi, Hobbes wrote a series of objections to the *Meditations*, which were answered by Descartes. Though the level of acrimony never reached the level that it did between Descartes and Gassendi, the exchanges were rather sharp. A central difference in the metaphysical realm was over materialism and the existence of incorporeal substance. Hobbes opposed Cartesian dualism with an uncompromising materialism. For Hobbes, incorporeal spirits were not only unnecessary: they were altogether incoherent. On Hobbes' view, "substance and body signify the same thing; and therefore, substance incorporeal are words which, when they are joined together, destroy one another, as if a man should say an incorporeal body."[63] Descartes, of course, presented his philosophy as an alternative to the dominant Aristotelianism of the schools. But Hobbes assimilates Cartesian dualism to the Aristotelian position on separated essences, rejecting both in one fell swoop.[64]

But there were differences at the level of physics as well. Descartes never wrote against Hobbes' physics, which was not published until after his death. But Hobbes clearly had Descartes' physics in mind when writing his own. Central to Descartes' physics is the way it is grounded in God. For Descartes, as for Hobbes, bodies are the objects of geometry made real, as everyone knows. But when they are made real, for Descartes, they are sustained by a god who keeps them in existence from moment to moment. Because of the way in which God does this, though, Cartesian bodies have certain properties that purely geometrical objects do not. In particular, real bodies satisfy a law of the conservation of quantity of motion, as well as series of laws of nature, that a body in motion remains in motion unless caused to change, that it tends to travel in a straight line, that it behaves in characteristic ways in collision, etc. This metaphysical foundation for physics was crucial for Descartes: one of the central reasons why he split with his erstwhile disciple Henricus Regius was precisely because Regius published a version of Descartes' physics in his *Fundamenta physices* (1646) that omitted the appropriate foundations in God.[65] Hobbes, too, recognized that bodies in motion will remain in motion, and that all

[62] See Garber, *Descartes' Metaphysical Physics*, pp. 136–143 and the references cited there.

[63] *Leviathan*, chap. 34.2.

[64] *Leviathan*, chap. 46.18.

[65] See, e.g., Descartes to Mersenne, 5 Oct. 1646, in Descartes, *Œuvres*, vol. IV, pp. 510f.

change happens through the collision with other bodies. But he very self-consciously *does not* call them laws, and *does not* ground them in the activity of a transcendent God. (It is, perhaps, also because he does not recognize the divine foundations of the laws that Hobbes does not recognize a conservation principle.) It is not altogether clear whether Hobbes was an atheist, but however one comes down on that, he explicitly argued that God has no role whatsoever to play in natural philosophy. In a way, one might say that Hobbes' physics is what becomes of Descartes' natural philosophy when God is no longer available.[66]

Descartes, Gassendi, and Hobbes were in sharp contention with one another. All of them set themselves against Aristotelian orthodoxy about the natural world. All three were attempting to present a vision of what a new natural philosophy should look like, something to replace the accepted world of prime matter and substantial form. But each also opposes the "new philosophies" proposed by their competitors. Though they all might agree that Aristotelian natural philosophy is wrong, and that everything in nature should be explained in terms of size, shape, and motion, they all pointedly disagree about a variety of other issues. And they are definitely unwilling to set aside these disagreements. The three represent what they would consider importantly different visions for the future of natural philosophy. In this way, there is nothing irenic about their programs: Descartes, Gassendi, and Hobbes direct their barbs as much against one another as they do against their common enemy, the philosophy of the schools. In this way, their intellectual programs depart in an important way from Boyle's mechanical philosophy. Though they may agree with the explanatory project Boyle proposes, they are quite at odds with his irenic project.

1.7 Novatores, Latitudinarians, and the Construction of the Mechanical Philosophy

The apparent conflict among Descartes, Gassendi, and Hobbes can be put into a larger intellectual context, one that goes a long way toward helping us to understand where the irenic perspective of Boyle's program may be coming from. Let me put the point in Kuhnian terms. In the first half of the seventeenth century there was a central paradigm for the investigation of the physical world: it was the Aristotelian

[66] In the physics, there is also an interesting exchange on creation. Descartes' physics proper begins with the creation of the world. On Descartes' account in the *Principia*, "hypothetical" because inconsistent with *Genesis*, God creates the world with particles of roughly equal size, all in motion; see *Principia* III.46. In *De corpore* 26.1, Hobbes sets aside the question of the initial state of the universe: "The questions therefore about the *Magnitude* and *Beginning* of the World, are not to be determined by Philosophers, but by those that are lawfully authorized to order the Worship of God." Which is to say, the question of the state of the world at creation is a theological question that lies outside of philosophy. Hobbes and Descartes also clashed on questions in optics after the publication of Descartes' *Dioptrics*. On this see Sabra, *Theories of Light from Descartes to Newton*, chap. 3.

natural philosophy, the philosophy as taught in the schools and learned by virtually every educated person. There was also opposition to the Aristotelian paradigm, a significant number of programs that opposed Aristotelian natural philosophy and offered alternative directions in which one could go. Those who opposed the Aristotelianism of the schools were generally called *novatores*, "innovators," by their contemporaries. Mersenne, for example, gives a number of different lists of such *novatores*. In the preface to his 1623 Genesis commentary it is Campanella, Bruno, Telesius, Kepler, Galileo, and Gilbert. Later in the Genesis commentary it is Bacon, Fludd, Hill, and Basso.[67] In *La vérité*, the list is: Patrizi, Basso, Gorlaeus, Bodin, Carpentier, Hill, Olive, "et plusieurs autres."[68] In a letter to Beeckman from 17 October 1630 Descartes groups together Telesius, Campanella, Bruno, Basso, Vanini, "novatores omnes."[69] On the other side of the aisle, among the Aristotelians there is Jean-Cécile Frey, a professor at the University of Paris in the 1620s. In 1625 Frey delivered a series of lectures against the new philosophers, which was later published by his students as the *Cribrum philosophorum qui Aristotelem superiore et hac aetate oppugnarunt*, "A Sieve for Philosophers Who Oppose Aristotle Both in Earlier Times and in Our Own". The figures he attacked were Campanella, Patrizi, Bacon, Telesio, Ramus, Chassins, Villon, Gassendi, Pomponazzi, and Valla. As late as 1651, Jean Bachout, the French translator of Jean D'Espagnet's alchemical work, *Enchyridion physicae restitutae*, sets out another such list, which includes Bacon, Fludd, Gorleus, and Taurellus. But to this list he adds a more recent name—René Descartes.[70] The overlaps between these different lists suggest that we have here a generally agreed category of thinkers. But what is interesting here is that though all opposed Aristotle in one way or another, it is difficult to see anything more than that that they all have in common. There is opposition to Aristotelianism, but there is no single paradigm around which these *novatores* rally.[71]

This, in a way, is the intellectual context in which to place the kind of activity that Descartes, Gassendi, and Hobbes were engaged in: each saw himself as providing an alternative to the Aristotelian natural philosophy, as well as an alternative to the others who were doing the same. Each of the three sees himself as the clear successor to the old orthodoxy. But it is important to realize that none of them is unique in holding such a view: there are a multitude of contenders for the new Aristotle, each of whom sees the choice as equally clear. And for each, the opponent is not

---

[67] Mersenne, *Quaestiones in Genesim*, col. 1838.

[68] Mersenne, *La vérité des sciences*, p. 109.

[69] Descartes, Œuvres, vol. I, p. 158.

[70] See [Espagnier, J. d'], *La philosophie naturelle restablie en sa pureté ...*, the (unpaginated) preliminary "Discours a la recommendation de la Philosophie ancienne restablie en sa pureté; Et sur le nom de son premier Author." (When the work was originally published in Latin in 1623, it was anonymous. In this brief introduction, the translator reveals the name of the author and puts him in the company of other *novatores*.)

[71] In her essay in this volume, Sophie Roux calls attention to largely the same lists of *novatores*, though she makes somewhat different use of them, seeing them as leading later in the century to a distinction between the ancients and the moderns. I will return to her point later in this essay.

only the Aristotelianism of the schools, but other *novatores* as well, who, in their different ways, are just as mistaken as the schoolmen. (It is worth pointing out, though, that from the point of view of the dominant Aristotelianism, they are all on a par, and all equally inadequate. As Mersenne, defending Aristotelianism, put it in 1625, "Aristotle is an eagle in philosophy, and the others are like chicks, who wish to fly before they have wings."[72])

Enter Boyle. I want to suggest that we can read Boyle's mechanical philosophy as a reaction to this context. That is, I suggest that the program for the mechanical philosophy that Boyle sets out in the *Origin of Forms and Qualities* and in other related texts can be read as a call to arms for a new program intended to respond to the chaos of anti-Aristotelian views fighting it out with one another in the learned world. Though many have read Boyle as another figure in a long, continuing, and well-established tradition of mechanical philosophers, the *Origin of Forms and Qualities* was in fact a kind of *manifesto*, the declaration of a *new* program, and not a simple description of a going program. It was, in essence, the construction of something new out of pre-existing elements. It was the articulation of a new paradigm for scientific activity, a paradigm that differed in significant ways from what had gone before. Boyle founded a club in which other, later thinkers could and did claim membership.

Boyle knew quite well the diversity among the adherents of the new anti-Aristotelian philosophies of the earlier part of the century, and the inconsistencies among the different systems. He quite clearly picked and chose among the different programs that were available. Sennert and Digby did not survive in his synthesis; Descartes and Gassendi did. Even though these figures may have seen themselves as engaged in distinct and competing intellectual projects, Boyle took these figures (and others as well), extracted what he wanted to claim were their central principles, and made them consistent with one another. In identifying certain doctrines as the *central* issues, and in setting aside all others over which thinkers had differed, Boyle created something new in the early 1660s, a new program, a new paradigm, one that turned out to be very successful.

Boyle's point here is irenic: he is trying to convince adherents of opposing positions that their agreement is more important than their disagreement. Earlier I talked about uniting the atomists like Gassendi and the Cartesians. But he wanted to go farther still and unite the mechanists and chemists. In the preface to *Some Specimens* he writes:

> And indeed, I freely confesse, that I shall think my self to have done no uselesse service to the Common-wealth of Learning, if I prove so fortunate, as by these, or any other Writings of mine to the like purpose, to beget a good understanding betwixt the Chymists and the Mechanical Philosophers, who have hitherto been too little acquainted with one anothers Learning.[73]

---

[72] *La vérité des sciences*, pp. 109f.

[73] Boyle, *Works*, vol. II, p. 90.

It is a central part of Boyle's project to undermine disagreement by focusing on fundamentals, indeed, to eliminate those points of disagreement from even being a proper part of scientific activity.

It is, perhaps, no accident that Boyle is trying to create this program at exactly this moment in exactly this place. In the last 30 or 40 years there has arisen an enormous literature about Latitudinarianism and the "Latitude-men" in Restoration England. In response to the bitter factionalism, zealotry, and enthusiasm that characterized religion during the English Civil War and the Commonwealth that followed, it is claimed that there arose a more moderate, more rational approach to religion. The adherents of this view, the Latitudinarians, were supposed to reject factional disputes about inessential doctrines, and concentrate on the heart and soul of Christianity, what really matters, and what all good and rational men can agree to. Furthermore, some claim that the "new science" of the Royal Society grew directly out of this attitude, a sort of post-Merton Merton thesis.[74] Others see this attitude as underlying the new emphasis on experiment and observation, something about which all good and rational men can agree as well.[75] There is certainly support for such a case. The two intellectual currents are explicitly linked in Simon Patrick's pamphlet *A brief account of the new sect of Latitude-Men, together with some reflections upon the New Philosophy* (1662). They also joined in Joseph Glanvill's continuation of Bacon's *New Atlantis*, titled *Anti-fanatical religion and free philosophy* (1676).[76] I do not want to make any grand claims of this sort. Some recent scholars have suggested that it is not entirely clear that there really was such a sect as the Latitude-men.[77] And it is even less clear that the existence of Latitudinarianism could explain the science of the Royal Society in general or experimental science in particular. The religious and scientific views of those associated with the Royal Society were sufficiently diverse that any such thesis is rather implausible.[78]

But, however the larger debate about Latitudinarianism and the "new science" might go, I think that something interesting and narrower can be said about the specific case of Boyle in this respect. Whether or not there was such a thing as the Latitudinarians, Boyle was certainly someone of an irenic temperament, someone "of a reconciling Disposition." In a letter to John Dury, 3 May 1647 Boyle wrote: "It is strange, that men should rather be quarrelling for a few trifling opinions, wherein they dissent, than to embrace one another for those many fundamental truths, wherein they agree."[79] While the comment is related specifically to religious sects, battling with one another, it is not unreasonable to see his creation of the mechanical

[74] See Jacob and Jacob, "The Anglican Origins of Modern Science" and "The Saints Embalmed. Scientists, Latitudinarians, and Society."

[75] See Shapin and Schaffer, *Leviathan and the Air-Pump*, chap. 7.

[76] This was published as essay 7 of Glanvill's *Essays on several important subjects*.

[77] See Spurr, "'Latitudinarianism' and the Restoration Church."

[78] See Hunter, "Latitudinarianism and the 'Ideology' of the Early Royal Society."

[79] Boyle, *Correspondence*, vol. I, p. 57.

philosophy in that spirit: helping his contemporaries to see that they should "embrace one another for those many fundamental truths, wherein they agree," rather than arguing uselessly about metaphysical issues such as atoms or the void. Whatever one might say about Latitudinarianism in general and its scientific influence, I think it is fair to say that in Boyle's mind, the attitude toward religion is directly connected with his advocacy of the mechanical philosophy. I am aware that the case I am suggesting needs further substantiation and argument. But it seems quite plausible that this broad irenic attitude toward religion will turn out to be what is behind the first formulation of the mechanical philosophy in England of the 1660s.

And so, even though Boyle appropriated Descartes and Gassendi for his new mechanical philosophy, and even though later commentators correctly see an affinity between those figures and Hobbes, it is misleading to include them (or any other earlier figures I know of) into Boyle's new program. If my reading is right, Boyle's mechanical philosophy was an irenic project, very much a part of his own particular intellectual context, but very foreign to anything that moved any of the three. Boyle's mechanical philosophy was a club to which they could have not nor would have belonged.

1.8 A Broader Conception of Mechanism?

I have argued that many of the figures who are often grouped as mechanical philosophers by later commentators do not really belong there, if we understand the mechanical philosophy as Boyle did. Bacon, Galileo, and Mersenne all fail for the obvious reason that they do not subscribe to the view that everything should be explained in terms of size, shape, and motion, what I have called Boyle's explanatory program. In the case of Descartes, Gassendi, and Hobbes, I have suggested that while they fit well under Boyle's explanatory program, the strong irenic spirit of Boyle's conception of the mechanical philosophy is quite foreign to anything that moved them.

But might this be too narrow a view? Might there be a broader conception of the mechanist program under which some, if not all of the figures from the first part of the seventeenth century that modern historians group as mechanical philosophers be joined? We might, for example, leave aside the irenic aspect of Boyle's program, and consider those who agree in explaining everything in terms of size, shape, and motion as a single and coherent group. This would allow us to group Descartes, Gassendi, and Hobbes at very least as mechanical philosophers in this broader sense. Or, we might be more liberal still. Recognizing the way in which the lever and, more generally, machines of other sorts were important to Galileo's reasoning in a variety of areas, we might include him as well.[80] (We should not forget here the fact that the

[80] This are the considerations that Machamer seems to suggest for considering Galileo as a mechanist in "Galileo's Machines, his Mathematics, and his Experiments." (Machamer and I have been disagreeing about whether or not Galileo counts as a mechanical philosopher for years.) Ursula Goldenbaum has also suggested in conversation that she thinks of Galileo as a mechanical philosopher.

Discorsi begins in the arsenal of Venice.) We might include other criteria as well to bring in the full range of figures that modern historians of science have treated as mechanical philosophers.

Now, I have no problem at all with applying the mechanical philosophy to understanding scientific activity in the second half of the seventeenth century. Once Boyle introduced the idea of the mechanical philosophy, there were certainly those who adopted it. Furthermore, there were certainly thinkers who called themselves mechanical philosophers and, more generally, mechanists from the 1650s on, many of whom would not necessarily have seen themselves as subscribing to Robert Boyle's narrower conception of the program in any strict sense.[81] We must be a bit careful here, though. As Sophie Roux has argued, the rather diverse and incoherent catch-all category of *novatores* that we were concerned with earlier in this paper, and earlier in the century, eventually gave way to a rather more coherent category of "new philosophers" or "moderns" sometime in the second half of the century.[82] While there were certainly those among the moderns who designated themselves as mechanists or mechanical philosophers, we cannot use the term "mechanical philosopher" to designate the moderns indiscriminately. Among the moderns there were also those who identified themselves as Cartesians, or as "experimental philosophers" or as mathematicians, emphasizing other aspects of their thought, their allegiance to the philosophy of Descartes, to the use of empirical methods, to the application of mathematics to nature. But even so, I have no trouble accepting the mechanical philosophy taken broadly as an important historiographical category for the second half of the century, an actor's category to which the historian is fully entitled.

But when we are talking about the first half of the century, in the years before Boyle (and no doubt others) made "mechanical" an adjective that could go naturally with "philosophy," I am somewhat more skeptical that it is appropriate to talk about mechanical philosophers, even in the broader sense at issue here.

First of all, I do not think that such a grouping constitutes a genuine actor's category. I certainly acknowledge that among the opponents of Aristotelianism in the first half of the seventeenth century are some figures that Boyle (and later historians) will later group together and call the mechanical philosophers. Boyle's grouping is not arbitrary, to be sure: there is a reason why Boyle grouped them together. And there is a reason why some later thinkers might have called themselves mechanists in a broader sense insofar as they accepted the principle that everything is explicable in terms of size, shape, and motion, or insofar as they saw the behavior of machines as a model for understanding nature, or some such thing. But it is also interesting and important that, so far as I can see, the group of earlier thinkers who may have shared these commitments did not see themselves as forming a coherent group, and as having a common core, that is, as subscribing to the same paradigm in the Kuhnian sense. Instead, I would suggest, they saw themselves as elements in this hurly-burly

[81] Alan Gabbey, though, points out how difficult it is to say what exactly it means to be a mechanist in this broader sense, and what the relation is to mechanics. See Gabbey, "What Was 'Mechanical' about 'The Mechanical Philosophy'?"

[82] See her essay in this volume.

of anti-Aristotelian philosophies, each fighting against one another to become the new direction for philosophy to take, the new paradigm to be adopted and to replace Aristotelianism. They certainly would have seen that they had some things in common with other figures, and would have differed in other ways. But there were many issues under contention, broad and narrow, from the use of mathematics and experiment in understanding nature, to issues about what are the appropriate concepts in terms of which natural phenomena are to be explained (form and matter versus size shape and motion versus appetite and desire, etc.) to specific doctrines such as the existence of atoms and the void, geocentrism, the nature of the elements, and so on. Particular subgroups from within the *novatores* might be defined by emphasizing some of these elements, and ignoring others. But I see no reason to believe that the features that were later to define the mechanists were seen by actors in the early seventeenth century as picking out a privileged group of thinkers who constituted a single coherent school. *We* can recognize the commonalities that unite thinkers like Descartes, Gassendi, and Hobbes in hindsight, given what was later to evolve in the intellectual community, but I am not convinced that they did. In this way, these earlier thinkers are not different variants of a common paradigm, but elements of a pre-paradigmatic stage in the development of the mechanical philosophy, part of its pre-history rather than its history.

But even if we recognize that the mechanical philosophy was not an actor's category for the early seventeenth century, might there be good reason for the historian to recognize these commonalities, and make use of the idea of the mechanical philosophy even before the idea (and the grouping) was available to thinkers like Descartes or Gassendi or Hobbes? Perhaps, though I would have to be convinced. On the other hand, even if we as historians decide to talk about the mechanical philosophy with respect to figures in this earlier period, it is very important to be very conscious of the fact that in doing so, we are dividing the world up in a way that might not have been fully intelligible to the figures in question. The early seventeenth century is a world of *novatores* and Aristotelians, I would argue. One can certainly give a history of their thought, but the history of the mechanical philosophy properly speaking only begins a few years later, with Boyle and his contemporaries.

Chapter 2
How Bacon Became Baconian

Guido Giglioni

Francis Bacon's metaphysics of material desires represents a major contribution to early-modern natural philosophy and theories of matter. By material desires, Bacon meant a limited set of primordial appetites deemed to govern all natural phenomena. He was convinced that through experimental trials natural philosophers could identify such basic appetites, classify them by means of increasingly comprehensive interpretative frameworks (inductions) and control them through direct manipulations (superinductions). Because of its focus on appetites—appetites within matter, but also appetites in men—Bacon's program of *inventio*, *inductio*, and *superinductio* of material desires can be described as an original model of natural-political inquiry.[1]

An accurate assessment of the nature of mechanical operations in Bacon's natural philosophy needs to focus on the interplay of these three levels of inquiry. By putting appetites at the center of his philosophical reflection, Bacon advanced a strikingly original solution to the question of the relationship between knowledge and action in nature, a solution that is markedly different from the one chosen by the major figures of seventeenth-century mechanical philosophy. By and large, mechanical philosophers of various creeds questioned any blunt assertion of ontological realism when explaining the nature of life and preferred to confine their investigations to a study of the external appearances of matter (what we might call a Kantian "as-if" solution) in addressing the question of how to account for intelligence in nature (or at least the recurrence of teleological patterns of action that seem to indicate the presence of some form of knowledge embedded in nature). For them, finalism in nature could no longer be explained through the unintentional teleology of

I would like to thank Sophie Roux, Daniel Garber, Michael Hunter, and one anonymous reviewer for their suggestions and comments.

[1] On Bacon's metaphysics of material desires, see Giglioni, *Francesco Bacone*, 59–95; "Mastering the Appetites of Matter: Francis Bacon's *Sylva Sylvarum*."

G. Giglioni (✉)
Warburg Institute, University of London, London, UK
e-mail: Guido.Giglioni@sas.ac.uk

D. Garber and S. Roux (eds.), *The Mechanization of Natural Philosophy*, Boston Studies in the Philosophy of Science 300, DOI 10.1007/978-94-007-4345-8_2,

Aristotelian kind. They interpreted design in nature either as the result of projections imposed onto nature by our mind or as the manifestation of an objective order originally established by God but not dependent on the very essence of nature. It was an argument that Ralph Cudworth described as a dilemma dividing "atomic atheists" from "bigotical religionists."[2]

Bacon opted for brutal realism: nature was a battlefield of ruthless and opposing appetites. The order of the cosmos did not result from intentional planning, but rather it remained a provisional settlement, highly unstable and flexibly compatible with a permanent state of tension in the very life of nature. Appetites were constantly in the process of overcoming the strenuous resistance—*antitypia*—offered by other appetites, which in turn reacted fiercely to any new attempt to be subjugated. The model followed by Bacon was a self-regulating mechanism based on the antiperistasis of appetites (that is, their tendency constantly to chase each other) rather than a teleological organization controlled by the irrepressible tendency to fulfill the very end of the process, be that end external or internal to the process. This means that in nature there was no real harmony, but a steady-state containment of conflicts subject to precarious balance and continuous readjustments.[3] Rather than relying on one of the possible models of apparent intentionality (the "as-if" intentionality) available at the time, and rather than falling back into an Aristotelian model of teleological realism (ends of nature as real and representing the highest level of ontological actualization), Bacon accepted a form of real but hidden (almost inscrutable) intentionality. The recurrent thesis that the universe—both natural and political—is divided between those who know and those who do not know is a distinguishing feature of Bacon's work as a whole. Those who know, know without others knowing that they know, which means that those who do not participate in knowledge act without knowing what they do. It is worth remembering that this rather complicated principle of knowing without being known represents the foundation of Bensalem's social organization in *New Atlantis*.[4]

Unlike mechanical philosophers, who assumed that nature and natural things do not know what they do because in fact they have no knowledge of their own (and therefore in the natural world there is no natural intention but also no natural activity), Bacon distinguished between beings that know because their appetite is driven by a higher level of knowledge and natural beings that do not know what they do because their perceptive power is feeble or because they are directed by beings that are endowed with superior knowledge. While the "as-if" solution presupposes a chasm between knowledge (whether coming from God or man) and absence of knowledge (nature), Bacon's solution postulates acts of complex mediation between

[2] Cudworth, *The True Intellectual System of the Universe*, II, p. 606.

[3] Clagett, *Giovanni Marliani and Late Medieval Physics*, p. 79: "Antiperistasis is defined as the supposed sudden increase of the intensity of a quality as a result of being surrounded by its contrary quality, for instance, the sudden heating of a warm body when surrounded by a cold."

[4] See Giglioni, "Fantasy Islands: *Utopia*, *The Tempest*, and *New Atlantis* as Places of Controlled Credulousness"; "The Material Foundations of Francis Bacon's Utopia."

different levels of knowledge (*perceptio* and *sensus*, *appetitus* and *voluntas*). Activity and order in nature are real, even though appetites are assumed to be held in check through the use of deception and ruse.

However much Bacon decided to explain the relationship between knowledge and matter by insisting on the primary role of the vital appetites of nature, this does not mean that he ignored the notion of mechanical action. Indeed, the phenomenology of mechanical operations in Bacon's work is quite rich (and this may explain why Bacon has often been taken—more or less *malgré lui*—as one of the fathers or heralds of the mechanical philosophy). In general, we can say that Bacon uses "mechanism" and related terms in three principal senses: to signify the arrangement of material and bodily parts, to denote the leveling and standardizing power of habit, and to emphasize the binding and compulsory nature of a true method once this is implemented in the act of knowledge. In other words, Bacon uses the words "mechanical" and "mechanicus" (1) when he describes the actions of both machines and organisms that depend on the specific disposition of their parts; (2) when he refers to patterns of action that are perfected by repetition and exercise (including the involuntary motions of the body); (3) and, finally, when he points to the inescapable production of knowledge resulting from the application of a specific set of rules.

In what follows, I will argue that the basic assumptions on which Bacon's natural philosophy rests have nothing in common with the principles underlying a mechanical philosophical project. I will then try to explain why, despite structural differences, English mechanical philosophers were able to appropriate Bacon's vital materialism. One of the principal aims of this essay is to alert the reader to the existence of various forms of competing Baconianisms throughout the seventeenth century and to the complex nature of their reception.

2.1 The Meaning of Mechanical Operation in Bacon's Oeuvre

Bacon has often been characterized as one of the tutelary deities of the modern notion of mechanical universe. In fact, his position is much more nuanced and elusive. As I hope will become apparent in the course of this paper, Bacon applies the terms "mechanical" and "mechanicus" in a broad range of semantic contexts, and yet the strict and technical sense of this term—the one that was to become the hallmark of seventeenth-century mechanical philosophies—remained largely foreign to his view of nature. As already anticipated, the words "mechanical" and "mechanicus" cover a wide spectrum of meanings in Bacon's work. However, they hardly include the cluster of meanings traditionally associated with the emerging of the new science of mechanics, such as the reduction of motion to a state ontologically identical to the state of rest, the indifference of any body to the opposite states of motion and rest, and the rejection of the belief in the existence of natural and violent motions in the physical universe.

One of the traditional meanings of the words "mechanical" and "mechanicus" acknowledged by Bacon is the one denoting manual labor as opposed to the supposedly spontaneous and creative work of the mind (*ingenium*). As was still common at the time, Bacon contrasts the domain of the liberal disciplines (that he often accuses of being barren of practical results and useful applications) with that of the mechanical arts, illuminated by the light of experience and devoted to the transformation of the natural bodies (*corporum naturalium praeparatio*).[5] It is true that mechanical arts and experiments play a very important part in Bacon's philosophy by forcing nature to disclose its secrets and by injecting new life in the development of human learning.[6] Purely intellectual occupations, Bacon argues, are dead and lifeless in their original principles, while mechanical arts—"that are founded on nature and the light of experience"—are like living organisms ("as if they were partaking of a certain breath of life, grow and get better by the day").[7] He also acknowledges the fact that technological changes in the form of new mechanical devices have caused momentous *rerum mutationes* that "no kingdom, no sect or star" has ever caused.[8] As is revealed by a page from the *Comentarius solutus*, a history of the mechanical arts was an integral part of Bacon's philosophical project:

> To procure an History mechanique to be compiled wth care and diligence and to professe it that is of the experim^ts and observations of all Mechanicall Arts. The places or thinges to be inquyred are; first the materialls, and their quantities and proportions; Next the Instrum^ts and Engins requisite; then the use and adoperation of every Instrum^t; then the woork it self and all the processe thereof wth the tymes and seasons of doing every part thereof.[9]

On the other hand, it cannot be denied that Bacon's overall opinion concerning both the mechanical consideration of nature and the science of mechanics is far from simply positive. Indeed, Bacon explains that in natural philosophy man's intellect has been corrupted by the observation of what happens in the domain of the mechanical arts to such an extent that some people are led to believe that the same processes of composition and separation that occur in men's artifacts also happen in nature. This misunderstanding has resulted in a number of philosophical dogmas, such as the fictional view that the four natural elements are the original constituents of the body.[10] Bacon is also prone to condemn accidental discoveries made by mechanics and the practice of guessing by trial and error (the *variatio quaedam experimentorum*).[11] Moreover, the experiments conducted by mechanics are confined

[5] Bacon, *Novum organum*, in *The Oxford Francis Bacon*, vol. XI, pp. 118, 134. In *De augmentis scientiarum*, Bacon calls this *mechanica* "operaria" (*Works*, vol. I, p. 572). See Weeks, "The Role of Mechanics in Francis Bacon's *Great Instauration*."

[6] Bacon, *Novum organum*, in *The Oxford Francis Bacon*, vol. XI, p. 156.

[7] Ibid., p. 13. See ibid, p. 118.

[8] Ibid., p. 195.

[9] Bacon, *Comentarius solutus sive pandecta, sive ancilla memoriae*, in *Works*, vol. XI, pp. 65f.

[10] Bacon, *Novum organum*, in *The Oxford Francis Bacon*, vol. XI, p. 102: "Inficitur autem intellectus humanus ex intuitu eorum, quae in artibus mechanicis fiunt, in quibus corpora per compositiones aut separationes ut plurimum alterantur, ut cogitet simile quiddam etiam in natura rerum universali fieri. Unde fluxit commentum illud elementorum, atque illorum concursus, ad constituenda corpora naturalia."

[11] Ibid., p. 116.

to the narrow domain of their very specific interests and usually they contribute very little to the real advancement of knowledge according to the wider project of *interpretatio naturae*.[12] In the domain of practical knowledge, mechanics (*mechanica*) is related to physics in the same way as magic is related to metaphysics in the theoretical domain, which means that *mechanica* and *magia* share the dubious honor of belonging to the sphere of ruse, cunning, and trickery.[13]

Even those passages that could be used to corroborate the view of Bacon as a mechanical philosopher *in nuce* need to be read in the context of his philosophy of material appetites. In introducing the notion of "solitary instance"—that is, one of those paradigmatic and privileged situations in nature where a specific natural property is manifested and brought into view in a clearer way than when it is displayed in other, more diverse natural contexts—Bacon characterizes the quality of color as a natural indicator that is not always particularly telling or trustworthy: "color has little to do with the intrinsic natures of any body, but only resides in the grosser and so to speak mechanical arrangement of the parts (*in positura partium crassiori et quasi mechanica*)."[14] Bacon is saying that the *quasi mechanica positura* of the parts of a body, however much it can account for such secondary qualities as color, has nothing very much to do with the "intrinsic nature" of a body. Here it is crucial to bear in mind that by *naturae intrinsecae corporis* Bacon means the primordial appetites of matter, which, to be sure, is something that cannot be described as mechanical. The mechanical properties of things are accidental and transient states of matter that affect only the so-called secondary qualities of the objects (they are, precisely, *crassa*), whereas for Bacon the primary qualities are vital propensities that cannot be erased or altered, but only controlled and manipulated. In this case, the difference between "mechanical" and "natural" is the same as the difference between "external" and "internal," "accidental" and "essential." In this context, mechanical simply means raw and coarse—what Bacon calls the level of "crass and tangible mass."[15]

Sometimes Bacon's use of "mechanical" is synonymous with "organic," in the traditional medical and biological sense of the word. In this case, it denotes a bodily part shaped in the form of a definite structure capable of being used by a "soul." Mechanical devices and organic bodies share the property of being "integral," in that their constitutive parts contribute collectively to the accomplishment of some common end (*conspiratio partium suarum*).[16] Bacon also uses the word *machina* in the sense of "diagram": "when you have a machine to hand (*astante machina*) a demonstration in mathematics is easy and transparent, but without one everything seems obscure and

[12] Ibid., pp. 156–158.

[13] Ibid., p. 214; *Sylva Sylvarum*, in *Works*, vol. II, p. 378. See *De augmentis scientiarum*, in *Works* vol. I, p. 571: "*Physica* siquidem et inquisitio *Causarum Efficientium* et *Materialium* producit *Mechanicam*; at *Metaphysica* et Inquisitio *Formarum* producit *Magiam*."

[14] Bacon, *Novum organum*, in *The Oxford Francis Bacon*, vol. XI, p. 274 (I, aph. 23).

[15] See also Bacon, *Historia vitae et mortis*, in *The Oxford Francis Bacon*, vol. XII, pp. 314–316.

[16] Bacon, *Abecedarium novum naturae*, in *The Oxford Francis Bacon*, vol. XIII, p. 186: "qualia sunt animata, et machinae, et organa mechanica, et huiusmodi."

more subtle than it actually is."[17] This brings us to what I think is the most genuinely Baconian meaning of "mechanical" and "mechanicus," namely, the one denoting the act of implementing the rules of a method, in such a way that acquisition and progress of knowledge are set in motion with inescapable necessity, independently of the intention or mental capability of the very user of the method. Method is, in a literal sense, a *regula* (ruler), that is, a measuring and leveling device:

> There remains but one way to health and sanity: to do the whole work of the mind all over again, and from the very outset to stop the mind being left to itself but to keep it under control, and make the matter run like clockwork (*res veluti per machinas conficiatur*). For if men really tackled work for machines (*opera mechanica*) with their bare hands, and without the help and force of instruments, in the same way as they have not hesitated to undertake work for the intellect (*opera intellectualia*) with little besides the naked force of the mind, there would have been very few things which they could have got going or mastered, even if they combined to use their best efforts.[18]

The renowned image of the clock, which "seems to mimic the heavens as they wheel round, and the pulse of animals as it beats in regular succession," is for Bacon a similitude that is most apt to explain the impersonal deployment of the method's rules rather than the inner working of natural processes. From this point of view, the notion of *machina* has epistemological rather than metaphysical connotations. It demonstrates less the great number and deep complexity of "men's original observations and nature's operation" than "the controlled motion of hand or instrument".[19] It may sound almost offensive to our delicate post-Romantic ears, but what for Bacon should ultimately become mechanical is the mind of man and not the activity of nature. The project outlined in the *Novum organum* is a program for taming the *ingenium*, mechanizing the mind, and controlling the appetites. No methodological blueprint for a future mechanization of nature is involved. If there is a real mechanical device in Bacon's universe, this is "the machine of the intellect" (*machina intellectus*), which is the "ladder" that leads to the knowledge of truth.[20]

Bacon's proposal to mechanize the mind sounds outrageous because it goes against the commonly held assumption that creativity relies on the freedom of the mind and represents the free expression of one's talent. It is also one of the most formidable blows dealt against the Renaissance notion of *ingenium*. Bacon is in favor of forcing the mind to follow a methodological path that has been successfully tested, where nothing is left to chance and everyone can reach the same results starting from the same mental capacities. Therefore, we can safely say that the common

---

[17] Bacon, *Novum organum*, in *The Oxford Francis Bacon*, vol. XI, pp. 42 f. (In this passage *machina* should be translated as "device" rather than "machine").

[18] Ibid., pp. 54f. See Stewart, "*Res, veluti per Machinas, Conficiatur*: Natural History and the 'Mechanical' Reform of Natural Philosophy", *Early Science and Medicine*, 17 (2012), pp. 87–111.

[19] Bacon, *Novum organum*, in *The Oxford Francis Bacon*, vol. XI, pp. 134f.

[20] Bacon, *Abecedarium novum naturae*, in *The Oxford Francis Bacon*, vol. XIII, p. 172: "Pertinet autem Abecedarium ad Instaurationis partem quartam, quae est scala sive machina intellectus." See Giglioni, "Reading Nature without Making a Book of It: Francis Bacon's *Novum Organum*."

denominator among Bacon's various uses of the words "mechanical" and "mechanicus" lies in the meaning of "being forced." The distinction between natural and mechanical motions corresponds to the distinction between original and induced motions, that is, between the primordial spontaneous appetites of matter and motions that are caused by force (*motus violentus sive mechanicus*).[21]

Violent and mechanical motion, says Bacon in *Cogitationes de natura rerum*, is in a way the motion in nature that everyone knows (*fere omnium motuum est vulgatissimus*) and is like "the life and soul" of "artillery, engines and the whole business of mechanics." Bacon's analysis of mechanical motion in projectiles, though, reveals that in the case of such like phenomena there is in fact no mechanical principle involved (once again, in the seventeenth-century sense of "mechanical"). Bacon's mechanical motion is a secondary vital reaction to a previous act of coercion:

> [T]he case is really this. The principal motion seems to be in the parts of the body projected, which being too subtle to be perceived by the eye, and men not being attentive enough but passing the matter by with a light observation, is not observed. But to an accurate observer it is manifest that hard bodies are most impatient of pressure, and have, as it were, a very acute perception thereof (*veluti sensum acutissimum habere*); so that when forced ever so little out of their natural position, they strive (*nitantur*) with great velocity to free themselves and return to their former state. And to do this, all the parts, commencing with the part struck, thrust and press one another forward, just like an external force; which produces a continuous and intense (though invisible) trepidation and commotion of the parts.[22]

As colors and other fleeting appearances in nature are—"so to speak" (*quasi*)—mechanical because they are in a way gross manifestations of the inner life of nature, so motions that superficially look like mechanical operations are in fact expressions of deeper appetites of matter. Appetite, in the form of unremitting trepidation, is the innermost motion in matter; mechanical motions are responses to external stimulations and provocations, as is apparent from the following excerpt from *Novum organum*:

> [T]hings do not come to *Penetration of dimensions*, and a *Vacuum*, except at the extremes of condensation and rarefaction, since these motions stop and leave off long before they get to that point, and are nothing more than the desires bodies have for conserving themselves in their proper consistencies (or, if people like, their own forms), and for not suddenly departing from them unless they are altered by gentle means and by consent. But it is far more necessary (for a great deal turns on it) to persuade men that violent motion (which I call Mechanical, and which Democritus, who in setting out his primary motions ranks even beneath run-of-the-mill philosophers, called motion of Percussion) is nothing other than motion of liberty, i.e. from compression to relaxation.[23]

---

[21] Bacon, *Cogitationes de natura rerum*, in *Works*, vol. III, p. 31.

[22] Ibid., pp. 29f.

[23] Bacon, *Novum organum*, in *The Oxford Francis Bacon*, vol. XI, pp. 386–387. See also *De principiis atque originibus*, in *The Oxford Francis Bacon*, vol. VI, p. 264: "omnis ille motus mechanicus ... nihil aliud est quam nixus partium corporis emissi ad se expediendum a compressione"; *Sylva sylvarum*, in *Works*, vol. II, p. 342 (the motion through which the smallest parts of a body respond to external pressures creating an "inward tumult" is "the chief root of all mechanical operations"), p. 382 ("the tumult in the parts of solid bodies when they are compressed, which is the cause of all flight of bodies through the air, and of other mechanical motions").

For Bacon, a mechanical motion is the same as a violent motion, with the proviso, though, that in the context of his metaphysics of natural motions understood as material appetites, violent motion means motion of liberty, that is, a motion whereby matter has the ability to react to what it perceives as a pressure or aggression from the outside. "In the term *motus mechanicus*," as the physician Francis Glisson would comment upon later in the seventeenth century, "Bacon seems to include in a concrete way both violent motion and motion of liberty, that is, that motion through which nature at once is acted upon and strives to defend itself from inflicted violence." The reality of such a reactive motion, Glisson concluded, testifies to the existence of a material substratum that is "a sufficient principle of natural life."[24]

A classic example in which Bacon's mechanical motion (as "motion of liberty") has been misinterpreted as mechanical motion in a mechanistic sense is Bacon's discussion of the nature of heat in *Novum organum*. Having arrived at a provisional assessment (the "first vintage") concerning the form of heat, Bacon argues that heat can be seen as a kind of motion (*natura, cuius limitatio est calor, videtur esse motus*). Interpreters have been keen to salute this definition as a remarkable anticipation of the mechanical definition of heat: heat, Bacon seems to say, is produced by particles in motion. In fact, when the sentence is read within its proper context, it becomes clear that what Bacon is arguing here is that heat is the result of struggling and conflicting appetites in matter. Heat, writes Bacon, is an expansive motion through which a body, whilst rising upwards, "strives towards self-dilatation, and takes up a greater sphere or dimension than it filled before." At the same time, the expansive motion "is impeded, repelled, and beaten back, in such a way that the body takes on a juddering motion, forever shaking, straining, and struggling, and unsettled by constant rebounding." Finally, the "motion of provocation or penetration" that triggers the appetite underlying the heating reaction has to be "rather rapid and not at all dull." This is the origin of "the rage of fire and heat."

Here we have an explanatory pattern that often recurs in Bacon's metaphysics of living matter. Nature works through conflicts of appetites. Every natural action involves reactions and repressions caused by the basic desires of matter. There is really no room for misinterpreting what Bacon means, for his is a language of natural desires, not of mechanical movements. The production of the form "heat" depends on the practitioner's ability to identify a specific set of conflicting appetites in matter, to arouse their power, and, finally, to "superinduce" the resulting form upon matter. If one is able to stimulate in a natural body "a motion of self-dilatation or expansion, and to repress the motion and turn it back on itself in such a way that the dilatation does not go forward smoothly but is now given its head and now forced to retreat," then he is able to "generate" heat.[25] In *New Atlantis*, Bacon recounts how the

[24] Glisson, *De natura substantiae energetica*, p. 340. See ibid., p. 375.

[25] Bacon, *Novum organum*, in *The Oxford Francis Bacon*, vol. XI, pp. 262–271. Robert Kargon is one of the very few who have acknowledged the non-mechanical nature of Bacon's explanation of heat (*Atomism in England*, pp. 51f.). On Bacon's discussion of the nature of heat, see Garber's chapter (Chap. 1) in this volume.

Bensalemites have reached such a high level of technological expertise that they can "generate heat only by motion." Further, they have "engine-houses, where are prepared engines and instruments for all sorts of motions."[26] And yet a careful analysis of the treatise and its basic assumptions reveals that Bacon's vision of a future engineering of natural motions is far from being mechanical. As the Father of Salomon's House explains at the beginning of his talk, the end of the island's "Foundation" is the knowledge of the "secret motions of things," and we know by now that by secret motions of things Bacon means the material desires of nature.[27] In the final analysis, we can say that for Bacon mechanical motions represent a stage in the appetitive life of matter. They signify the moment in which the natural propensity of matter, its *nisus* and *conatus*, meets with a source of resistance and tries hard to overcome it.

If appetites dominate matter, matter cannot be seen as an inert, unresponsive substratum. A key passage from *Sylva Sylvarum* provides incontrovertible evidence that in Bacon's universe natural motions cannot be explained according to the laws of mechanics.

> It is certain that all bodies whatsoever, though they have no sense, yet they have perception: for when one body is applied to another, there is a kind of election to embrace that which is agreeable, and to exclude or expel that which is ingrate: and whether the body be alterant or altered, evermore a perception precedeth operation; for else all bodies would be alike one to another. And sometimes this perception, in some kind of bodies, is far more subtle than the sense; so that the sense is but a dull thing in comparison of it.[28]

Appetites are characterized by tendencies to follow what is conducive to their immediate preservation and to reject what may destruct their energy. Bacon calls these tendencies perceptions and distinguishes them from sense perceptions in that they are devoid of sense awareness. The sentence "else all bodies would be alike one to another" provides a decisive clue for interpreting Bacon's notion of motion. Bacon is saying that precisely to avoid a mechanistic outcome—that is, the absolute homogeneity and indifference of matter and motion—the existence of some form of perception in matter must be postulated. Without tendencies (appetites) there would be no real change and no real bodies. Motions are propensities through which bodies are able to feel and discriminate. In this way, they establish bodily and perceptual identities in the otherwise undifferentiated continuum of matter. Unlike the representation of the material substratum given by mechanical philosophers, where material differences are transient *modi* in a neutral continuum of matter and motion, Bacon's matter throbs with a primal tendency to establish relationships of identity and otherness. Sense, the defining characteristic of animals and human beings, is just an effete form of the original power of natural perception embedded in matter.

One might object that Bacon still distinguishes between animate and inanimate matter. In *Historia vitae et mortis*, for instance, he clearly assumes a difference

[26] Bacon, *New Atlantis*, in *The Major Works*, pp. 484–486.

[27] Ibid., p. 480.

[28] Bacon, *Sylva Sylvarum*, in *Works*, vol. II, p. 602.

between animate and inanimate beings. However, as is evidenced by the following passage, this does not mean that inanimate matter has no perception or appetite:

> It is certain that all bodies are endowed with a desire to assimilate what is contiguous to them. Subtle and pneumatic bodies, such as flame, spirit and air, do so in a generous and eager way; on the contrary, bodies that have a crass and tangible mass, in quite a feeble way because that desire to assimilate is bound by a stronger desire for rest and immobility.[29]

Desire affects reality in its entirety, both animate and inanimate. In Bacon's metaphysics of matter, motion, understood as obscure feeling, is better defined as desire. The difference between inanimate and animate matter, therefore, is the result of a distinction of degree and not of kind. Inanimate matter (which is alive because intrinsically pervaded by primordial appetites) becomes animate, that is, vegetative or sentient, when it undergoes a process of organization. The drive to organization, however, is implanted in matter and is an integral part of its essence. What is even more important to note here is that inanimate matter, which is inherently affected by the motions of desire, contains life in a more original way than animate matter. The same reasoning applies if we look at matter from the point of view of the division between tangible and pneumatic matter. In every tangible body there is spirit, which means that sluggish matter encloses another type of matter that is more active, and that such a difference in degrees of activity is responsible for all the transformations in matter.[30] As for the distinction between vital and dead spirits, this, again, does not presuppose any original gap between inert and living matter. Like the difference between animate and inanimate matter, the distinction between mortal spirits (*mortuales spiritus*, in the sense of deadly, actively mortal) and vital spirits (*vitales spiritus*) is not an essential and qualitative distinction. *Spiritus mortuales* are present in all bodies, regardless of whether animate and inanimate. Despite the term used by

[29] Bacon, *Historia vitae et mortis*, in *The Oxford Francis Bacon*, vol. XII, pp. 314–316. See also *Novum organum*, in *The Oxford Francis Bacon*, vol. XI, p. 290 (II, aph. 27): on account of the analogical instances (*instantiae conformes*), "organa sensuum et corpora quae pariunt reflexiones ad sensus, esse similis naturae. Rursus ex hoc ipso admonitus intellectus, non aegre insurgit ad axioma quoddam altius et nobilius. Hoc nimirum: Nihil interesse inter consensus, sive sympathias corporum sensu praeditorum, et inanimatorum sine sensu, nisi quod in illis accedat spiritus animalis ad corpus ita dispositum, in his autem absit. Adeo ut quot sint consensus in corporibus animatis, tot possint esse sensus in animalibus, si essent perforationes in corpore animato, ad discursum spiritus animalis in membrum rite dispositum, tanquam in organum idoneum. Et rursus, quot sint sensus in animalibus, tot sint proculdubio motus in corpore inanimato, ubi spiritus animalis abfuerit; licet necesse sit multo plures esse motus in corporibus inanimatis, quam sensus in animatis, propter paucitatem organorum sensus. Atque huius rei ostendit se exemplum valde manifestum in doloribus. Etenim quum sint plura genera doloris in animalibus, et tanquam varii illius characteres (veluti alius est dolor ustionis, alius frigoris intensi, alius puncturae, alius compressionis, alius extensionis, et similium) certissimum est omnia illa, quoad motum, inesse corporibus inanimatis; veluti ligno, aut lapidi, cum uritur, aut per gelu constringitur, aut pungitur, aut scinditur, aut flectitur, aut tunditur, et sic de aliis; licet non subintrent sensus, propter absentiam spiritus animalis."

[30] Bacon, *Historia vitae et mortis*, in *The Oxford Francis Bacon*, vol. XII, p. 346 "Inest Omni Tangibili *Spiritus*, Corpore crassiore obtectus, et obsessus."

Bacon—*mortualis*—such spirits are not devoid of life, that is, lifeless; on the contrary, they are alive in the highest degree, but they bring death and destruction to the hosting body. Instead of "lifeless spirits," they should more properly be called "spirits of death." In their desire to multiply and escape the body in which they feel they are trapped, they bring death to the body that hosts them.[31] *Animatus* and *inanimatus* mean respectively "with" or "without" soul (or, to express the same concept with different words, made up of more or less organized matter); *mortualis* and *vitalis* mean "bearer of death" and "bearer of life."

The texts presented so far elucidate the meaning of such key words as *naturalis* and *mechanicus*, *sensus* and *perceptio*, *animatus* and *inanimatus* in Bacon's writings. When dealing with his metaphysics of matter, it is of paramount importance to keep in mind that every time Bacon is speaking of the ultimate appetites of matter—that is, its motions—, he is addressing metaphysical issues, that is, he is pointing to a level of speculation that transcends the outward and sensible aspects of the material bodies. Bacon characterizes the distinction between the two levels of knowledge—experience and metaphysics—as a distinction between concrete and abstract physics.[32] The level of the motions of matter is deeper than the one characterized by the distinction into "tangible" and "pneumatic" matter; the latter refers to natural bodies, the former to reality as such. As Bacon makes clear in the above-mentioned passages from *Sylva Sylvarum* and *Historia vitae et mortis*, "[i]t is certain that all bodies whatsoever, though they have no sense, yet they have perception," and "[it] is certain that all bodies are endowed with a desire to assimilate what is contiguous to them."[33] Natural bodies are collections of original forms resulting from various combinations of original motions; motions are propensities resulting from appetites. As a consequence, bodies are transient manifestations of original appetites. Appetites—that is, the primordial motions of matter (of matter taken as a whole, both tangible and pneumatic)—are the "letters" that form the alphabet of nature.

On the basis of this metaphysics of material appetites, we need to reassess the meaning of such key words as "mechanical," "empirical," and "experimental" when they are used in relation to Bacon's oeuvre. But this also means that we should revisit Bacon's relationship with seventeenth-century natural philosophy as a whole (mechanical corpuscularianism, empiricism, and experimental philosophy) by pursuing a more complex and nuanced approach.[34] The following are some of the questions that historians of Bacon's natural philosophy cannot shirk any longer: How did it happen that such a wide gap opened up between Bacon's own philosophy and its later reception? Why did the gap originate so soon, almost immediately after Bacon's death? Finally, how did it come that those who decided to expand on Bacon's notion

[31] Ibid., pp. 348–354.

[32] Bacon, *De augmentis scientiarum*, in *Works*, vol. I, p. 566.

[33] Bacon, *Sylva Sylvarum*, in *Works*, vol. II, p. 602; *Historia vitae et mortis*, in *The Oxford Francis Bacon*, vol. XII, pp. 314–316.

[34] On this point, see Giglioni, "*Historia* and *Materia*: The Philosophical Implications of Francis Bacon's Natural History."

of material desires appear to have been so few and soon forgotten? Can this be taken as evidence that, for all Bacon's declarations of vital materialism, the true fulfillment of his natural philosophy lies in any case in the interpretation that was given of it by mechanical and corpuscularian philosophers, regardless of the written record and Bacon's own original intentions? In the rest of this essay, I shall try to provide some preliminary and tentative answers to these questions. Inevitably, the essay will have some of the haphazard and impressionistic tones that are typical of all initial attempts. For reasons of both space and argumentative focus, I will confine my excursus to seventeenth-century England.

2.2 Mechanical and Vital Readings of Bacon's Natural Philosophy in Seventeenth-Century England

Generally speaking, we can say that both historical events and ideological constraints influenced the way people read Bacon during the seventeenth century. Among the historical events, one cannot play down the significance of the court disgrace. Dugald Stewart, to provide only one example, mentioned the case of George Hakewill's *An Apologie of the Powers and Providence of God in the Government of the World*, a book published in 1627. Though the language is heavily indebted to Bacon, the name of the Chancellor barely recurs in the whole book as a result of the political misfortunes that afflicted the last years of his life.[35] Another crucial event that helped to alter the perception of Bacon's natural philosophy among his contemporaries was the establishment of the Royal Society and its growing success. It is well known that, to counter the metaphysical and epistemological threats coming from Hobbes' philosophy, Boyle and other members of the Society who were particularly committed to convincing the public of the new philosophy of nature went to great lengths to emphasize the experimental aspects of Bacon's natural philosophy.[36] A third historical development was the gradual establishment of a distinctively British tradition in philosophy, characterized by a peculiar impatience with the dogmatism of reason. Despite the fact that—at least in regards to the philosophy of mind and matter—Boyle and Locke were more Cartesian than Baconian, it became increasingly common to link these thinkers with Bacon, perhaps

[35] Stewart, *Dissertation First Exhibiting the Progress of Metaphysical, Ethical and Political Philosophy since the Revival of Letters* (1817 and 1822), cited by Penrose, *The Reputation and Influence of Francis Bacon in the Seventeenth Century*, p. 14. On the cultural meanings associated with "Baconianism" since the seventeenth century, see Rossi, "Formiche, ragni, epistemologia"; Pérez-Ramos, *Francis Bacon's Idea of Science*, pp. 7–31.

[36] Malherbe, "L'induction baconienne," p. 194: "Le baconisme des expérimentalistes est probablement, en partie, une arme dirigée contre le concept hobbien de la science." On the use of experimental philosophy in an anti-Hobbesian fashion, see the classic study by Shapin and Shaffer, *Leviathan and the Air-Pump*.

in order to produce the appearance of a British philosophical pedigree that seemed nobler than the one based on influences from continental rationalism (a facile historiographic device that seems to work perfectly well even today).

The ideological constraints that prevented and still prevent a more straightforward reading of Bacon's works are intertwined with the historical events mentioned above and are of epistemological, nationalistic, and historiographic nature. What happened in all these cases was the expunction of the metaphysics of material desires from the various accounts of Bacon's philosophy of nature and the gradual transformation of a form of fully-fledged materialism into a new genre, what we might call "epistemological parenetics," made up of methodological slogans and rhetorical exhortations. Due to this combination of ideology and propaganda, and depending on the chosen viewpoint, Bacon became a subject of both enraptured eulogies and scathing denigrations. At the turn of every century, new ideological layers were added, and at the end of the nineteenth century, views on Bacon were often just interpretations of interpretations of famous authoritative interpreters.

Although an increasingly standardized view of Bacon established itself during the centuries, the early stages in the reception of his work (in the first half of the seventeenth century) were still characterized by a certain degree of hermeneutical fluidity. Given the fact that these appropriations took place in times of intense social, religious, and political unrest, in a way it was almost inevitable that the reception became "politicized" and biased towards various agendas. One possible answer to the question of why Bacon's metaphysics of material desires was loathed, rejected, avoided, ignored, or forgotten may simply lie in Bacon's very attempt to advance an overtly materialistic system that programmatically did not rely on any form of clearly theorized dualism or on a theological solution. Other philosophers at the time, who, in a similar manner, were endorsing the idea of a material universe ruled by physical forces, relied on the dualism of mind and matter or assigned to God the task of imposing laws on the material universe. On the contrary, intentional planning with respect to the minds of both God and man had no explanatory value in Bacon's cosmos. This point represented a radical departure from previous positions and made Bacon's metaphysics original and innovative. Unfortunately, insisting on the indifferent and arbitrary desires of matter, as Bacon did, was not a good way of advertising natural philosophy at the beginning of the seventeenth century. Closely related to this point is the recurrent, more or less veiled accusation that Bacon was in fact a disguised atheist.

The years after Bacon's death saw the posthumous publication of a number of treatises that helped keep Bacon's fame alive. Here we can only mention a few examples. William Rawley, Bacon's chaplain and secretary, the author of a very influential biography of the Lord Chancellor, published two collections of Baconian writings in his possession, the *Opuscula varia posthuma* in 1638 and the *Resuscitatio* in 1657. Gilbert Watts, fellow of Lincoln College, Oxford, published a translation of Bacon's *De augmentis scientiarum* in 1640. In 1653, Isaac Gruter, rector of the Erasmus Latin School at Rotterdam and avid collector of books and manuscripts, came into possession of the papers of William Boswell, ambassador to the Netherlands from 1632 to 1650 and Bacon's literary executor. From the writings in his possession, he published *De principiis atque originibus* (written by Bacon around 1612) and *Cogitationes de rerum natura* (written before 1605).[37] As argued by Robert Kargon, the 1640s and 1650s were characterized by a growing interest in atomism and the decision to publish these treatises written by Bacon during the first decade of the seventeenth century "contributed to the acceptance of atomism in England, despite's Bacon's avowed rejection of atomistic solutions in his later theory of matter, in the *Novum organum* (1620), the *Historia vitae et mortis* (1623) and the posthumous *Sylva Sylvarum* (1627)."[38] Judging from all this editorial and publishing fervor, one might think that Bacon's work was undergoing a process of intense exegesis. In fact, the picture is more complicated.

Despite open declarations of intents, engagement with Bacon's natural philosophy followed a more circuitous path. One has only to read Bacon's *Sylva Sylvarum* to be immediately confronted with the paradox that the purported founding father of modern science was still perfectly at ease with sympathies and antipathies, sentient spirits, appetites, and desires in matter. Alexander Ross, an unapologetically staunch defender of scholastic Aristotelianism, spoke ironically about Bacon's *Sylva*—"my Lords new Philosophy"—as a real forest where "a young Scholar may quickly lose himselfe, and shall encounter with many bryers and brambles."[39] Even the eulogizing Thomas Sprat admitted that "[Bacon's] Rules were admirable: yet his *History* not so faithful."[40] Seth Ward, Savilian professor of astronomy at Oxford and bishop of Exeter, found that Bacon's insistence on induction could in fact be detrimental to the very progress of knowledge. "It was a misfortune to the world," he wrote, "that my Lord Bacon was not skilled in mathematics, which made him jealous of their assistance in naturall enquiries; when the operations of nature shall be followed up to their staticall (and mechanicall) causes, the use of induction will cease, and sylogisme succeed in the place of it."[41] Ward presented the experimental work in which he and his fellow students of nature were involved at the time as a project "to salve mechanically, and statically the phenomena of nature."[42] To save the phenomena of nature mechanically, though, was exactly the meaning of "mechanical"—the genuinely mathematical and mechanistic meaning—which interested Bacon least, as we have seen.

One of the most successful attempts to appropriate Baconian ideas was made by Samuel Hartlib and his associates in the 1630s. Hartlib, "the Great Intelligencer of Europe," found in Bacon's work a powerful source of inspiration for his own projects of fostering empirical knowledge, intellectual regeneration, and religious enlightenment. Charles Webster has shown how Bacon became an influential source of inspiration in Cambridge at the end of the 1620s for people like Joseph Mede,

[37] Rees, "Introduction" to Bacon, *The Oxford Francis Bacon*, vol. VI, pp. lxx–lxxxv.

[38] Kargon, *Atomism in England*, p. 53.

[39] Ross, *Arcana Microcosmi*, p. 263.

[40] Sprat, *History of the Royal Society*, p. 36.

[41] [Ward and Wilkins], *Vindiciae academiarum*, p. 25.

[42] Ibid., p. 36.

John Worthington, John Milton, and Henry More.[43] In the hands of Hartlib, Bacon's *New Atlantis* was transformed into a model of religious utopia. The core of Hartlib's, Dury's, and Comenius' reform program was religious, expressing a willingness to extend forms of industrious piety, worldly asceticism and ecclesiastical pacification through learning and social progress. In other words, Bacon's philosophy contained instructions for laying the foundations of a spiritual kingdom on Earth, a task that Bacon had vividly described in the "fable" of Bensalem in *New Atlantis*. In Hartlib's interpretation of Bacon's natural philosophy, the dimension of practical usefulness was a key factor. However, if it is true that, generally speaking, the theoretical aspects of the Baconian worldview were relevant insofar as they could foster such practical activities as husbandry, mining, metallurgy, chemistry, and navigation, this did not prevent the members of the Hartlib circle from transforming the Baconian great renewal of learning into a pansophical enterprise. In *Considerations Tending to the Happy Accomplishment of Englands Reformation*, Hartlib insisted on the importance of having a comprehensive and far-reaching reading of Bacon's theories. "[I]n Matters of Humane Sciences," he wrote, "the End of his Negotiation should be, 1. To put in Practice the Lord *Verulams* Designations, *De Augmentis Scientiarum*, amongst the Learned. 2. To help to perfit Mr *Comenius* Undertakings, chiefly in the Method of Teaching, Languages, Sciences, and of Ordering Schooles for all Ages and Qualities of Scholars."[44] Hartlib's most characteristic Baconian trait lay in his use of induction as an argumentative mechanism for expanding knowledge.

Once the utopian and pansophical components that could be easily attributed to or projected onto Bacon's work began to be integrated into programs of educational reform, the radical implications of the "great instauration" were inevitably brought to the fore. In John Webster's notorious *Academiarum examen* (1653), Bacon was presented as one of the chief authorities in a general plea for a universal reform of learning. In Webster's opinion, a practical knowledge of nature would lead to the disclosing of "all mechanick operations," with a wide range of applications made possible in the arts of navigation and war, engineering and commerce. The discovery of the "mechanick" operations of nature was to have momentous consequences for merchants and astronomers, mariners and mechanics. By "mechanick operations," Webster meant direct and manual involvement with the operations of nature ("laborious tryals, manual operations, assiduous observations"). Here Webster is very close to one of Bacon's uses of the term "mechanic," the one related to physical and manual labor. The reason why, for instance, the Aristotelians preferred the use of syllogisms to that of inductive reasonings, Webster argued, was that inductions were "too mechanical and painful" to carry on, for they required a greater mental effort and larger amounts of physical energy. In keeping with the original meaning of Bacon's method, Webster interpreted Baconian induction as a way of "consorting or sympathizing with nature it self."[45] Not far from the research model outlined by

[43] Webster, *Samuel Hartlib and the Advancement of Learning*, p. 6.

[44] Hartlib, *Considerations Tending to the Happy Accomplishment of Englands Reformation*, p. 132.

[45] Webster, *Academiarum examen*, pp. 19f., pp. 66–69.

Bacon in *Sylva Sylvarum*, Webster advocated a set of practices centered around natural magic, such as chemistry, a reformed kind of medicine, physiognomy (in all its varieties: anthropological, celestial, and subcelestial), oneiromancy, magnetism, and atomism. The natural magic he defended was "that noble and laudable science"—and here Webster felt legitimated to introduce Bacon's own words from *De augmentis scientiarum*—"*Quae cognitionem formarum abditarum ad opera miranda deducat, atque activa passivis conjungendo, magnalia naturae manifestet*; which lead the cognition of occult forms into wonderful works, and by conjoining actives to passives doth manifest the grand secrets of nature."[46]

In Webster's hands, Bacon's program for dissecting and anatomizing nature had turned into a Fluddian project of "mystical" anatomy, capable of laying bare "the true Schematism or signature of that invisible *archeus* or *spiritus mechanicus*."[47] Ward's response, especially to this last point, was quick and sharp: "How little trust there is in villanous man! he that even now was for the way of strict and accurate induction, is fallen into the mysticall way of the cabala, and numbers formall: there are not two waies in the whole world more opposite, then those of the L. Verulam and D. Fludd, the one founded upon experiment, the other upon mysticall ideal reasons."[48] Ward's reaction is understandable. In Webster's appropriation of Baconian themes, Bacon's naturalism had come full circle, from the materialism of the basic appetites of nature to the Platonizing spiritualism of Fludd's Rosicrucianism. And yet, if one bears in mind certain passages from the *Novum organum* (especially book 2) and *Sylva Sylvarum*, then Webster's statement sounds less capricious and fanciful. It would be not so preposterous to argue that Webster was closer to the original inspiration of Bacon's program to reform learning by promoting an up-to-date form of magic than—to give just one example—Boyle's experimental philosophy based on a mechanical understanding of nature. In all likelihood, what Bacon would have vehemently rejected in Webster's appropriation of his own ideas was the spiritualistic tones, but not the thesis of an irritable "archeus" (read appetite) embedded in matter.

Of all the speculative readings of Bacon's natural philosophy (pansophical, Rosicrucian, and panpsychistic), Francis Glisson's hylozoistic interpretation is probably the most metaphysically-oriented and systematic. Glisson acknowledged Bacon's materialism with philological accuracy and intellectual acumen. In 1672, in his barely noticed *De natura substantiae energetica*, Glisson presented Bacon as one of the most important, if not the most important source to support a radical view of living matter. Already in *Anatomia hepatis* (published in 1654), Glisson had applied the Baconian notion of predatory vital spirits to a physiological account of bodily fluids to explain the way in which the blood assimilated the nutritive juice.[49]

---

[46] Ibid., p. 69.

[47] Ibid., p. 74.

[48] [Ward and Wilkins], *Vindiciae academiarum*, p. 46.

[49] Glisson, *Anatomia hepatis*, pp. 375–379. On Glisson's anatomical use of Bacon's theory of vital spirits, see Giglioni, *The Genesis of Francis Glisson's Philosophy of Life*, pp. 86–88.

It is in the philosophical treatise on the energetic nature of substance, however, that Bacon's theory of living and appetitive matter (originally combined with Francisco Suarez's doctrine of distinctions) became a metaphysical turning point. In Glisson's philosophical scheme, matter is characterized as a pliable substratum endowed with two general perpetual motions, Bacon's *motus antitypiae* and *motus nexus*, that is, matter's primal motions of resistance and connection.[50] For all its philological correctness, though, Glisson's use of Bacon's metaphysics of living matter apparently left no trace behind. Indeed, Glisson's case is historically relevant in a sort of photo-negative effect, being probably one of the very few instances in which someone decided to embark in an actual Baconian research project, that is, an experimental investigation into the basic desires of matter. (In passing, it is also worth pointing out that such an investigation was not completely unsuccessful if we think that it led Glisson to elaborate his theory of irritability).[51]

Glisson's plan for a systematic research concerning the fundamental motions that pervade inanimate beings is Baconian in spirit and style. In keeping with the ideas outlined in *Novum organum*, he looks at the primordial motions of matter as the cause of natural forms and divides the "variety and entangled complexity" of such motions according to five principal classes of forms. The first class includes natural motions and forms leading to the schematisms of matter (thickness and thinness, density and subtlety, etc.). The second class contains motions and forms that provide the particles of matter with their respective degrees of tension (continuity and divisibility, tenacity and friability, etc.). The third class includes motions and forms that cause rest and restlessness (*inquietudo*) among the particles of matter and give a body a determinate measure of solidity (*consistentia*) (delimitation and fluidity, suppleness and hardness, etc.). The fourth class comprises motions and forms that determine the position (*situs*) of the bodies (similarity and dissimilarity, sympathy and antipathy, etc.). Finally, the fifth class includes motions and forms that have an immediate bearing on the very life of nature (motion and rest, heat and cold, etc.). Each class refers to a specific characteristic of matter: the first to the actual extension of matter, the second to its divisibility, the third to the agitation of the particles of matter, the fourth to the position of the body with respect to the benefit or harm that such a position can procure to the body, the fifth to the activity of the particles.[52] If we compare this classification with the division of simple natures outlined by Bacon in the *Abecedarium novum naturae*,[53] we can understand the level of commitment displayed by Glisson in his attempt to identify the original passions of matter according to Bacon's directions.

[50] Glisson, *De natura substantiae energetica*, pp. 86, 246, 340, 353, 354–357, 375f., 408f., 418f., 425.

[51] On Glisson's sfortuna, see Giglioni, "What Ever Happened to Francis Glisson? Albrecht Haller and the Fate of Eighteenth-Century Irritability."

[52] Ibid., pp. 359–363.

[53] Bacon, *Abecedarium novum naturae*, in *The Oxford Francis Bacon*, vol. XIII, pp. 172–224.

There is little doubt that aphorism 48 of the second book of *Novum organum* is at the heart of Glisson's account of matter. That section of Bacon's work deals with the "instances of wrestling" (*instantiae luctae*) and encapsulates the most speculative part of Bacon's metaphysics of matter. The state of neglect in which this important part of the work lies speaks volumes about the early as well as the later reception of Bacon's oeuvre. The *Novum Organum Epitomiz'd*, an anonymous abridged translation published in 1676, condensed the 6,000 words of the section into one sentence: "The Four and Twentieth sort are instances of Predominancy." The translator is aware of the difficulties in abridging Bacon's text and he apologizes for that: "I am forced to cut short, and abbreviate many excellent directions, and to pass over several weighty observations because I am limited. However this abbreviation may give the Reader a taste of the whole."[54] In the original version, aphorism 48 is a thorough account of 19 original motions of matter, presented under the heading of the "instances of wrestling," which Bacon also calls "instances of predominance" (*instantiae praedominantiae*). They indicate

> the ascendancy of virtues over each other or their submission to each other, and which of them is the stronger and gets the upper hand and which the weaker and goes under. For the motions and exertions of bodies are no less composed, decomposed and intermixed than the bodies themselves.[55]

This is the semantic area in Bacon's natural philosophy that one should explore to make sense of Baconian atomism. Bacon espouses a view of matter in which its irrepressible primary drives are the real atomic entities that rule nature. Bodies and motions are tangles of a limited number of primitive tendencies. Each body is a momentary concretion of conflicting impulses. Bacon's nature is not a pacified state of affairs, always poised to actualize the best possible arrangement. Appetites are blind, arbitrary and moved by domineering tendencies. There is no mechanical necessity, nor teleological fate in Bacon's natural world. Bodies result from the energy produced by conflicts without resolution that constantly occur in matter. From this point of view, aphorism 48 can be seen as the actual manifesto of Bacon's vital materialism.

To be sure, a view of matter in which the field of transformations is dominated by tendencies of "fight," "predominance," and "submission" could hardly be less reassuring. Indeed, it is even more frightening than Hobbes' state of nature, where at least the will to survive and the tendency to self-preservation point to rational patterns of behavior. It does not come as much of a surprise, therefore, that almost no one engaged intellectually with Bacon's theory of material appetites. The truth is that Machiavelli was already sufficiently discredited in political science to be adopted as a model of investigations in the field of natural philosophy. Even Glisson, who embraced Bacon's metaphysical model of matter, did not go so far as to accept the Chancellor's thesis that appetite prevails over knowledge in the realm of nature.

---

[54] *The Novum Organum Epitomiz'd*, p. 32. On the importance of aphorism 48, see Rees, "Introduction" to Bacon, *Novum organum*, in *The Oxford Francis Bacon*, vol. XI, pp. lxxxvii–lxxxviii.

[55] *Novum organum*, in *The Oxford Francis Bacon*, vol. XI, p. 383.

In his work on the anatomy of the stomach and the intestines, published just before his death in 1677, Glisson acknowledged again his debt to Bacon's materialism; in particular, he showed how the Baconian notion of responsive matter could be used as a speculative foundation for a theory of irritability. However, he rejected unswervingly Bacon's concept of blind appetite as logically untenable.[56]

In the 1670s, Glisson's reappraisal of vital Baconianism appears to have been quite an isolated episode. At the time, the tensions underlying the appropriation of the meaning of "Baconian" seem to have reached some sort of stable equilibrium. Thomas Tenison, the Archbishop of Canterbury, published *Baconiana, or Certain Genuine Remains of Sir Francis Bacon* in 1679. A quotation from *Baconiana* outlines the principal divisions in philosophical lineage according to a scheme that was to last for a long time. "It is true," said Tanison,

> there lived in part of the last, and this, century, many memorable advancers of philosophical knowledg. I mean not here such as Patricius, or Telesius, Brunus, Severinus the Dane, or Campanella. They only spun new cobwebs, where they had brush'd down the old. Nay, I intend not, in this place, either de Chart, or Gassendi. They were certainly great men, but they appeared somewhat later, and descended into the depths of philosophy, after the ice had been broken by others. And those I take to have been chiefly Copernicus, Father Paul the Venetian, Galileo, Harvey, Gilbert, and the Philosopher before-remembered, Sir Francis Bacon, who if all his Circumstances be duly weighed, may seem to excel them all.[57]

Around 1680, the divisions are clearly delineated: on the one hand, we have the obsolete weavers of philosophical cobwebs; on the other, the heralds of a new understanding of nature. In this picture, Bacon does not seem to have anything to do any longer with his fellow Renaissance philosophers of nature.

Tenison's admiration for Bacon may have been transmitted to him by his kinsman Thomas Browne. It is safe to characterize Browne as another example of "Baconian" investigator of nature, but his case, too, is particularly complex and elusive. The nature of Browne's loyalty to Bacon's ideas in natural philosophy is difficult to assess. Alexander Ross put Bacon in the same company as William Harvey, Thomas Browne, and Jan Amos Comenius as instances of an incorrect attitude towards the study of nature, and one must admit that the way he arrived at this list is not completely arbitrary. The fact is that, apart from Comenius (who enthusiastically drew on Bacon's philosophy), Harvey and Browne, despite their sharing many ideas with the Lord Chancellor, were not so keen on acknowledging their borrowings. However, unlike Harvey, who seems to have disliked Bacon quite openly, Browne's attitude is more nuanced. Thomas Browne's natural philosophy stands quite apart in the panorama of seventeenth-century views on nature. He leans towards pantheism as much as Bacon towards atheism. What for Bacon is superstition (the reality of witchcraft, the intermediate nature of spirits and demons, etc.) is for Browne foundational evidence and material for scientific investigations. If it were not for his idiosyncrasies, Browne

---

[56] Glisson, *Tractatus de ventriculo*, pp. 366f.

[57] Tenison, in *Baconiana*, pp. 5–6. The passage is a clear echo of the beginning of Bacon's *Historia naturalis et experimentalis ad condendam philosophiam*, published in 1622, in *Works*, vol. II, p. 13.

could be taken as the missing link between Bacon and Cambridge Platonism on the one hand, and Bacon and Boyle on the other. The contradiction is only apparent because Browne's understanding of nature is supple enough to allow both mechanical and vital readings of natural phenomena. It is above all in matters of theology that Browne differs from Bacon the most: the Platonic chain of being, the argument from design, the parallelism between ideas in the mind of God and ideas active in nature, angels and demons—these are all tenets that are of momentous importance for Browne and rejected by Bacon.

It should not come as too much of a surprise, therefore, to find out that Cudworth was particularly uneasy with Bacon's philosophy, in which he suspected atheistic "insinuations". In particular, he did not hesitate to denounce Bacon's characterization of final causes as a form of *idola*—"to use that affected language"—as "the very spirit of Atheism and infidelity." Cudworth welcomed Bacon's strictures about the "delusion" of "reducing natural actions to the mode of human"; however, he warned against Bacon's tendency to deny the existence of an "intellectual platform" behind nature.[58] In Cudworth's opinion, Bacon's critique of anthropomorphism was good in that it reaffirmed the autonomy of nature; to make appetite the root of nature's life, though, was tantamount to abdicating the very rationality of the created world in favor of a view of power as opaque and secret agency. Cudworth's nature could be seen as mechanical insofar as its plastic power was viewed as an instrument in the hands of a superior source of knowledge; but it was vital insofar as its potential was directed by levels of more or less transparent intentionality. Bacon was not openly mentioned in the *True Intellectual System*, not just out of reverence, but because the literary device of not naming names was part of the very game characterizing the whole work, namely, to address contemporary authors through ancient doubles. In a letter to Boyle, however, Cudworth broke his usual restraint and told him: "You have much outdone Sir Francis Bacon in your Natural Experiments; and you have not insinuated any thing, as he is thought to have done, tending to Irreligion; but the contrary."[59]

Boyle appreciated Bacon for his use of quantitative methods,[60] a number of methodological suggestions[61] and the role attributed to experiments.[62] In *Certain Physiological Essays*, he explained that he started collecting a whole series of "particulars" and experimental materials in order to write "a continuation of the lord *Verulam's Sylva Sylvarum*, or Natural History." He even added that his "intended centuries" were supposed to resemble the ones written by Bacon and "to be annex'd"

[58] Cudworth, *The True Intellectual System of the Universe*, II, p. 608. See also ibid., III, pp. 101f.

[59] Cudworth to Boyle, 16 October 1684, in Boyle, *Correspondence*, vol. VI, p. 48.

[60] Boyle, *New Experiments Physico-Mechanical touching the Spring of Air*, in *Works*, vol. I, p. 261.

[61] Boyle, *Certain Physiological Essays*, in *Works*, vol. II, p. 14.

[62] Boyle, *A Defence of the Doctrine touching the Spring and Weight of the Air*, in *Works*, vol. III, pp. 13, 50. On Boyle's Baconianism, see Hunter, "Robert Boyle and the Early Royal Society"; Knight and Hunter, "Robert Boyle's *Memoirs for the Natural History of Human Blood* (1684)"; Anstey and Hunter, "Robert Boyle's *Designe about Natural History*."

to them. Boyle adopted Bacon's language and typical turns of phrase. For instance, the reason why natural philosophers including Aristotle and Tommaso Campanella failed to accomplish their philosophical endeavors was that "they have too hastily, and either upon a few Observations, or at least without a competent number of experiments, presum'd to establish principles, and deliver Axioms." "[S]peculative devisers of new Hypotheses," such as Bernardino Telesio, have been "more ingenious than fruitful."[63] If someone hopes to find an extensive treatment of Bacon's natural philosophy in Boyle's works, though, he will be utterly disappointed. An attitude of respect and reverence is evident in his treatment of Bacon. If we looked at Telesio's and Bacon's philosophical views on matter and spirits as historians of philosophy, we would easily realize that the difference between them is not considerable, and yet, in Boyle's eyes, the former is a fanciful naturalist while the latter is a serious researcher. A very eloquent hermeneutical case in point is the way Boyle introduces his investigation on the nature of heat and cold. While the characteristically Baconian phraseology is used as an ornamental device, the Telesian root is completely ignored (despite the fact that Telesio is the one who re-thought in a very original manner the parallelism of heat and cold as the two sentient forces of nature and, above all, insisted on looking at cold as a positive and substantial power):

> And certainly Cold, and Heat, especially when imploy'd by turns, are the two grand Instruments by which Nature performs so many of her Operations here below, that our great Verulam did not speak so inconsiderately, when he called Heat the Right hand of Nature, and Cold her Left.[64]

(A further indication that this is just a decorative emblem is the fact that nature is personified, a view that is completely foreign to Boyle.)[65]

But it is not merely a matter of style and belletrism; indeed Bacon rarely appears in Boyle's writings. It would not be too much of an exaggeration to say that, while Boyle lavishes effusive praise on Bacon's figure, his natural philosophy lacks a real engagement with the most characteristically Baconian tenets on matter. Indeed, on the specific question of heat and cold, if we compare the way Boyle dealt with Bacon's ideas to the almost contemporary use of them made by Comenius, we find ourselves once again in the historiographically unsettling situation that Comenius is, in fact, a more faithful interpreter of Bacon's natural philosophy than Boyle.

In *Disquisitiones de caloris et frigoris natura* (1659), Comenius describes heat and cold as forms of motion. To clarify this definition, he sets out a number of principles concerning matter that are reminiscent of the way Bacon had already described the vital nature of matter in *Novum organum*, part 2, aphorism 48: "The matter of the universe needs a place to expand itself," "Matter loves so much to maintain its condition of self-continuity that it cannot be broken up," "Matter loves spaces that

---

[63] Boyle, *Certain Physiological Essays*, in *Works*, vol. II, pp. 13, 24f.

[64] Boyle, *New Experiments and Observations touching Cold*, in *Works*, vol. IV, p. 208.

[65] On this point, see Davis and Hunter, "Introduction" to Boyle, *A Free Enquiry into the Vulgarly Received Notion of Nature*, in *Works*, vol. X, and Hunter and Davis, "The Making of Robert Boyle's *Free Enquiry*."

may correspond to its quantity," "There is no penetration of bodies," "Matter is like a sponge."[66] To stress even more the close relationship between matter and motion, Comenius distinguishes different types of motion according to spatial dimensions. He writes that motion can be motion in length, in length and width, and in length, width and depth. Motion can also be within itself and outside itself. These various directions provide the basic coordinates for all possible types of vital motions: stretching in (*intensio*), stretching out (*protensio*), contraction (*contractio*), extension and expansion (*extensio seu expansio*), folding and wrinkling (*complicatio seu corrugatio*), distension and amplification (*distensio seu dispansio*), concentration (*contrusio*), compression (*compressio*), coagulation (*coagulatio*), and thickening (*spissatio*). Here Comenius is clearly following the Baconian "fold of matter," that is, nature's inherent tendency to expand and contract. If motion as active energy lies at the core of the comprehensive definition of heat and cold, a more specific description characterizes the twin powers in terms of "violent torsions" in the parts of matter. Comenius compares heat and cold to natural weavers (*textores*), which, by means of their power, produce the natural threads that make the universal fabric of bodies. The whole world of nature is a work of ceaseless weaving. In the theatre of nature, the succession of heat and cold, of day and night, of summer and winter is like the alternation of warp and woof in the great loom of the creation.

A typically Baconian point addressed by Comenius is the asymmetry between heat and cold in nature. While it is rather easy to point to various sources of heat throughout nature, it is not as easy to identify the material origin of cold. Distancing himself from Telesio (who had characterized the earth as an active cause of cold) and from the astrologers in general (who attribute the origin of cold to cold stars and planets such as Saturn), Comenius maintains that cold and heat come from the same source, "by virtue of the balance resulting from the tension of the opposites."[67] He explains the disproportion between heat and cold already noticed by Bacon in terms of natural propensities embedded in matter: every thing in the material universe has a natural tendency to cohere and persist in its own situation because of its innate impulses to self-love. To maintain the general balance of the universe, God endowed matter with a natural tendency towards condensation and provided an external powerful source of heat—the sun—to counteract the otherwise unstoppable drive to self-implosion ingrained in matter. This means that, of the two natural tendencies of heat and cold, cold is in fact the deeper and more "natural" one, for it coincides with the tendency towards self-aggregation and self-preservation. Cold is original, heat accidental.

Comenius' use of Baconian concepts is a further reminder of the ideological and epistemological biases that mar the view that a supposedly mechanical and corpuscularian understanding of heat can be found in Bacon. Taking Bacon as his mentor—and definitely taking him at face value—Comenius says that heat and cold are the result of an active motion whereby bodies are variously altered; more specifically,

---

[66] Comenius, *Disquisitiones de caloris et frigoris natura*, in *Opera omnia*, vol. XII, p. 278.

[67] Ibid., p. 272.

they manifest a motion that weaves bodies providing them with a specific "texture": "the specific effect of cold is to interweave (*contexere*), that of heat is to unweave (*retexere*); the effect of both is to weave (*texere*)."[68] It is worth noticing that for Comenius (like for Bacon) the word *textura* has no mechanical resonances, but rather indicates a condition of matter that is the result of a primordial "weaving" impulse in matter. The best evidence that heat and cold act as natural "weavers" is the formation of ice arabesques on windowpanes in the cold days of winter. The colder air outside and the warmer air inside struggle against each other, each twisting in different directions (the *motus torsivus* fighting against the *motus distorsivus*), drawing whirlpools of ice crystals by interweaving the warp (*stamen*) with the woof (*trama*) of matter.[69] As in Bacon, the model of action and reaction rests on a delicate antiperistatic balance of appetites and tendencies.

In *Novum organum*, Bacon had posited the "appetite" of trepidation as one of the innermost motions in matter. Comenius follows suit and he strengthens the role of trepidation in the universe by characterizing the basic powers of heat and cold as "torsions" of matter. A whole range of titillations and irritations displayed by a large number of natural phenomena mark the underlying, persistent *torsio* of matter. As in Bacon's view of nature, action and reaction dominate Comenius' cosmos. Both happen in the form of a constant state of natural, insensible "wriggling." What at first sight looks like action (heat trying to invade the domain ruled by cold) and response to action (cold protecting itself against the onslaught of heat) is, in fact, the response (heat as a vital spur introduced by God to maintain the continuity of the life of the universe) to the original action (cold as the tendency to self-love and rest). It is on this theological note that Comenius parts company with Bacon. While heat and cold are the weights established by God to keep the equipoise of the *machina mundi*, in Bacon's natural philosophy the tendency towards rest and death are inherent in the very essence of the material universe, and—what is more alarming—they are the result of blind and arbitrary appetites.[70]

Of the two contemporary readings of Bacon's theory of heat and cold, the one by Boyle and the other by Comenius, the former was certainly the more successful one. The association of Bacon's name with the rise of experimental science happened very early. Historians have long pointed to the ideological uses of Bacon's philosophy of nature. Through the good offices of such apologists as Thomas Sprat, Joseph Glanvill, and Abraham Cowley, both the Royal Society and the College of Physicians could claim for themselves the title of legitimate heirs of Bacon's research program.[71] Cowley saluted Bacon as the new Moses who had

---

[68] Ibid., p. 273.

[69] Ibid., p. 282. For a comparative analysis of Bacon's and Comenius' theories of matter, see Giglioni, "The Darkness of Matter and the Light of Nature."

[70] On this point, see Giglioni, "The Hidden Life of Matter."

[71] Webster, "The College of Physicians," p. 410; Webster, *The Great Instauration*, pp. 96, 315; Hunter, *Science and Society in Restoration England*, p. 29.

dispelled the "Scar-crow" of authority from the enterprise of learning.[72] Sprat claimed that Bacon had promoted experimental philosophy with such admirable rules and cogent arguments that some of his writings should have been used as a preface to the *History of the Royal Society*.[73] Baconian catchphrases echo throughout the apologetic work: the experimental philosophy "opens our eyes to perceive all the realities of things"; the old philosophies "fix'd, and determin'd their judgements, on general conclusions too soon"; "[t]he true philosophy must be first of all begun, on a scrupulous, and severe examination of particulars … from experimenting, to demonstrating, and from demonstrating, to experimenting again"; the "bare knowledge of things" is opposed to "the artifice of words."[74] The language is definitely Baconian, but the referent has changed. The level of rhetorical and methodological theorizations (*interpretatio naturae*) is kept separate from the level of natural philosophy (*materia plica*).

Hugh Trevor-Roper famously spoke of "vulgar Baconianism." He wrote that the 1640s in particular should be described as "the heyday of vulgar Baconianism." He believed that the name of Bacon, after having been "devalued" by "millenarian visionaries" and by "utilitarian gadgeteers," was restored by the group of the Oxford natural philosophers. This was also the interpretation underpinning Margery Purver's study of the Royal Society.[75] Against Purver's account, Michael Hunter has argued that "[I]t is undoubtedly simple-minded to postulate a straightforward Baconian orthodoxy in late seventeenth-century England, or to see this as the only key to intellectual advance."[76] Given the presence of so many forms of creative uptake of Baconian ideas in the years between the 1630s and the 1680s, we cannot dismiss any of the different strands of Baconianism as irrelevant or "vulgar." Experimental, metaphysical, social, and utopian Baconianisms were not just various historically legitimate forms of Baconian exegesis. They also underwent a series of interesting crossovers.

In the final analysis, the Baconian notion of matter, with its emphasis on both natural appetites and "mechanick" operations, and the underlying assumption that "labor" characterizes both the striving of the appetites and the work of the mechanical arts, explain why at a certain point it became possible and plausible for the upholders of mechanical corpuscularianism to appropriate Bacon's naturalism. Indeed, the shift from "mechanick" to "mechanical" philosophy occurred almost insensibly in the early phases of the reception of Bacon's philosophy. It transformed a metaphysics of nature that Bacon had conceived of as a precondition for taking control of the natural appetites of matter (its "secret motions") into an investigation seen as a preliminary stage in a general program for mechanizing qualities and

[72] Cowley, "To the Royal Society," in Sprat, *The History of the Royal Society*, sig. B^{v}.

[73] Sprat, *History of the Royal Society*, pp. 35f.

[74] Ibid., pp. 26, 30f., 40.

[75] Trevor-Roper, "Introduction" to Purver, *The Royal Society*, p. xv.

[76] Hunter, *Science and Society in Restoration England*, p. 33.

reducing natural phenomena into patterns of corpuscularian motions "for the effecting of all things." The insistence on the "mechanick operations" (manual labor, practical applications, concrete results, the betterment of the human condition) helped to shift the emphasis from the brutal realism of the appetites of matter to a philosophy of natural appearances, from mechanical *art* to mechanical *philosophy*. In his response to Webster's *Examination of Academies*, Ward reproached radical "pamphleteers" for having abused Bacon's characteristic and powerful tirades against bookish learning in order to ruin "our academicall institution," namely, the claim "that instead of verball exercises, we should set upon experiments and observations, that we should lay aside our disputations, declamations, and publick lectures, and betake our selves to agriculture, mechanicks, chymistry, and the like." From this passage it is apparent that at the time the typical combination of Baconian motifs—the experimental approach to nature, a public and institutional basis for natural philosophy, and the emphasis on the productivity of the mechanical arts—could be used as a flexible tool of propaganda, to be applied in very different contexts. Ward, for instance, did not accept the educational implications of the Baconian program with their universal and radical outcomes, like Comenius and Webster had done, because universities were still the place for the education of gentlemen and noblemen—"a more generall and comprehensive institution," where they "may become rationall and gracefull speakers, and be of an acceptable behaviour in their countries." "Which of the nobility or gentry," Ward continued, "desire when they send their sonnes hither, that they should be set to chymistry, or agriculture, or mechanicks?"[77] Like Ward, many at the time became aware that the introduction of the Baconian encyclopaedia into the university curriculum would have overstressed the importance of mechanical arts and technology, with disruptive effects on the social establishment. A better solution was therefore to keep experimentations and investigations on nature and matter confined to restricted circles of inquirers and to maintain the direct involvement with matter, characteristic of mechanical disciplines, separated from the theoretical investigation of nature.

In Bacon's comprehensive philosophical program, control over the appetites of matter was crucial. It had several major consequences: epistemological (the universal leveling of the *ingenium*), ethical (self-control of one's own specific and defining clusters of appetites), political (manipulation of blind appetites by the few in control the many), and social (the rehabilitation of manual labor). Bacon was fully aware of these implications. The Baconians who historically proved to be successful in their attempt to transform Bacon into the "father" of experimental science, mechanical philosophy, and empiricism made the momentous choice to abandon the far-reaching scope of Bacon's philosophical project, with its distinctive penchant for realism and materialism, to focus on the methodological and epistemological aspects of his metaphysics of matter.

---

[77] [Ward and Wilkins], *Vindiciae academiarum*, pp. 49f.

2.3 Conclusion

Judging from this brief excursus into the early reception of Bacon among English natural philosophers, one can safely say that the great majority of his immediate followers preferred to dismiss, more or less tacitly, the Lord Chancellor's characterization of matter as a continuous and supple material substratum riddled with original appetites. One exception was Glisson, but his interpretation, while philologically correct, had barely any impact on seventeenth-century English natural philosophers. At the other end of the spectrum, Boyle's position, hermeneutically reticent as it was, proved to be enormously successful. As is often the case in history, such a discrepancy between philological correctness and hermeneutical creativity turned into a source of momentous cultural change.

Needless to say, a considerable part of this interpretative output was, in fact, propaganda. To explain the extraordinary success of seventeenth-century Baconianism as a mere result of propagandistic strategies, however, would be untenable. It is more correct to say that, throughout the century, Bacon's natural philosophy met the favorable response of various "universes of expectations." Most of all, his work provided a new language to voice concerns on natural knowledge at the time. To be sure, this was not a minor reason behind the popularity of Bacon's philosophy. Such a language could supply a conceptual armory to address matters of methodological and scientific correctness; it could provide a natural philosophical worldview suitable for attitudes of worldly asceticism and spiritual regeneration; finally, it could give natural philosophers a specific vocabulary to describe the life of material appetites. Hartlib and Dury used Bacon's philosophy to justify their efforts to achieve comprehensive educational and social reform. Their rhetorical treatment of Bacon's ideas largely focused on the establishment of international "fraternities" of practitioners and scholars, on the free exchange of information from all the corners of the world, and on a program of public funding for scientific enterprises. "Mechanical" meant for them "mechanic." Boyle and other practitioners interested in creating a new style of doing research on nature focused on the natural appearances of matter (phenomena) and dismissed both the desires of matter and the passions of men as belonging to the inscrutable province of knowledge. Whereas Hartlib's rhetorical adaptation of Baconian themes was eminently political and religious, Boyle tried to limit his endeavor strictly to the field of natural philosophy. Of course, as seen in Hartlib and Boyle, tracing the line between the domains of political and natural philosophy is a very delicate task, but there is no question that Boyle adopted a more diluted version of Baconian philosophy, both metaphysically and politically.

In examining the reputation of Bacon in the seventeenth century, Stephen Penrose has argued that such reputation was also built up by "hearsay, propaganda, or imagination divorced from a reading of Bacon himself." And he continues: "There was a genuine interest in his works, and it is probable that he was read more at this period than at a later time when he had become a much more imposing, though legendary figure under the ministrations of the French Encyclopedists."[78] I suspect that the actual reading of Bacon's works, especially his *Novum organum*, stopped very early. (By "reading," I mean of course a transformative experience in which the text works as a productive source of inspiration.) A glaring example is the fortunes (or misfortunes) of the *Novum organum*. Very few editions and no complete translation into English appeared in Britain throughout the seventeenth century. The anonymous translator of the already-mentioned epitomized version of the *Novum organum* characterized Bacon's original as an "obscure but useful book."[79] The concepts outlined by Bacon in the second half of the *Novum organum* to investigate the ultimate structures of matter found no concrete application and his map of the natural desires of matter remained a dead letter. Can Glisson really be the only exception? One can answer this question only after further research has been conducted. Graham Rees, in the introduction to his edition of the *Novum organum*, rightly pointed out that, despite the fact that Bacon's discussion of prerogative instances accounts for almost three-quarters of Book II, "the commentators to their discredit have found few occasions for exerting themselves in that and none for celebration."[80]

We can conclude, then, by trying to provide an answer, however tentative, to the question with which we opened this essay: Why did Bacon's materialism of natural appetite, which represents the core of his natural philosophy, pass almost unnoticed from the very beginning? The question is even more intriguing because the solution which has been offered many times—namely, that after all Bacon separated the discourse on metaphysics from the discourse on method—is not satisfactory. Even *Novum organum* cannot be taken as a mere work of methodology and epistemology, for it is filled with examples of what Bacon meant by metaphysics of matter and appetite. In other words, his methodological machine does not work without metaphysical fuel. Indeed, method and metaphysics cannot really be separated from one another because Bacon's notion of induction is highly speculative and the revelation of matter's appetites can only happen through the method he devised.[81]

As we said at the beginning, both historical events and ideological constraints contributed to the demise of Bacon's metaphysics of material appetites early in the seventeenth century. However, there also are reasons internal to the very metaphysics elaborated by Bacon that may explain why Bacon became the Baconian we all know. His philosophy revolved around a series of pivotal arguments that were of

[78] Penrose, *The Reputation and Influence of Francis Bacon in the Seventeenth Century*, pp. 21f.

[79] *Novum organum Epitomiz'd*, sig. A2.

[80] Rees, "Introduction" to Bacon, *Novum organum*, in *The Oxford Francis Bacon*, vol. XI, p. lxvii. Recent studies have emphasized the need to re-examine the reception of Bacon's natural philosophy during the seventeenth and eighteenth centuries. Here I am referring the reader to some of these studies: C. Pastorino, "Francis Bacon"; Anstey, "Locke, Bacon and Natural History"; Jalobeanu, "Bacon's Brotherhood and its Classical Sources"; Jalobeanu, "The Fascination of Solomon's House in Seventeenth Century England"; Hoquet, *Buffon: Histoire naturelle et philosophie*; Peterschmitt, "Bacon et la chimie"; Yeo, "Between Memory and Paperbooks."

[81] Giglioni, "Reading Nature without Making a Book of It: Francis Bacon's *Novum Organum*."

formidable audacity: that matter is inherently alive; that the nature of matter's life is constitutively appetitive; that the appetites of matter are many and conflicting; and finally, that matter's appetites are blind. These four assumptions—materialism, centrality of appetite, its plurality and arbitrary causation—were daring enough to incur severe censure. By and large, the fortunes of vital philosophies of nature were rapidly declining at the time, after the interlude of Elizabethan and Jacobean naturalism (one can think of Giordano Bruno, John Dee, William Gilbert, William Harvey, to mention only a few names). On the other hand, though, Bacon was too famous and respected to be openly criticized and rejected. It was much better, then, to keep exploiting his fame and to transform him into an icon of the new science and philosophy, defusing in the meantime the "darker" components of his philosophy.

This process was the result of both deliberate choices and unintended consequences. Historical changes of a cultural nature are characterized by discontinuities, sudden disappearances, and silent sinking. Continuities may be noticed by contemporary observers who manage to remove themselves from the hustle and bustle of the ideological fracas, or these continuities can be restored by later interpreters who cautiously and sagaciously avail themselves of the benefit of hindsight. It would not be historiographically too hazardous to say that Bacon's *appetitus* resurfaced as *conatus* in the philosophies of Hobbes and Spinoza. Indeed, it could even be the case that the most genuinely Baconian legacy is to be sought precisely in that direction. But this is another story, and it will be the subject of another essay.

Chapter 3
An Empire Divided: French Natural Philosophy (1670–1690)

Sophie Roux

3.1 Introduction

For some 30 years, historians of science and philosophy working on the seventeenth century have endeavored to blur the edges of an opposition that had until then been generally accepted, namely, between the "new philosophy" on the one hand and an "old philosophy" on the other. In a nutshell, this generally received opposition was based on the idea that there had been a Scientific Revolution in the seventeenth century that consisted not only of the appearance of new sciences and of the discovery of new results in old sciences, but also of a change in the very principles of natural philosophy. From an ontological viewpoint in particular, the "new philosophers" had the right idea of substituting the clear principles of matter and motion for the old obscure entities that were the various occult qualities, substantial forms and virtues of all sorts. And thanks to these new ontological principles, which earned the new philosophers who defended them the title of "mechanical philosophers," it became possible to begin truly to explain natural phenomena and the triumphal road to modern physics was opened.

There were several ways to question the opposition of old and new, as well as the idea that, because of the intrinsic weakness of their principles, ancient philosophers

Before my arrival in Grenoble, I presented this chapter in Paris, Oxford and Aix-en-Provence; I wish to thank the organizers of the corresponding seminars and conferences for giving me the opportunity to make my point more precisely, and the audiences for having obliged me to do so.
Unless otherwise indicated, translations are mine.

S. Roux (✉)
Université Grenoble II/Institut universitaire de France,
Grenoble, France
e-mail: Sophie.Roux@upmf-grenoble.fr

D. Garber and S. Roux (eds.), *The Mechanization of Natural Philosophy*, Boston Studies in the Philosophy of Science 300, DOI 10.1007/978-94-007-4345-8_3,

were doomed to disappear when modern philosophers arose.[1] These ways have more or less all been used in the historiography of these last 30 years. It has been noted that the clarity of the mechanical explanations was at times insufficient and at times illusory, in particular because they turned out to be tautological; more radically even, it was argued that clarity and obscurity are relative epistemic criteria, depending on accepted etiologies and ontologies.[2] Learned studies have shown that the great authors owed more to the Scholastics than they had been willing to admit, and for the lesser authors, that they had at times taken such complex positions that the division between the old and the new became as such practically impossible.[3] Other studies pointed out the autonomy of scientific practices: it is now clear that the step was ultimately rarely made from new scientific practices to the formulation of alternative principles of natural philosophy, even in this philosophical age.[4] In particular, it was entirely possible to participate in the new scientific practices while remaining either silent on the question of the principles or an Aristotelian of the strictest obedience, and conversely, to be an innovator in principles, while lacking initiative or coherence when dealing with science. Lastly, while mechanics and astronomy were placed at the heart of the Scientific Revolution, disciplines or ways of thinking once considered as peripheral, such as the hermeticism and naturalism of the Renaissance, alchemy or chemistry, and medicine, were put back on the table.[5] In short, because of all

---

[1] In the words of Brockliss, "Descartes, Gassendi, and the Reception of the Mechanical Philosophy," p. 465: "Twenty years ago, when historians accepted the propaganda of mechanist philosophers, this [the question "why in the course of the half century or so after 1660 some form of the mechanical philosophy replaced scholastic Aristotelianism as the predominant physical philosophy all over western and central Europe"] was an easy question to answer: seventeenth-century Aristotelians were obscurantists, philosophical dinosaurs condemned to extinction in the new world of the Scientific Revolution. Today, … it is a question that seems to become more and more difficult to solve." Brockliss wrote this in 1995.

[2] See for example, on the first point, Chalmers, "The Lack of Excellency of Boyle's Mechanical Philosophy"; Gabbey, "Mechanical Philosophies and their Explanations"; on the second point, Hutchison, "Dormitive Virtues, Scholastic Qualities and the New Philosophies."

[3] In this respect, Des Chene, *Physiologia*, and Leijenhorst, *The Mechanisation of Aristotelianism*, are exemplary studies. About the complexity of *minores*, among a flourishing literature, see for example Henry, "Occult Qualities and Experimental Philosophy"; id., "Robert Hooke, the Incongruous Mechanist"; Lüthy, "Thoughts and Circumstances of Sébastien Basson"; id., "An Aristotelian Watchdog as Avant-Garde Physicist. Julius Caesar Scaliger"; id., "David Gorlaeus' Atomism."

[4] On the autonomy of scientific practices, see for example Bennett, "The Mechanics' Philosophy and the Mechanical Philosophy." On the specificity of the seventeenth century with regard to the "vexing problem of the nature of natural philosophy," see Lüthy, "What to Do With Seventeenth-Century Natural Philosophy?" pp. 165–167, 174–177.

[5] See for example Yates, Giordano Bruno and the Hermetic Tradition; Rhigini Bonelli and Shea (eds.), *Reason, Experiment and Mysticism in the Scientific Revolution*; Westman and McGuire (eds.), *Hermeticism and the Scientific Revolution*; Newman, *Gehennical Fire: The Lives of George Starkey*; id., *Atoms and Alchemy*; Principe, *The Aspiring Adept: Robert Boyle and its Alchemical Quest*; Clericuzio, *Elements, Principles and Corpuscles*; Maclean, *Logic, Signs and Nature in the Renaissance*.

these works, but also, undoubtedly, because of a general historiographic trend to privilege local and singular case studies over sweeping panoramas, the categories of old and new philosophy are today under the constant suspicion of being both conceptually ill defined and without any real historical pertinence. It is particularly feared that they merely represent a bad partition of seventeenth-century natural philosophies, ill suited for drawing out their inexhaustible richness, and poorly distinguished one from the other.

It is not bad for us intellectual historians to lose a bit of our innocence—in less metaphorical terms, to notice that the categories of ancient and modern philosophers are not natural givens, and to ask to what we commit ourselves to when we use them. And yet, fleeing to local and singular case studies is no more satisfying than indulging in sweeping panoramas: panoramas may erase the differences, but case studies may be so particular that it is not even possible to see in which respect they differ from other particular case studies. More specifically, at least two reasons could allow us to restrain ourselves from wholly banning the opposition between these two categories, or, what comes down to almost the same thing, from using them with the intellectual tweezers of quotation marks. As we will see, the first reason is just a hackneyed version of the idea that there had been a Scientific Revolution in the seventeenth century; the second reason, however, will lead us to the heart of this paper.

The first reason may come from the spontaneous feeling that there is a family resemblance between Descartes, Hobbes or Gassendi, and that this resemblance is stronger than the resemblance between any two of these authors and, for example, Thomas Aquinas, Ockham, Bradwardine or Oresme. To illustrate this feeling by a comparison taken from a common experience, when you look at pictures of early twentieth-century French bourgeois, you do not differentiate them according to the kind of professional life they led; what strikes you is rather their common historical look: before being doctors or traders, the ancestors of our doctors and of our traders, you see them as men of their time, a time separated from ours by some profound changes. But does this comparison hold, and can we find in such a feeling some good reasons to maintain the opposition of the old and the new? I doubt it.

Taking previous studies as a starting point, in particular those of Edward Grant and Charles B. Schmitt, one can contrast an essentialist conception of the Aristotelian tradition, according to which Aristotelianism has a hard doctrinal core that cannot be touched, with a Wittgensteinian conception, according to which the fundamental plurality and diversity of Aristotelians should be accepted. According to the Wittgensteinian conception, Aristotelians form a family loosely linked by resemblances.[6] Such a move could indeed be taken in most areas in intellectual history. When we seek to constitute historically pertinent categories, the problem is never the establishment of the essential

[6] Thijssen, "Some Reflections on Continuity and Transformation of Aristotelianism in Medieval (and Renaissance) Natural Philosophy." This paper actually finds in Lovejoy and Lakatos the first glimmers of a third way, that I shall not discuss here.

identity of two works, or even of two propositions, a task that would be impossible. Rather, it is to look for resemblances between certain works, and then to propose a weighing of these similarities so as to give substance to the conviction that there is indeed something like a family resemblance between one work and another. *Prima facie*, Wittgenstein's metaphor of family resemblance has thus the advantage of allowing a looser use of categories like "Aristotelian," "mechanical philosopher," "seventeenth-century philosopher," and the like. But the problem lies in this very looseness: since we have no proper means of knowing what kind of resemblances should be taken into account and how the various similarities should be weighed, it is highly probable that, when we speak of "the modern philosophers (or scientists)" with no further specification, we are actually relying on the tacit assumption that a Scientific Revolution happened in the seventeenth century, and on some other tacit assumptions about the nature of this Scientific Revolution.

Thus, the feeling that there is a stronger family resemblance between Descartes, Hobbes, and Gassendi than between any two of these authors and, for example, Thomas Aquinas, Ockham, Bradwardine or Oresme, as long as it is not more precisely substantiated, might turn out to be the reheated cabbage of the old Scientific Revolution. That is why I prefer to flee to a safer ground and to insist on another reason for not banning the opposition between the old philosophers and the new philosophy. This second reason is that philosophers of the seventeenth century themselves considered their intellectual enterprise in terms of a battle between two camps, the ancient philosophers and the modern philosophers. It is possible that in so doing, the philosophers, both ancient and modern, were under an illusion as to their own historical situation, that they were dominated by their polemic spirit rather than governed by a concern for precision and intellectual rigor. Nevertheless, this stance was significant to them, and, as such, it had some real effects that can be the object of a second-order historical analysis. In other words, and to use once again a comparison, even if the categories of old and new are not legal tender, they are nonetheless a currency that, good or bad, actually circulated. Hence, it might be worth studying how a work of historical categorization and demarcation between sides, of ordering and partitioning the empire of natural philosophers, was carried out in the seventeenth century. This is precisely what is at stake in this paper.

3.2 A Debate on Natural Philosophy

Yet, such an enterprise immediately calls for a series of remarks that will help me to focus on a specific debate. To begin with the most obvious point: the result of the demarcation process between the old and the new philosophers differed according to time and place. It is not only that those that are called "new" are by definition always changing; it is also that the very format of the categorization changed. Indeed, until the middle of the seventeenth century, the lists of innovators are very long, and they include the authors who had attempted a reform of Aristotelianism from the end of seventeenth century on, mostly, but not only, in Italy; but after 1660, only a few great names remain, including, at the forefront, Descartes. Let us give a few references to substantiate this assertion.

In an *Apologie* written in 1625 to defend great men charged with magic, Gabriel Naudé, a physician and the librarian of the Président de Mesmes, mentions a "swarm of innovators that grows from one day to the next, under the guidance of Telesius [Telesio], Patrice [Patrizi], Campanella, Verulamio [Bacon], Jordan Brun [Giordano Bruno] and Basson, who have no other plan than to elbow aside this philosophy and ruin this great edifice that Aristotle and more than 12,000 others who have interpreted him worked to build over so many years." In his *Advis pour dresser une bibliothèque*, written 2 years later, the swarm has indeed swelled since he adds to the previous *novatores* "Gilbert, Gassendi, Gomesius [Gómez Pereira], Charpentier [Carpenter], Gorlaeus," who, according to him, intend to inspire in all minds "a thousand [*milliace*] openings and new conceptions."[7] In his *Quaestiones ad Genesim* (1623), Mersenne draws up several lists of *novatores*: Campanella, Bruno, Telesio, Kepler, Galileo and Gilbert are mentioned in the *Preface* and, in the body of the book, Bacon, Fludd, Hill, and Basso are to be met.[8] A similar list is presented in *L'impiété des déistes* (1624), where Mersenne outlines his plan to write an Encyclopedia to refute all kinds of lies, and more particularly those of "Gorlee [Gorlaeus], Charpentier [Carpenter], Basso [Basson], Hill, Campanella, Brun [Bruno], Vanin [Vanini] and several others."[9] As for *La vérité des sciences* (1625), it places Patrizi among "Basson, Gorlaeus, Bodin, Carpentier [Carpenter], Hill, Olive" and other chicks who set their hearts on flying before their wings have grown, whereas Aristotle is compared to an eagle.[10] In the Preface of his *Cribrum philosophorum qui Aristotelem superiore et hac aetate oppugnarunt*, composed in 1628, published in 1646, Jean-Cécile Frey, teacher of philosophy at various colleges of the University of Paris, explains that his intention is to refute all those who challenged

[7] Naudé, *Apologie pour tous les grands hommes qui ont esté accusez de magie*, chap. 13, p. 240, and id., *Advis pour dresser une bibliothèque*, chap. 7, p. 133, quoted by Pintard, *Le libertinage érudit dans la première moitié du XVIIe siècle*, pp. 451f. In secondary literature, Antonio Gómez Pereira (1500?–1560?) is generally known because of his *Antoniana margarita*, which is supposed to have anticipated the Cartesian theory of animal machines; see also Navarro's chapter (Chap. 4) in this volume. Here, however, Naudé was more likely thinking of Pereira's *Nova veraque medicina experimentis et evidentibus rationibus comprobata*. Nathanael Carpenter (1589–1638) published under the pseudonym of N.C. Cosmopolitanus the anti-Aristotelian treatise *Philosophia libera triplici exercitationum decade proposita*. The other names that Naudé mentions are familiar enough to us today.

[8] Mersenne, *Quaestiones in Genesim*, resp. Praefatio, n.p., and *Paralipolema et observationes*, col. 1838. Nicolas Hill (1570?–1610?) published a *Philosophia epicuraea, democritiana, theophrastica*. Daniel Garber mentioned this text to me; in his essay in this volume, he calls attention to largely the same lists of *novatores*, though he makes a somewhat different use of them, contrasting the earlier *novatores* with the later mechanical philosophers.

[9] Mersenne, *L'impiété des déistes, athées et libertins de ce temps*, I, pp. 238f.

[10] Mersenne, *La vérité des sciences contre les sceptiques ou pyrrhoniens*, I, 9, p. 109. I was not able to identify Olive.

Aristotelian doctrines, Campanella, Patrizi, Bacon, Telesio, Chassinus, Peter Ramus and recently the "Vile Villon."[11] In 1630, Descartes explains that he owes nothing to his first mentor Beeckman nor to any philosopher, whether old or new; *novatores* are for him at this point Telesio, Campanella, Bruno, Basson and Vanini.[12]

We could think that nothing had changed over the next 40 years when we read a letter from Leibniz to his former teacher Thomasius, in which he's already defending a reformist program relying on a reconciliation of the ancients and the moderns.[13] Indeed, Leibniz makes a long list of *moderni*, whom, he wrote, Thomasius should censure as he censured the now-unknown Michel Baghemihn, a municipal magistrate in Stettin. With all the erudition required when addressing one's former teacher, Leibniz gathers at this point at least three generations of innovators: he begins quite traditionally with the Aristotelian naturalists of sixteenth-century southern Europe (Patrizi, Telesio, Campanella, Bodin, Nizolio, Frascator, Cardano), then he goes on with philosophers of the early seventeenth century to whom we would today attribute different orders of greatness (Galileo, Bacon, Gassendi, Hobbes, Descartes, Basson, Digby, Sennert), and finally he ends up with Protestant partisans of atoms from Thomasius' own generation, whom almost nobody remembers today (Sperling, Derodon, Deusing).[14] And yet I think that Leibniz was here the exception that confirms the rule, the general tendency being at the time to simplify the story and to cut down on the catalogue of innovators.

For example, in the *Advertisement* of his voluminous *Physica*, published in 1669–1671, Honoré Fabri criticizes the *recentiores* who, confounding physics and geometry with forging hypotheses at leisure, are as ridiculous as Democritus, who thought it necessary to tear out his eyes in order to better understand corporeal things.

[11] This passage is quoted and commented by Ariew and Garber, in their introduction to Frey, *Cribrum philosophorum*. Geoffroy Chassins [15??–16??] is the author of a *De natura sive de mundo*. Antoine de Villon advanced in 1624, together with Étienne de Clave and Jean Bitaud, 24 anti-Aristotelian theses, that were condemned by the Parlement of Paris, at the request of the Faculty of Theology; the main documents of this affair are published in Kahn, "La condamnation des thèses d'Antoine de Villon et Étienne de Clave."

[12] Descartes to Beeckman, 17 October 1630, in Descartes, *Œuvres*, vol. I, p. 158.

[13] Leibniz to Thomasius, 30 April 1669, *Leibniz-Thomasius, Correspondance*, p. 97. According to Bodéüs, ibid., p. 119, in a dissertation from 1665, Thomasius identified as innovators Cardano, Campanella, Fludd, Gilbert, Comenius, Hobbes and Digby.

[14] Johann Sperling (1603–1658), professor of medicine at Wittenberg, was so prolific an author that it is not easy to pin down the book to which Leibniz refers; note however that he published the *Antiparasceve. Pro traduce*, in which, following his mentor Daniel Sennert, he rejected the Aristotelian theory on the origin of forms. Anton Deusing (1612–1666), professor of medicine in Groeningen, is the author of the *Disquisitio physico-mathematica gemina de vacuo et de attractione*. As for David Derodon (1600?–1664), Leibniz may here be thinking either of his *Disputatio de libertate et atomis*, or of his more popular *Philosophia contracta*. This triplet confirms the studies arguing that there was an affinity between Protestantism and physical atomism, see for example Heyd, *Between Orthodoxy and Enlightenment: Jean-Robert Chouet and the Introduction of Cartesian Science in the Academy of Geneva*; Leijenhorst and Lüthy, "The Erosion of Aristotelianism"; Brockliss, "Pierre Gautruche et l'enseignement de la philosophie de la nature," pp. 207f.

The allusion is crystal clear: he holds a grudge against Cartesians.[15] It is perhaps Daniel who best expressed the tendency I am trying to capture. In 1691, he notes that, except in schools, Thomists, Scotists and Nominalists are not distinguished anymore: "We put them all in the same category, and in the same party, which we call ancient philosophy, to which we contrast the philosophy of Descartes, or the new philosophy. You [Descartes] have even had the good fortune to erase, in a way, everything that appeared from the new philosophers at the same time as you." Just as it was customary in sixteenth-century Spain to call all heretics Lutherans, whatever their particular sect, it is now usual, Daniel adds, to call "Cartesians" all those who try to elaborate more sophisticated opinions in physics. Even Gassendi happens to be put among Descartes' disciples, although he is older than Descartes.[16] In sum, at this moment, the innovators were no longer the sixteenth-century swarm of prolific Renaissance Aristotelians, but the disciplined and undifferentiated cohort of followers of Descartes, among whom Gassendists are to be found.

Lists of innovators of the years 1670–1690 confirm Daniel's diagnosis, whether they were drawn up by old or new philosophers. On the side of the old philosophers, Jean-Baptiste de La Grange (1678) states explicitly in the title of his book that his target is "the new philosophers, Descartes, Rohault, Gassendi, Father Maignan"[17]; in the body of his book, he names in addition Regius and criticizes Malebranche's doctrine of vision in God.[18] When Louis Le Valois, in his book published in 1680 under the pseudonym of Louis de la Ville, wants to establish that, according to every Cartesian, the essence of matter is extension (Maignan and Cordemoy being the exceptions that confirm the rule since they do not back such a thesis), he successively examines those who are, according to him, Cartesian authors: Clerselier, de la Forge, Rohault, the author of *Discours sur les Entretiens de M. Rohault*, the author of *Recherche de la vérité* (Malebranche), Cally, Bernier, Gadroys, Antoine

[15] "[U]nde est, quod novas & inanes quorundam recentiorum physicas hypotheses reiicas, qui ut novam physicam facerent, novam sibi naturam, novumque mundum fabricarunt; nec tam apud illos invenias, quid re ipsa sit, quàm quid ab iis fictum sit; novum sibi mundum finxerunt, nova, seu potiùs nulla elementa; non coelestes globos, at subtilis materiae turbines, seu vortices; Democritico certè suo haud absimiles, esse voluerunt, qui ut res corporeas, sensibiles, & visibiles meliùs intelligeret, oculos, ut aiunt, eruendos sibi esse putavit" (Fabri, *Physica*, Auctor lectori, § 3, n.p.). In Fabri's works, Descartes is the Democritean leader, *democritici* being those who explain everything through three principles, *moles, figura, motus*, see for example ibid., § 33; id., *Epistolae tres de sua hypothesi philosophica*, I § 23, § 27, § 30, resp. p. 52, p. 58, p. 62; id., *Œuvres de Fabri*, vol. II, ff. 75–77; vol. IV, f. 102r, ff. 207v–209r.

[16] Daniel, *Voyage du monde de Descartes*, pp. 184f. Daniel is not completely consistent when he mentions, p. 113, "les chefs des nouvelles Sectes de Philosophie, entre lesquels nous réconnûmes aisément M. Descartes, M. Gassendi, le P. Maignan, & plusieurs autres."

[17] Emmanuel Maignan (1601–1676) was a Minim teaching philosophy in Toulouse; his *Cursus philosophicus* is famous for rejecting substantial forms and being committed to atoms and void, but it also contains a mechanical explanation of transubstantiation, reproduced in Maignan's *Philosophia sacra*, and commented on by Rohault, *Entretiens sur la philosophie*, pp. 120–122.

[18] La Grange, *Les principes contre les nouveaux philosophes*, Preface, §§ 33–39, pp. 30–38; chap. 4, §§ 5–13, pp. 77–87.

Le Grand, Ambroise Victor, the author of *Critique de la Critique de la Recherche de la vérité* (Robert Desgabets), and the author of *De l'éducation des dames* (François Poullain de la Barre).[19]

Ancients and moderns disagreed on many things, but they at least did agree on who was what. The *Requeste des maîtres es arts* (1671) counts as innovators scientists from the previous generation (Blaise Pascal, Gilles Personne de Roberval) or from the Académie des sciences (Jean Picard, Adrien Auzout, Jean-Dominique Cassini, Jean Pecquet), but in the final *Arrêt* "Cartists and Gassendists" are mentioned, the longest list including "Gassendy [sic], Descartes, Rohault, Denis, Cordemois [sic], Clercelier [sic], Delaunay [sic] and their followers" and the incriminated *libelles* being Rohault's *Physics*, Port-Royal *Logic*, and Gassendi's *Exercitationes*.[20] When Malebranche responds to Le Valois, who decided at some point to include Bernier among Cartesians, notwithstanding the fact that he professes to be Gassendist, he opposes Cartesians and Gassendists to the new Aristotelians and takes over his opponent's classification, mentioning the author of *Recherche*, Clerselier, Cally,

[19] Le Valois, *Sentimens de M. Descartes touchant l'essence et les proprietez des corps*, I, chap. 4, pp. 61–86. The *Discours sur les Entretiens de M. Rohault* mentioned here is probably the *Discours sur les sujets traités dans les Entretiens de Mr. Rohault* that replaced Rohault's Preface in the 3rd edition of his *Entretiens sur la philosophie* (1673). The priest Pierre Cally (1630–1709) is, with Rohault and Malebranche, the author the most frequently quoted by Le Valois: this does not result from the intrinsic quality of his *Institutio philosophiae*, or of his *Universae philosophiae institutio*, but rather from his being Le Valois' local rival. Like Le Valois, Cally was at the time a professor of philosophy in Caen, albeit at the secular college du Bois, not at the Jesuit college: he was introduced to Cartesianism by Pierre-Daniel Huet in the 1960s, defended physical atomism, but objected to animal-machines and to plant-machines; according to Bayle, *Nouvelles de la République des Lettres*, January 1687, pp. 96f., he was dismissed from Caen with two of his fellows for Jansenism and Cartesianism. Claude Gadroys (1642?–1678), secretary of the quartermaster of the Armée d'Allemagne and in charge of the military hospital in Metz earned his reputation as a Cartesian because of his *Discours sur les influences des astres selon les principes de M. Descartes*. The Franciscan Antoine Le Grand (1629–1699) first taught philosophy in Douai, then in London, where he published in 1675 his *Institutio philosophiae, secundum principia Renati Descartes*; Le Valois may also have heard of his *Apologia pro Renato Descartes, contra Samuelem Parkerum*, first published in 1679. The Oratorian Ambroise Victor (1621–1695) was professor of theology and philosophy in Saumur; he published five volumes of *Philosophia cartesiana* in Paris in 1667; a sixth volume defending the thesis of animal-machines was published in Saumur in 1671. *La critique de la critique de la Recherche de la vérité* is the answer by Dom Robert Desgabets (1610–1678) to Simon Foucher's *Critique de la Recherche de la vérité*. The author of *De l'éducation des dames pour la conduite de l'esprit dans les sciences et dans les mœurs*, Paris, J. du Puis, 1674, is François Poullain de La Barre (1647–1725).

[20] Boileau et al. *Requeste des maîtres es arts*, pp. 236f. In the many editions of the *Requeste* that were published in the late seventeenth century and early eighteenth century, Malebranchists and Pourchotists were added to Cartists and Gassendists. Gilles de Launay gave popular Gassendist conferences in Paris as early as the mid-1650s; among other work, he published *Les essais physiques*, *Dissertation de la philosophie*, and *Les essais métaphysiques du sieur de Launay*. Jean-Baptiste Denis (1640–1704) was ordinary consulting physician to Louis XIV; he is famous for having performed the transfusion of blood in dogs and in men in 1667–1668. Denis is the author of *Recueil des mémoires et conférences qui ont esté presentées à Monseigneur le Dauphin*.

Bernier.[21] Thanks to a well-known parable that begins with an analogy between nature and an opera, Fontenelle lets his Marquise of the *Entretiens sur la pluralité des mondes* (1685) note how philosophy became mechanical. When she inquires about the actors of this transformation, he unswervingly answers: "Descartes and a few other Moderns."[22]

Many examples of similar lists could be given. In all of them, Renaissance Aristotelians have disappeared, Cartesians and Gassendists are united against unnamed philosophers of the School, and, sometimes but not always, Gassendists appear as a subset of the Cartesian set. This partition of the empire of natural philosophy presents in my eyes two characteristics that constitute two good reasons to explore it more thoroughly.

First, this partition succeeded in becoming stable beyond the particular moment of its inception. Even today we continue to use it when we refer to Descartes as the archetype of the early modern natural philosopher, the absolute reference in novelty, the benchmark by which the works of other natural philosophers of the seventeenth century fall either on the side of the old or the new (or are even forgotten). This alone would be sufficient to focus on this division, but a second characteristic gives further support.

To state once again the obvious: if a demarcation between two camps is to be studied, it is important not unduly to privilege one of them. In the case at hand, this is not always easy, not only because the court of history has judged in favor of the moderns, but also because the ancients, who, by definition, lived before the moderns, did not have the opportunity to refute them: you can indeed contrast the views of Thomas Aquinas and Descartes on a given topic, but you find no refutation of Descartes by Thomas Aquinas to comment on. In fact, most of the works which try to re-evaluate scholastic works written in the late sixteenth or early seventeenth centuries (rarely those of the Aristotelians of the late seventeenth century) presuppose that since the Scholastics are *per definitionem* conservative, they did not evolve.[23] And yet, these Aristotelians did not vanish into thin air when Gassendi, Descartes, Mersenne, or Hobbes appeared, and they most likely had successors capable of reacting to the new philosophers, whether to argue against them, to adopt some of their views, or to formulate more or less promising compromises. In other words, if we wish to apply an elementary principle of symmetry to philosophers, we must find moments when they actually exchanged arguments with each other. From this point of view as well, the partition of natural philosophy into the old and the new in France between 1670 and 1690 is interesting to study.

The circumstances in which this confrontation took place are well known. In the middle of the 1660s, Claude Clerselier, who was at the time in charge of editing Descartes' correspondence, and, to put it bluntly, was the general of the Cartesian

[21] Le Valois, *Sentimens de M. Descartes touchant l'essence et les proprietez des corps*, § 47, p. 83: "M. Bernier, quoyqu'il fasse profession d'estre tout Gassendiste, & nullement Cartesien"; Malebranche, *Œuvres complètes*, vol. XVII–1, pp. 486, 488.

[22] Fontenelle, *Entretiens sur la pluralité des mondes*, in *Œuvres complètes*, vol. II, p. 21.

[23] Here also, there are some exceptions that confirm the rule, see for example the confrontation between Descartes and Schoock, as now presented in Theo Verbeek, *La querelle d'Utrecht*; Blackwell and Kusukawa, *Philosophy in the Sixteenth and Seventeenth Centuries*.

army, launched a campaign among theologians to win support for the explanations of the Eucharist that Descartes had proposed in response to objections from Arnauld and, later on, in two letters to Mesland. Clerselier was soon assisted in his efforts by Dom Robert Desgabets, who imagined explanations of this mystery of his own devising.[24] Through a series of events, the details of which do not need to retain our attention here, their propaganda actually led to condemnations of Cartesian philosophy pronounced by the King, the University and the teaching orders. Descartes' works had already been put on the Index of Prohibited Books *donec corrigantur* in Rome in 1663.[25] In 1671, the King of France sent to the University of Paris, via the Archbishop, the order to teach "no other doctrine than the one brought forth by the rules and statutes of the University," and this to "prevent the course of an opinion that could bring some confusion in our mysteries"[26] (1671). Following a 2-year conflict between some Oratorians of Angers, including Bernard Lamy and the rector of the University of the same city, an *Arrêt* of the Conseil d'État explicitly forbade them to teach Cartesian philosophy in any way or manner whatsoever, ordered their works, including lecture notes and theses, to be examined by the rector, and reasserted the power of the Faculty of Theology over the Faculty of Philosophy (1675). In Caen, where Pierre Cally was teaching, the Faculty of Theology prohibited the teaching of Descartes' philosophy, on pain of being refused access to any academic grade (1677). The Chapter General of the Benedictines of St. Maur and of the Genofevins condemned Jansenism and Cartesianism in a single stroke (1675 and 1678 respectively). More importantly, to comply with the Jesuits, and after several local conflicts, the General Assembly of the Oratorians issued a decree spelling out what should be taught and what not (1678). The professors of the Parisian Faculty of Arts finally followed their colleagues of the Faculties of Theology and of Medicine in their ban on Cartesianism, and made a list of prohibited propositions (1691). The last noteworthy condemnation in the period that we consider is by the General Assembly of the Jesuits in 1706.[27]

[24] Descartes' and Desgabets' explanations are studied in Armogathe, *Theologia cartesiana*; see as well Ariew, *Descartes and the Last Scholastics*, chap. 7, pp. 140–154.

[25] Documents concerning this condemnation are published in Armogathe and Carraud, "La première condamnation des *Œuvres* de Descartes, d'après des documents inédits des Archives du Saint-Office."

[26] I follow Ariew's translation, "Damned if You Do," pp. 257f. The order of the King referred to the condemnation of 1624 mentioned *supra*, note 12.

[27] The main facts and documents are to be found in Babin, *Journal*; Bayle, *Recueil de quelques pièces curieuses*; Duhamel, *Philosophia universalis*, vol. V, Appendix, pp. 1–45. For recent accounts, see Cousin, "De la persécution du cartésianisme en France," in *Fragments philosophiques*, vol. III, pp. 297–332; Bouillier, *Histoire de la philosophie cartésienne*, vol. I, pp. 466–485; McClaughin, "Censorship and Defenders of the Cartesian Faith"; Ariew, "Quelques condamnations du cartésianisme"; id., "Damned If You Do"; id., "Bernier et les doctrines gassendistes et cartésiennes de l'espace" (a French version of the previous text); id., *Descartes and the Last Scholastics*, chap. 9 (another English version), pp. 172–187; Azouvi, *Descartes et la France*, pp. 19–28, 38–47. The motivations behind censorship are variously interpreted, but I shall not discuss this question here.

Until 1691, the focus of these condemnations, when explicit, is not on the dangers of doubting, as it will be later on, but on the danger that Cartesian physics constitutes for the mysteries of the (Catholic) faith, namely the mystery of Eucharist as officially defined in 1551 by the Council of Trent. Namely, the Cartesian thesis by which the essence of the body is extension and its consequence that the sensible qualities we perceive in bodies, without really belonging to them, result from certain configurations and motions of the extended substance, are judged incompatible with two aspects of the Eucharist. First, according to this mystery, the whole body of Jesus Christ is genuinely, and not symbolically, present in every host; hence, if we admit that the essence of the body is extension and that all extension is local, the body of Jesus Christ has the same local extension as the host. This is however absurd: the extension of the host is too small to contain the body of Jesus Christ; at a given time, there are several hosts in different places of the world, but only one Jesus Christ. Second, according to the dogma of transubstantiation, the species of wine and bread are real accidents, not false appearances, and they are distinct from the corporeal substance, so that they can subsist when the host is changed to Jesus Christ; but, in the Cartesian natural philosophy, all sensible qualities are appearances that necessarily result from certain configurations and motions of the corporeal substance. It is thus impossible in Cartesian physics to have sensible accidents that do not result from the corporeal substance, yet this is precisely what happens in transubstantiation.

To be sure, this kind of reasoning takes up many pages in the books that I consider here and the entirety of some other works that I will not be considering.[28] Moreover, it can reasonably be argued that a general ongoing question was at stake: how should one solve a conflict between a truth of the faith (the mystery of Eucharist) and a truth of the reason (the Cartesian theses on matter), or, to put the matter in institutional terms, between theology and philosophy? According to the new philosophers, reason should not be used to judge the mysteries of the faith, and, reciprocally, faith should not be used to decide matters of reason: this amounts to giving reason a definite autonomy with respect to faith and to assert that one can, as a simple physicist, talk about things as they are in their natural state.[29] According to

[28] See, among the ancients, La Grange, *Les principes contre les nouveaux philosophes*, Preface, §§ 2–4, and chap. 6–8, resp. pp. 2–6 and pp. 99–135; Rochon, *Lettre d'un philosophe à un cartésien de ses amis*, §§ 2–9 et §§ 26–31, resp. pp. 4–17 and pp. 66–81; Vincent, *Discussio peripatetica*, II, sect. 5–6, pp. 55–60; Le Valois, *Sentimens de M. Descartes touchant l'essence et les proprietez des corps*, II and III, pp. 99–317; Charles-Joseph de Troyes, *La philosophie de Monsieur Descartes contraire à la foi catholique*; Huet, *Censura philosophiae cartesianae*, chap. 8, § 3, iv, pp. 173–185; and, among the moderns, Rohault, *Entretiens sur la philosophie*, pp. 119–137; Malebranche, *De la recherche de la vérité*, III, II, chap. 8, § 2, pp. 360–363, and the texts of the polemic with Le Valois, in Malebranche, *Œuvres*, XVII-1, pp. 477–531; Arnauld, *Examen du traité de l'essence des corps*, pp. 29–63; Régis, *Réponse à Huet*, chap. 8, pp. 312–325.

[29] Arnauld, *Plusieurs raisons pour empêcher la censure ou la condamnation de la philosophie de Descartes*, §§ 6–7, pp. 310–313; Rohault, *Traité de physique*, I, chap. 7, § 9, chap. 5, § 12–14, vol. I, pp. 35 and 41, passim; in *Entretiens sur la philosophie*, p. 111, Rohault refers to these declarations; Clerselier, Preface, in Rohault, *Œuvres posthumes*; Malebranche, *De la recherche de la vérité*, III, II, chap. 8, § 2, p. 360.

the old philosophers, however, the relation is not reciprocal and precedence should be given to faith over reason: to be sure, reason should not be used to judge the mysteries of the faith, but faith has a right of inspection with respect to reason, simply because faith is the rule of truth.[30] To sum up, according to the new philosophers, the competing claims of faith and reason should be adjudicated by a separation of their respective domains of jurisdiction; the old philosophers argue, quite traditionally, for a submission of reason to faith. At this point, although he is not one of the authors we study in this paper, one cannot help thinking about a famous sentence by Bossuet, in his letter to one of Malebranche's disciples on 21 May 1687:

> I see … a great combat being prepared against the Church under the name of Cartesian philosophy … under the pretext that one can only accept that which one understands clearly … there is introduced … a liberty to judge that means that, without regard to tradition, one boldly advances that which one thinks.[31]

In the following however, I would like to leave aside books (or parts of books) concerned exclusively with the mystery of Eucharist or with the relation of faith and reason, and to focus on the norms to be adopted in natural philosophy. Namely, it is noteworthy that the official censures and condemnations pronounced in the name of theology, as numerous and radical as they were, were not considered to be sufficient. They were assisted by numerous polemical works, the audience of which was not an assembly of narrow-minded regents and boisterous students but, rather, learned companies of courteous *honnêtes gens*, and the object of which was not (or not only) to reassert the rights of faith, but to defend a certain way of proceeding in physics. In the intellectual context of the time, it is somehow natural that official censorship was not felt to be a sufficient measure to eradicate Cartesianism and that, somewhat belatedly, old philosophers tried to get rid of their image as stubborn pedants, to become more popular among *honnêtes gens* and to respond to the rise of the new natural philosophy.[32]

---

[30] La Grange, *Les principes contre les nouveaux philosophes*, chap. 5, § 1, pp. 90 and 100; Le Valois, *Sentimens de M. Descartes touchant l'essence et les proprietez des corps*, II, chap. 3, art. 2 and art. 4, chap. 4, pp. 119–139, 147–149, 183–220; Babin, *Journal*, p. 42; Charles-Joseph de Troyes, *La philosophie de Monsieur Descartes contraire à la foi catholique*, Préface, p. iij; Huet, *Censura philosophiae cartesianae*, chap. 8, § 3, iv, pp. 173f.; Daniel, *Suite du voyage du monde de Descartes*, p. 100.

[31] Bossuet to ***, 21 May 1687, in Bossuet, *Correspondance*, vol. III, pp. 372f.

[32] La Grange, *Les principes contre les nouveaux philosophes*, Preface, § 44, p. 40, declares this delay surprising: "Il est assez surprenant qu'il n'y a pas encore eu personne depuis vingt ans que la Philosophie de Descartes fait bruit, qui ait tâché de prouver la fausseté de ses opinions, & la verité des principes de la philosophie commune"; Vincent, *Discussio peripatetica*, Dedicace, n.p., explains however that there is nothing surprising here, since 20 years earlier (he published the first volume of his *Cursus philosophicus* in 1658), nobody knew of Descartes: "Cum liberiori calamo coepi spargere in lucem publicam Cursum Philosophicum, nominare eum nec libuit ne oportuit; quia per id temporis latebat etiamnum ferme Anonymus, Et si jam alicuis nominis inter Arctoos, nullius erat in Gallia, vel ad summum tantilli."

In the 1660s, an intense Cartesian propaganda emerged, which involved personal patronage, even family networks, and ostentatious manifestations such as the reburial of Descartes at Saint-Étienne-du-Mont in 1667.[33] As far as publications are concerned, the *Discours de la méthode*, first published in the Netherlands, was reissued in Paris (1658, 1660), while numerous editions of Descartes' other works were released. His unedited works were posthumously published under the supervision of his followers: Claude Clerselier took care of the *Letters* in three volumes (1657–1667) and of *Le monde* (1664), Nicolas Poisson of mechanics and music (1668), Louis de La Forge of *De l'homme* (1664). Many books were issued to defend and illustrate the Cartesian cause: Du Roure's *La physique expliquée suivant les sentiments des anciens et nouveaux philosophes; et principalement de Descartes* (1653) being a special case, I am here rather alluding to works from Gérauld de Cordemoy, Nicolas Poisson, Louis de La Forge, Jacques Rohault, Antoine Arnauld, Nicolas Malebranche, Pierre-Sylvain Régis.[34] The ideas of Descartes were discussed in salons, *conférences*, or private academies; the aristocracy had Cartesian preceptors—the most eminent Cartesian aristocrats of the period are the duc de Luynes, prince Louis II de Condé, the marquis de Liancourt, the prince de Conti, the cardinal de Retz, the marquis de Pomponne and, of course, women like the comtesse de Grignan, the marquise de Sablé or the duchesse du Maine.[35] In a word, the diffusion of Cartesianism

[33] On familial networks involved in the defense of Cartesianism, see Azouvi, *Descartes et la France*, pp. 29–32; van Damme, *Descartes*, pp. 48f.

[34] On the diffusion of Cartesianism through books, see Bouillier, *Histoire de la philosophie cartésienne*, vol. I, pp. 430–437; Martin, *Livre, pouvoirs et société*, vol. II, pp. 874–883; Azouvi, *Descartes et la France*, pp. 32–35; van Damme, *Descartes*, pp. 29–37. Cartesians established a sort of editorial continuity between the master's work and their own, the first edition of *Le monde* being for example published together with a *Discours de l'action des corps* by Cordemoy and a *Discours des fièvres* by Rohault.

[35] On the social fashion of Cartesianism, see Bouillier, *Histoire de la philosophie cartésienne*, vol. I, pp. 437–447; Cousin, "Le cardinal de Retz cartésien," in *Fragments philosophiques*, vol. III, pp. 140–296; Azouvi, *Descartes et la France*, pp. 32–35, 59–64; van Damme, *Descartes*, pp. 47–61. See however the disillusioned observations of Rochon, *Lettre d'un philosophe à un cartésien de ses amis*, § 91, p. 215: "Ils se flattent d'avoir dans leur party de grans Genies & des personnes de la plus haute qualité. Mais ils se méprennent assurément, & ne font pas assez de reflexion que ce sont deux choses que d'entendre Descartes, & d'estre Cartesien. Ces grans esprits & ces personnes illustres en naissance prennent plaisir à sçavoir ce qu'on dit.... Mais ils sont bien au-dessus de ces bassesses et de ces emportements"; and of Daniel, *Voyage du monde de Descartes*, p. 142: "[C] ertains jeunes Abbez, Cavaliers, Avocats, Médecins,... se disent Cartésiens dans les compagnies, pour avoir un titre de bel esprit, qu'ils obtiennent quelquefois par la seule hardiesse de parler à tort & à travers de matiére subtile, d'Automates, de phénomenes, sans sçavoir autre chose, que ces termes." Allusions to oral discussions of Cartesianism in *salons* and *conférences* are to be found in Rohault, *Traité de physique*, Preface, n.p.; Rohault, *Entretiens sur la philosophie*, pp. 110, 138, 150; Rochon, *Lettre d'un philosophe à un cartésien de ses amis*, § 9, § 50, § 75, resp. pp. 19, 129, 172; Le Valois, *Sentimens de M. Descartes touchant l'essence et les proprietez des corps*, I, chap. 4, § 5, p. 61; Daniel, *Suite du voyage du monde de Descartes*, p. 48.

was not confined to teaching institutions; from a social point of view, the philosopher was not only a professor, but also an author, and even a polished and worldly conversationalist.[36] Consequently, if this diffusion was really to be stopped, the publication of lists of propositions was clearly not enough, because these were taught in relatively obscure and confined places such as colleges and universities. The *honnêtes gens*, learnèd but not specialized, good Christians but not furious theologians, were the ones to be convinced.[37] That the audience of *honnêtes gens* was the target not only of the moderns, but of some of the ancients as well, is clear by more than one allusion.[38] This goes along with two characteristics of this debate.

First, concerning the format of their books, the philosophers I am concerned with often gave up heavy tomes for small in 12° works that could fit in a hand or in a pocket.[39] Most of the time, they deserted Latin for French. La Grange is the more explicit in this respect:

> I first composed the greater part of this work in Latin, with the idea I had that the Latin language would better suit the matter with which I was dealing than would the French language. I also believed that reading the work should be allowed only to the learned [*gens d'estudes*]. But, in addition to the fact that most books of our new philosophers are in French, and that it is about undeceiving those that may study their books and follow their feelings, I do not know how it has happened that the most knowledgeable people prefer French books to those in Latin; it seems that Latin is being neglected, because it is too old,

[36] For a long-term study of the various social figures of the philosopher, see Ribard, "Philosophe ou écrivain?"

[37] Pascal famously said of the *honnête homme* that he could not be qualified as a mathematician, as a preacher, as being eloquent, but only as being an *honnête homme*; this is exactly what is at stake in the description of the public of Rohault's conferences given by Clerselier in the Preface he wrote for Rohault's *Œuvres posthumes*, n.p.: "des personnes de toutes sortes de qualitez & conditions, Prélats, Abbez, Courtisans, Docteurs, Médecins, Philosophes, Géometres, Régens, Escoliers, Provinciaux, Estrangers, Artisans, en un mot des personnes de tout âge, de tout sexe, & de toute profession." Vincent, *Discussio peripatetica*, IV, sect. 526, p. 521, has a typical reaction of an Aristotelian professor in front of the new Cartesian public: "[E]vulgant ad omne hominum genus, pueros, adultos, viros, foeminas, doctrinam illam, quae in scholis ad paucos, solosque viros extenderetur." Vincent was teaching in Toulouse, where Régis had been sent in the mid-1960s by Rohault with the mission to propagate Cartesianism; Fontenelle, *Éloge de M. Régis*, in *Œuvres complètes*, vol. VI, pp. 143f., explains that Régis' success was such that the magistrates of the city awarded him a pension.

[38] *Honnêtes gens* are explicitly presented as the best judges in matter of philosophy in Rohault, *Entretiens sur la philosophie*, pp. 146, 154; Rochon, *Lettre d'un philosophe à un cartésien de ses amis*, § 52, § 76, § 78, § 82, resp. pp. 131, 176f., 183 and 192f.; Rapin, *Réflexions sur la philosophie*, Preface, pp. 329, 332. Ladies are mentioned and sometimes opposed to old professors by Rohault, *Entretiens sur la philosophie*, p. 152; Rochon, *Lettre d'un philosophe à un cartésien de ses amis*, § 78 and § 81, resp. pp. 183 and 188; Daniel, *Voyage du monde de Descartes*, pp. 140, 161; Huet, *Nouveaux mémoires pour servir à l'histoire du cartésianisme*, p. 49.

[39] As noted by Brockliss, "Pierre Gautruche et l'enseignement de la philosophie de la nature," pp. 189f., plenty of manuals were published after 1650, but, except for Gautruche's *Philosophiae ac mathematicae totius institutio*, not by Jesuits.

and this language will share the same fate as ordinary philosophy, whose antiquity appears unpleasant. But what primarily obliged me to write in French, is the little the learnèd [*gens d'estudes*] make of philosophy or Scholastic theology.[40]

Moreover, some of these authors did not think dry philosophical arguments or threatening comparisons with Calvinists sufficient for their purposes. In certain cases, they invested mundane genres like the dialog, the conversation or the letter to a friend, and they imagined pleasant fantasies or picturesque narratives, sometimes manipulating with ease the polemic tools of satire and parody. To Fontenelle's *Entretiens sur la pluralité des mondes* (1686), where a Cartesian narrator instructs a marquise in natural philosophy, Father Daniel responds with *Voyage du monde de Descartes* (1690), where the narrator, disconcerted by the various accounts he has heard about the newly discovered Cartesian world, sets his mind to visiting it, as well as other worlds forged by philosophers, which he succeeds in doing thanks to an old man who knows Descartes' recipe to separate one's soul from one's body through the inhalation of some special tobacco.[41] In a similar way, to the *Requeste des maîtres es arts* imagined by Boileau, Bernier and Racine (1671) Daniel responds with *Histoire de la conjuration faite à Stockholm contre Monsieur Descartes* 1693, added as an appendix to his *Suite du voyage*. Written in a pedantic legal language the *Requeste* is a parody of what the teachers of the University of Paris might have said to the parliament in 1671 against the introduction of new ideas in physics, here the Court of Parnassus, followed by the *arrêté* pronounced by this Court: among many things, each one more absurd than the other, it prohibits blood from circulating in the human body, sends to the heart the injunction to continue being the principle of the nerves, reestablishes into their rights entities, identities, petreities and polycarpeities.[42] Now, in the *Histoire de la conjuration*, all the scholastic entities are gathered in Stockholm to sue Descartes, who is accused of having denied their existence.

[40] La Grange, *Les principes contre les nouveaux philosophes*, Preface, § 47, pp. 43f. Azouvi, *Descartes et la France*, pp. 55–57, notes moreover that Cartesians like Cordemoy (*Discours physique de la parole*, 1668), Louis Le Laboureur (*Avantage de la langue françoise sur la langue latine*, 1669) and François Charpentier (*Deffence de la langue françoise*, 1676) explicitly argued that French was the language of philosophy *par excellence*.

[41] The prefaces of these works show the similarity of their motivations, see Fontenelle, *Entretiens sur la pluralité des mondes*, Preface, pp. 9f.: "J'ai voulu traiter la Philosophie d'une manière qui ne fût point philosophique; j'ai tâché de l'amener à un point où elle ne fut ni trop sèche pour les gens du monde, ni trop badine pour les Savans"; Daniel, *Voyage du monde de Descartes*, Avis, p. *4: "J'ai tâché de varier, & d'égaïer un sujet aussi mélancolique, et aussi sec, que le peuvent être des matiéres de Philosophie, tant par la diversité des incidents ..., & même par quelques conversations assez animées de gens, qu'on ne sera pas fâché d'entendre parler." To these works, one might be tempted to add Huet's *Nouveaux mémoires pour servir à l'histoire du cartésianisme*: in his letter to Bossuet of 5 April 1692, in Bossuet, *Correspondance*, vol. V, pp. 108f., Huet writes that his book was inspired by Daniel's *Voyage du monde de Descartes*. However, contrary to Daniel's work, his *Nouveaux mémoires* is a fanciful biography with no explicit philosophical argument, in which Descartes, disappointed by Queen Christine, stages a fake death, travels here and there, and finally ends up teaching philosophy to the Lapps, to whom he bears great resemblance because of his small stature, big head, dark hair and swarthy complexion. In this respect, it should rather be read as an answer to Baillet's biography.

[42] Boileau et al. *Requeste des maîtres es arts*, pp. 237f.

At the beginning of the trial, motion plays the role of advocate for Descartes, but when he understands that, according to his client, he is only a mode, he resigns. Being definitively isolated, Descartes is condemned by the whole assembly as a novateur and chief of a sect, a rebel with regard to the laws of the old and true philosophy, enemy of all kinds of scholastic entities, and finally sentenced to death.[43] I have summarized only the most coherent and consequent writings, but, as we will see, this jesting mood animates most of the works here at stake, even those that were still written in Latin and occupied heavy tomes.

Second, as far as the content is concerned, there was a back and forth between natural philosophy and theology. As already noted, the new philosophers ventured to move to somewhat slippery theological ground when they proposed mechanical explanations of the Eucharist. Moreover, in reaction to the accusations of defending opinions dangerous for religion, they sometimes argued that substantial forms lead to licentiousness or paganism.[44] Conversely, however, the defenders of the old philosophy did not always entrench themselves in theological citadels. To be sure, some of them, in particular Louis Le Valois and Charles-Joseph de Troyes, considered that it is sufficient to recall that faith is the rule of truth and what the truths of the faith are. Others however did not content themselves with asserting the precedence of truths of faith over truths of reason; they set to refute the Cartesian theses *qua* theses of natural philosophy as well, whatever their competence in this domain.[45] Many physical questions were thus examined, from the question of what it is possible for God to create, to the adequacy of the Cartesian explanations of light, magnetism or tides, through the plurality of worlds or the genesis of the three elements.

In the following sections of this chapter, I shall examine neither questions pertaining to metaphysics, nor questions pertaining to particular or special natural philosophy, but rather concentrate on two correlated questions pertaining to what was at this time called general natural philosophy, namely, the question of what kind of ontological entities are necessary for the establishment of a good physics, and the correlated question of what norms should be adopted in natural philosophy. The first question will lead me to focus on substantial forms in general and animal souls in particular, these scholastic entities *par excellence*. To use once again Father Daniel

[43] "Et les voix ayant été recueillies, ce Philosophe infortuné fut déclaré Novateur et Chef de secte, Rebelle aux Lois de l'ancienne et veritable Philosophie; perturbateur de l'ordre des *Categories*: Ennemi des *Vertus & Facultez occultes*; des *Accidens absolus, & non absolus*; des *Qualitez premieres, & secondes*; des *Formes, des Elemens* et des *Mixtes;* des *Ames materielles*, soit *vegetatives*, ou *sensitives*; des *Instincts, substances incompletes*, et generalement de toutes les *Formes* tant *substantielles* qu'*accidentelles*" (Daniel, *Suite du voyage du monde de Descartes*, p. 245).

[44] See *infra*, Conclusions.

[45] Thus, they responded to the criticism of Rohault, *Entretiens sur la philosophie*, pp. 106f.: "[C]ette Doctrine de M. Des Cartes, a esté depuis peu attaquée, par des personnes qui croyent avoir droit, sous pretexte de nouveauté, de s'opposer à tout ce que l'esprit peut découvrir dans les choses Naturelles. Ces personnes ne la combattent point par des raisonnemens philosophiques, et ils s'imaginent qu'il est bien plus aisé de la décrediter en exposant simplement qu'elle est contraire à la Religion" and his following injunction to examine Descartes' doctrine from a purely philosophical point of view.

as a guide, “the article on substantial forms is the one that caused the most uproar and division between the two parties.”[46] In the France of this period, you belonged to the camp of old philosophers if you thought that substantial forms are necessary in physics, and to the camp of new philosophers if you thought that they are not.[47] But be aware that this ontological question cannot be dissociated from the second one: with respect to which kind of norms were these entities excluded or, on the contrary, defended?

These questions are not totally unprecedented. In a series of seminal papers, Keith Hutchison systematically confronted the old Scholastics with the new Mechanical Philosophers, arguing that the incentive of their disagreement was not the methodological concern of the latter to avoid the vacuous explanations of the former, but their competing worldviews, in particular with respect to the questions of what should be considered as a cause and what should be considered as a physical entity.[48] I shall not discuss here the details of Hutchison's argument. I do agree with his methodological attempt to render justice to the old philosophers, rather than to put them away on a back shelf. But I do not think that we have to decide whether the bone of contention was whether certain explanations were vacuous or whether they were false: both assertions were at stake. Moreover, as explained, I think that it might be fruitful to study an actual confrontation, in other words to consider a set of books that explicitly refer and respond to one another. In order to understand this confrontation, I will now review the fairly well-known arguments that were formulated by new philosophers against the ancients and thus make precise what exactly the latter were responding to. Only then will I isolate the generally lesser-known arguments of the ancients. Finally I will reflect on what we can conclude from this confrontation.

3.3 On the Side of the New Philosophers

According to the longest lists mentioned above, Descartes, Gassendi, Maignan, Cordemoy, Rohault, La Forge, Lamy, Bernier, Cally, Malebranche, Arnauld and Régis were considered new philosophers in the France of the years 1670–1690. In this paper, I shall exclude Descartes, Gassendi, Maignan, Bernier and Cally and restrict myself to studying Cartesians who published during the years 1670–1690,

[46] Daniel, *Voyage du monde de Descartes*, p. 143. Daniel has a deep concern with the characterization of philosophical sects; as we will note *infra* note 102, he characterizes more specifically the Cartesians by the exclusion of animal souls, and it was indeed a point on which they disagree with their fellows in innovation, the Gassendists.

[47] There is no such statement with exceptions that confirm the rule; the most interesting exceptions are Fabri in the camp of old philosophers (on his requalification of substantial forms, see Roux, “La philosophie naturelle d'Honoré Fabri,” pp. 87–90) and Leibniz in the camp of new philosophers.

[48] Hutchison, “What Happened to Occult Qualities in the Scientific Revolution,” “Dormitive Virtues, Scholastic Qualities and the New Philosophies” and “Individualism, Causal Location and the Eclipse of Scholastic Philosophy.”

including minor figures like Nicolas Poisson, Bernard Lamy or Claude Gadroys.[49] My aim is indirect and negative: I do not want to examine the ontological entities and physical norms defended by the new philosophers as such, but rather, the criticisms they addressed to the ontological entities and physical norms of the old philosophers; and I do not want to determine if these criticisms were correct or pertinent, but rather, to understand the kind of natural philosophy their proponents were opposed to.

It is to be noted that, when they speak of old philosophers, new philosophers use generic designations, as if dealing with a litter of young animals, among which it is neither important to make differences nor to set precedence; a position is neither specified by nor associated with the name of a philosopher, the title of a book or the number of a paragraph, it is presented as a generic position defended in the "ordinary" or "common" philosophy.[50] This type of designation suggests that the new philosophers do not take into account the differences that were significant to the eyes of the old ones, and that in so doing, they may oversimplify some real conceptual problems. Indeed, the new philosophers systematically characterize substantial forms as entities separated from matter, whereas, to the old philosophers, this was only one option among several available.[51] Hence, one sees that it is one thing to establish that, according to the new philosophers, substantial forms were in an illegitimate way credited with a separate existence, and quite another to establish that it was indeed the case that substantial forms had a separate existence. In what follows, it is only the first of these that interests me.

But even if one keeps this clarification in mind, the criticism of the old philosophers often takes a form so grotesque that one gets the feeling of a paradigm shift, as if the new philosophers could not understand the natural philosophy of the ancients anymore, this philosophy being truly incommensurable to theirs in its objectives and in the means it used to achieve them. One can nonetheless identify three factors that enter into this critique, not clearly separated from each other: the entities of the old philosophers are unnecessary, the explanations they offer are trivial or even vacuous, the words they use are unclear or even meaningless. I shall consider these three reasons in turn. Overall, the charge made by the modern philosophers

[49] Their biographies are known thanks to the impressive work realized by Pierre Clair and François Girbal.

[50] Rochon, *Lettre d'un philosophe à un cartésien de ses amis*, § 77, pp. 178–181, reproaches the new philosophers precisely for speaking of the School in general, and he insists that some distinctions should be introduced between different Aristotelianisms.

[51] Rohault, *Traité de physique*, I, chap. 18, § 2, vol. I, pp. 141f.; Malebranche, *De la recherche de la vérité*, VI, II, chap. 2, p. 642; Arnauld and Nicole, *La logique*, III, chap. 18, p. 240; La Forge, *Traité de l'esprit*, chap. 13, p. 206. According to Gilson, *Études sur le rôle de la pensée médiévale dans la formation du système cartésien*, pp. 162–163, the criticism does not concern essential forms as such (which can be designed, but not exist in a separate state), but essential forms as they are apprehended within the framework of Cartesian noetics (for which any true idea is the idea of a substance); for a detailed examination of separate forms in Scholasticism, see Des Chene, *Physiologia*, pp. 53f., 65, *passim.*

is serious enough to make one wonder what, according to them, forced the old philosophers to adopt such a way of thinking infused with redundancy, triviality and vacuity. We shall see that the new philosophers also had an answer to this second-order question.

3.3.1 The Methodology of Ontology: Beings Should Not Be Multiplied Without Necessity

The heart of the new philosophers' argument is here that the entities relied on by the old philosophers are superfluous. Once one accepts matter and its modes, figure and motion, substantial forms are no longer necessary to account for natural phenomena. According to the new philosophers, the old philosophers thus made use of superfluous entities, violating the principle of ontological economy by which one must never unnecessarily multiply beings. Here is a concise formulation of this idea by Cordemoy: "To reasonable people who know that one must not multiply beings without necessity, it is enough, in order to believe that it [a watch] has none [substantial form] to see that everything that it does can be explained by the body."[52] Insofar as the body having such and such material configuration is sufficient to explain all the properties of a watch, one need not refer to a form that would add to them in order to understand these properties. One may well nominally call "form" the body as it has such and such material configuration, but this does not justify adding a new entity. The bogeyman of the unnecessary multiplication of entities does not only concern substantial forms, but is also used at several steps of the analysis. Thus, to each newly discovered property of a body, the old philosophers would have matched a new being, and thus increased the entities, if not unnecessarily, at least without restriction.[53] In this way they derided the way the Aristotelians would have analyzed the concrete changes, even trivial, of the bodies that surround us: a new being corresponding to each new state of the body. To explain a trivial transformation, one must constantly bring about a multitude of new beings that follow one another like characters in an enchanted ballet, when, for example, wheat becomes bread or when a dog passes from life to death.[54]

---

[52] Cordemoy, *Six discours sur la distinction et l'union du corps et de l'âme*, pp. 122f. The same example is used by Malebranche, *De la recherche de la vérité*, III, II, chap. 8, § 2, pp. 357f.

[53] Malebranche, *De la recherche de la vérité*, III, II, chap. 8, § 1, p. 355: "[V]oici ce qui arrive ordinairement aux philosophes. Ils voient quelque effet nouveau: ils imaginent aussitôt une entité nouvelle pour le produire. … [I]ls donnent libéralement au feu autant de facultés ou de qualités réelles, qu'il est capable de produire d'effets différents"; ibid., VI, II, chap. 2, p. 642: "Les philosophes … prétendent qu'il y a … une infinité de petits êtres…: et ils en supposent d'ordinaire autant qu'ils ont de différentes sensations des corps, et qu'ils pensent que ces corps produisent d'effets différents."

[54] For the transformation of wheat, see Cordemoy, *Six discours sur la distinction et l'union du corps et de l'âme*, pp. 113f.; Malebranche, *De la recherche de la vérité*, VI, II, chap. 2, p. 639; for the transformation of a dog, see ibid., I, chap. 16, § 4, p. 127.

3.3.2 *The Way of Physics: Physics Should Explain Phenomena, Namely, Give Efficient Causes*

The new philosophers also accuse the old philosophers of being all talk. The latter would rely on general principles, vague and indeterminate, unable to yield any new knowledge, and in the end, unable really to explain the phenomena. Here is, for example, what Cordemoy says about elasticity: "The entire School says that this is done by an *elastic* virtue, that is to say, in common language, that there is something that has the power or virtue of making a spring: but this does not explain this thing."[55] The idea is well known: the explanations of old philosophers are trivial, vacuous and circular, because the *explanans* (e.g., elasticity) is only a general term derived from the *explanandum* (e.g., the fact that some bodies are elastic, i.e. have the capacity to come back to their original state after a deformation). Generality here is bad generality, because one associates with the particular phenomenon in need of an explanation a principle whose only characteristic is that it is the cause of this phenomenon. It is precisely the criticism Arnaud and Nicole address to Aristotle's physics:

> [T]he main fault one can find there is not that it is false, but that, on the contrary, it is too true and that it teaches us only things that it is impossible not to know. … [A]fter learning all these things, it does not seem that one has learned anything new, nor that one is better able to reason as to any of the effects of nature.[56]

Contrary to how it was in this physics, one must "descend to the particular."[57] Obviously, the question at this point is to know what it means to "give a reason for a natural effect," "descend to the particular"—in other terms, to know what the new philosophers expect from an explanation. According to them, explaining a natural phenomenon means in general being able to find its efficient cause, to explain by what means and after what manner it comes about; for mechanical philosophers in

[55] Cordemoy, Six discours sur la distinction et l'union du corps et de l'âme, p. 123.

[56] Arnaud and Nicole, *La logique*, Discourse II, p. 33. See as well La Forge, *Remarques*, in Descartes, *L'homme*, pp. 183f.: "De dire aussi que c'est une Qualité du Corps ou une proprieté de l'Ame [that makes the heart beat], cela ne sert de rien pour expliquer ce que c'est; non plus que si demandant ce que c'est qu'un Elephant, on me répondoit que c'est un Animal d'Affrique"; ibid., p. 217–219; Malebranche, *De la recherche de la vérité*, VI, II, chap. 2, pp. 640f.: "[L]es scolastiques parlent si généralement, qu'ils ne se hasardent pas beaucoup. Une qualité est ce qui fait qu'on appelle une chose d'un tel nom, on ne peut le nier…."

[57] "[O]n ne s'arreste souvent qu'à des questions si abstraites & si generales, que quand bien mesme tous les Philosophes seroient de mesme avis sur chacune, cela ne pourroit servir à expliquer en particulier le moiundre effet de la Nature; Cependant une science d'usage doit bien-tost descendre dans le particulier…. Car enfin raisonner toujours, & ne raisonner que sur des choses aussi generales que celles sur lesquelles on raisonne ordinairement, sans descendre à rien de particulier, ce n'est pas le moyen d'acquerir des connoisssances" (Rohault, *Traité de physique*, Preface, n.p.); "si nous voulons … dire quelque chose de plus que le commun, il faut se resoudre à descendre dans le particulier, nonobstant la coûtume des Philosophes, qui n'y descendent presque jamais, & qui se contentent pour l'ordinaire de proposer plusieurs questions fort vagues, & qui peuvent même passer pour superfluës, en ce qu'on n'en peut tirer aucune utilité" (ibid., I, chap. 18, § 1, vol. I, p. 141).

particular, it amounts to showing what microscopic configuration of corpuscles could produce this phenomenon.[58] Ancient principles are thus criticized for not being able to produce a specific natural phenomenon. To this criticism are linked two considerations, one relating to the abusive way in which the old philosophers would have used final causes, the other relating to their incapacity to predict some future effects.

First, old philosophers are accused of being satisfied with final causes. The new philosophers that I study here do not exclude them because of a metaphysical criticism on the type of assumptions one makes about the Creator when one speaks of final causes; more simply, they declare final causes to be insufficient to explain the phenomena. To do so, they use witty analogies: to be satisfied with final causes would be as ridiculous as to say that wood comes to Paris for fear of cold, or to explain the motion of a ship by its navigator's destination.[59]

Secondly, the new philosophers say that their mechanical explanations are predictive. To know how the phenomenon is produced is to know which microscopic corpuscular configuration can be its cause, so that he who knows this configuration will be able to predict yet unknown effects. When in this case, Rohault says, one is dealing with experiences that are most useful for the physicist, these are experiences that reasoning "anticipates."[60] On the contrary, the old philosophers, insofar as their principles would only restate the *explanandum* in general terms, not only do not explain effects, but also are unable to anticipate some still unknown effects. Here is for example what Malebranche writes in this respect:

> Even if one knows that there is in fire a substantial form accompanied by a million faculties like those of heating, of dilating, of melting gold, silver and all metals, of lighting, of burning, of cooking, if one proposed to me this problem to solve, namely, if fire can harden mud and soften wax, the ideas of substantial forms and the faculties to produce heat, rarefaction, fluidity, etc. would do me know good to know if fire would be able to harden mud and soften wax …. The same goes for all general ideas: they are thus totally useless for resolving any question.[61]

---

[58] On mechanical explanations, see McMullin, "Structural explanations"; Clarke, *Occult Powers and Hypotheses*, pp. 164–191; Gabbey, "Mechanical Philosophies and their Explanations."

[59] "[D]e quelque façon qu'ils la [the expression "crainte du vide"] prennent, ils ne satisfont pas à ce que l'on demande; non plus que feroit un homme, qui estant interrogé comment le bois vient à Paris des Provinces éloignées, répondroit qu'il y vient par la crainte du froid: car ce n'est pas là répondre à la question; puisque c'est apporter la cause finale, au lieu de l'efficiente que l'on demande" (Rohault, *Traité de physique*, I, chap. 12, § 3, vol. I, pp. 78f.); "Seroit-ce assez (par exemple) pour expliquer le mouvement d'un vaisseau qui seroit porté tantost en Syrie, & tantost en Affrique, de dire que le Pilote qui est dedans a dessein d'y aller, et qu'il a connoissance de la route qu'il doit tenir, ne faudroit-il pas outre cela qu'il sçeust parfaitement bien l'usage des instruments du Vaisseau, & qu'il eust l'adresse de s'en bien servir en vrai Pilote & le pouvoir bien conduire" (La Forge, *Traité de l'esprit*, chap. 4, p. 122).

[60] "[L]es experiences de la troisiéme sorte sont celles que le raisonnement previent, qui servent à justifier ensuite s'il est faux, ou s'il est juste; Ce qui arrive, lorsqu'aprés avoir consideré les effets ordinaires d'un certain sujet, & formé une certaine idée de sa Nature …, nous venons par raisonnement à connoistre que si ce que nous croyons de sa Nature est veritable, il faut necessairement qu'en le disposant d'une certaine maniere, il en arrive un nouvel effet, auquel nous n'avions pas encore pensé" (Rohault, *Traité de physique*, Preface, n.p.).

[61] Malebranche, *De la recherche de la vérité*, VI, II, chap. 2, p. 641.

3.3.3 Ontological Categories: The Bipartition Between Body and Soul Should Be Respected

Finally the new philosophers argue that while the old philosophers use obscure and meaningless terms, the entities they relate to cannot be conceived. These are two different criticisms, one about language and the other about ideas, but the new philosophers confound them *de facto*.[62] The interlacing of these criticisms is such that obscurity and inconceivability are both reduced to an ontological question: according to the new philosophers, it is because the entities used by the old philosophers do not exist that the words they use are obscure and that they do not correspond to any conceivable idea.[63] In a number of texts, it is difficult to know whether obscurity and inconceivability are considered as signs of an incorrect ontology or, conversely, assuming the correct ontology is known, they infer that ancient philosophers use obscure terms and have no conceivable idea.[64] In any case, the fact is that the new philosophers constantly refer to simple categorizations of beings: a being is either substance or accident, and then, either spiritual or corporeal. For the sake of brevity I henceforth only focus on the second categorization. The categorization of beings as soul or body is a strict partition: everything belongs either to the category of the soul, or to the category of the body. The reproach addressed to the old philosophers is thus that they do not respect this partition: they do not understand that these two categories cover the totality of beings (in other words, they affirm that some beings depend neither on body nor on soul), and they do not realize that this is a strict partition (in other words, they relate certain beings to both body and soul).

The first type of error with respect to the ontological bipartition, asserting the existence of beings that depend neither on the body nor on the soul, is denounced by La Forge, not directly in relation with his ontology, but indeed in relation to his theory of ideas—the presupposition being that something of which we cannot have any idea does not exist:

> [U]ndoubtedly terms must seem obscure when no idea corresponds in one's mind to the meaning given to them. Yet we have the idea of only two sorts of beings, generally speaking, namely that which is extended, which we call *body*, and that which thinks, which we

[62] Rohault, *Traité de physique*, I, chap. 12, § 1, vol. I, pp. 77f.: "[D]es paroles qui ne signifient rien que l'on puisse concevoir"; Malebranche, *De la recherche de la vérité*, VI, II, chap. 2, p. 640: "Aristote … propose et résout toutes choses par ces beaux mots de *genre*, d'*espèce*, d'*acte*, de *puissance*, de *nature*, de *forme*, de *facultés*, de *qualités*, de *cause par soi*, de *cause par accident*. Ses sectateurs ont bien de la peine à comprendre que ces mots ne signifient rien"; ibid., Elucidation 12, p. 948: "C'est principalement dans les matières de physique qu' … on se sert de termes qui ne signifient rien.... Ce sont des termes vides de sens."

[63] Ibid., p. 942: "Tout ce qui existe se réduisant à l'être ou aux manières d'être, tout terme qui ne signifie aucune de ces deux choses ne signifie rien; et tout terme qui ne signifie aucune de ces deux choses distinctement et en particulier, ne signifie rien de distinct."

[64] Clarke, *Occult Powers and Hypotheses*, pp. 167f., chooses the first option and concludes that this criticism is doomed to failure, because it is relative to the ontological framework in which it is formulated.

> call *soul*. And starting from when one speaks of other beings, that can be related neither to one nor the other,... such as those beings that one calls substantial forms of bodies, real, impressed, intentional, occult, sympathetic, or specific quality, concoctive, retentive or expulsive faculty etc., it is impossible for any idea to correspond to them in the mind, or that what one then says has any meaning one can conceive.[65]

The second type of error with respect to this ontological bipartition is, according to the new philosophers, to assert the existence of entities that are both body and soul, or at least to treat something as if it belongs to one of these categories when in fact it belongs to the other. In the exemplary case of sensible qualities, which soon came to be called secondary qualities, the old philosophers are claimed to have the mistaken belief that the feeling that a body causes in us belongs to this body. A passage from Arnauld and Nicole illustrates this idea:

> The soul, which saw that it was not by its will that its feelings were excited in it, but that it had feelings only when occasioned by certain bodies, as when it felt heat when approaching fire, was not content to judge that there was something outside itself that was the cause for its having these feelings, about which it would not have been mistaken, but it went beyond this, believing that that which was in these objects was entirely the same as the feelings or ideas that had been occasioned. And from these judgments it formed ideas, by transporting these feelings of heat, color, etc. into the very things that are outside of it. And these are these obscure and confused ideas that we have of sensible qualities, the soul having added its false judgments to what nature let it know.[66]

Thus, in a recurrent and insistent way the old philosophers project onto bodies sensible qualities that, in fact, exist only in mind.[67]

In the case of qualities such as gravity, *horror vacui*, or the capacity of a plant to grow, the criticism is different and, so to speak, reversed. The old philosophers are not accused of projecting onto bodies that which exists only in mind, but of interpreting properties of the body as if they were properties of the mind. Arnauld and Nicole describe this error when they reconstruct the way the old philosophers elaborated their concept of gravity. The starting point consists in two true ideas—there is something that falls, and there is something that is causing this fall. But following a hurried judgment arises a proposition that is not true: the cause of the fall is in the stone. Given the axiom that no body can move by itself, this proposition suggests that the cause of the fall is a being that exists in the stone while being distinct from it. Hence, the stone includes matter, which receives the motion, and a substantial form, which gives the motion and which is the cause of the fall.[68] At this point, however, the old philosophers, assuming they accept the ontological bipartition

---

[65] La Forge, *Traité de l'esprit*, Preface, p. 77. See as well Régis, *Système de philosophie*, Physique, VII, II, chap. 17, vol. II, p. 631.

[66] Arnaud and Nicole, *La logique*, I, chap. 9, pp. 71f.

[67] See Rohault, *Traité de physique*, I, chap. 23–27, §§ 1–2, vol. I, resp. pp. 217, 249, 264, 270, and 291; Malebranche, *De la recherche de la vérité*, I, chap. 10–19, and VI, II, chap. 2, resp. pp. 89–139 and 636–639.

[68] Arnaud and Nicole, *La logique*, I, chap. 9, p. 77.

between body and mind, would think of the substantial form as a mind: if they thought of it as matter, the substantial form would be part of the matter of the stone; therefore one would be right back in the situation one wished to avoid, namely the situation where a body moves by itself.[69] Whether they accept this or not, the ancient philosophers are therefore led to think of gravity as a mind acting on the body, and thus once again to commit an error of ontological categorization.

3.3.4 The Social Twist

The new philosophers also had an answer to the question of what could possibly have made the old philosophers adopt such a system of thought replete with redundancy, triviality and emptiness: they denounce the learned contention of the old philosophers and their relation to the authority of Aristotle. If the old philosophers came to lose the taste for things themselves, to use empty words, to believe that they know something new when they use a new name, or that they had resolved a problem when they had multiplied entities, it is first because they wanted to cover up their ignorance, to appear more knowledgeable than they actually were, and to distance themselves from those they call ignorant or common people. Rohault, for instance, criticizes the tendency for them to

> be accustomed to saying in general that this effect is produced by a quality. For from this custom comes that of giving words as if they were reasons, and the foolish vanity of believing that one knows more than common people, when one knows words the common people don't know.... Indeed, what difference can there be between the responses of a peasant and a philosopher, if when each is asked, for example, why a magnet attracts iron, one says that he does not know the cause, and the other that it takes place by a virtue and an occult quality? Is this not in good French saying the same thing with different words? And is it not visible that the entire difference between the one and the other is that one has enough sincerity to avow his ignorance, and the other enough vanity to want to hide it?[70]

The other source of the faults of the old philosophers is the respect they have for Aristotle. In fact, the new philosophers describe the relationship of the old philosophers

---

[69] Ibid., IV, chap. 7, p. 322.

[70] Rohault, *Traité de physique*, Preface, n.p. The knowledge of peasants and the knowledge of ancient philosophers are constantly compared by Rohault, see for example *Entretiens sur la philosophie*, pp. 104 and 115. See as well Clerselier, in Descartes, *L'homme*, Preface, p. ii–ij; Arnaud and Nicole, *La logique*, III, chap. 19, § 3, pp. 246–247: "[I]ls s'imaginent en être plus savans pour avoir trouvé ce mot." See as well Gadroys, *Discours sur les influences des astres selon les principes de M. Descartes*, Preface, n.p.: "[L]a Philosophie commune a des principes si foibles, qu'on n'en peut tirer aucune conclusion.... C'est une science de mots; ce n'est pas une science de choses; elle remplit la bouche & elle laisse l'esprit vuide"; Malebranche, *De la recherche de la vérité*, III, II, chap. 8, p. 355: "[O]n s'imagine savoir mieux que les autres, ce que toutefois on sait beaucoup moins"; ibid., VI, II, chap. 2, p. 640: "[O]n n'est pas plus savant qu'auparavant"; ibid., Elucidation 9, pp. 916–917; ibid., Elucidation 12, p. 948.

to Aristotle in two complementary ways. It is first a relationship of dependence: the old philosophers are criticized for having an undue respect for Aristotle and for being his friends rather than friends of the truth.[71] Nonetheless, the relationship from the old philosophers to Aristotle is also described as a betrayal: according to the new philosophers, the ancients betrayed Aristotle by preferring his scholastic followers to him. The problem with the Scholastics is not only that they pale in significance beside Aristotle, but more essentially that they misunderstand the status of his physics. Aristotle had conceived physics as a logic or a dialectic, and, as such, it was valid, but his scholastic commentators took for physical reality what was merely logical or dialectical.[72] His Arab commentators are moreover accused of having introduced unnecessary subtleties in physics.[73] This strategy of dissociating Aristotle from his followers does not call into question the principle of authority; it is rather a way of capturing Aristotle's prestige to the detriment of the old philosophers. If the old philosophers misunderstand Aristotle, their theses may well be incompatible with the theses of the new philosophers, while the same does not apply to Aristotle.[74]

The argumentation of these new philosophers is neither remarkable for its novelty nor its sophistication. On the contrary, one can easily show that regarding the substance of the matter, they merely adopt arguments, ideas or examples that could be found in the older generation, most notably in Gassendi and Descartes and beyond, in some Renaissance discussions. To conceptual sophistication and detailed discussion, they obviously prefer polemics, including swift satire and exaggerated simplification, possibly inspired by their performances in salons, *conférences* and

[71] Arnauld and Nicole, *La logique*, III, chap. 20, b6, pp. 282–283; Rohault, *Traité de physique*, Preface, n.p.; Malebranche, *De la recherche de la vérité*, II, II, chap. 3, pp. 210–214.

[72] Poisson, *Commentaire*, V, p. 166: "[Matter and form] ne sont que des termes, qui de soy sont tres bons & tres explicatifs des diverses manieres dot [sic] on peut considerer cet or; mais qui n'ont jamais dû estre employez comme si c'estoit des choses réellement & substantiellement existentes, n'ayant esté pris au commencement que comme des façons de parler"; Malebranche, *De la recherche de la vérité*, III, II, chap. 8, § 1, p. 356: "Si les philosophes ordinaires se contentaient de donner leur physique simplement comme une logique, qui fournirait des termes propres pour parler des choses de la nature…, on ne trouverait rien à reprendre dans leur conduite.… Mais … ils veulent absolument que la Physique de leur maître Aristote soit une véritable Physique, qui explique le fond des choses, et non pas simplement une logique."

[73] Lamy, *Entretiens sur les sciences*, pp. 253f.: "Pour ce qu'il [Aristotle] dit de la Phisique en general, c'est plûtôt une Dialectique ou maniere de parler des choses naturelles, qu'une veritable Physique. Aussi il n'y a rien de mieux dit, ni de plus vrai que ce qu'il en écrit, quand on prend bien sa pensée. Ce qu'il dit de la matiere & de la forme n'est que pour marquer précisément que ce que l'on entend par ces noms.… Il en est de même de la quantité et des qualitez, dont il ne fait qu'expliquer ce que leurs noms signifient dans l'usage de la Langue.… Aujourd'huy…, ce n'est point proprement sa Philosophie qui règne dans les Ecoles, c'est celle des Arabes." See as well Arnauld, *Plusieurs raisons pour empêcher la censure ou la condamnation de la philosophie de Descartes*, § 6, pp. 310f.; Rohault, *Entretiens sur la philosophie*, pp. 115 and 152f. Both Arnauld and Rohault refer explicitly to Rapin, on which see *infra*, note 118.

[74] Rohault, *Traité de physique*, Preface, n.p., notes that his disagreement is not with Aristotle, but with his commentators.

private academies. Thus we might well wonder about the old philosophers being targeted in this way. In particular, we wonder if some of them should not have been a bit more subtle than Moliere's physicians invoking the "dormative virtue" of opium, if they could hear all of these critics without even trying to answer to them, and if they were all so firmly attached to the Aristotelian camp that they needed to hear yet again what had already been said 50 years earlier.

3.4 On the Side of the Old Philosophers

We may be at a loss when asking ourselves who exactly should be included in this group. To include those, who were classified as ancients by the moderns will not do; as already noted, whereas the ancients explicitly name the moderns that they wish to refute, moderns use generic designations when they castigate the "ordinary" or "common" philosophy. In the following, to delineate the hard core of ancients that I shall study, I use two criteria, one that pertains to the format (the typical book I am dealing with is a small format volume written in French for *honnêtes gens*) and another one which pertains to the content (the book should aim at refuting new philosophers). These two criteria do not always coincide during the period in question: on the one hand, there had been Aristotelian manuals written in French since the early seventeenth century; on the other hand, some refutations of new philosophers are written in Latin. Here I shall not consider the former, represented for example since the beginning of the century by the manuals and introductions to philosophy written by Scipion Dupleix, Pierre Du Moulin, Théophraste Bouju, Léonard de Marandé, Charles Sorel, Gilles de Launay or Louis de Lesclache. As a rule, I shall also exclude the typical products of the teaching institutions, such as manuals and theses, but also systematic refutations of Descartes written in Latin by professors who had published a manual earlier on, such as Pierre Godart or Jean Vincent, superior of the Congregation of Christian Doctrine in Toulouse.[75]

The hard core of the corpus obtained by applying these two criteria is composed of works written by Jesuits who would later obtain eminent positions in Paris: Ignace-Gaston Pardies, Antoine Rochon, Louis Le Valois, Gabriel Daniel, René Rapin.[76] Considering the singular position of Honoré Fabri in the Jesuit institution

[75] Godart's diptych is composed of his *Totius philosophiae summa* and his *Dissertatio in qua egregie ludicra Cartesii revelantur*. Vincent's diptych consists of *Cursus philosophicus* and his *Discussio peripatetica in qua philosophiae cartesianae Principia … examinantur.*

[76] To take them by alphabetic order, Gabriel Daniel (1649–1728) was the librarian, then the Superior, of the Parisian Jesuit *maison professe* and historiographer of France; Louis Le Valois (1639–1700) taught philosophy first at the Collège de Clermont in Paris (from 1682 on, collège Louis-Le-Grand), then in Caen for 10 years, and was finally appointed confessor to Louis XIV's grandsons; Ignace-Gaston Pardies (1636–1673) taught philosophy and mathematics at the Collège de Clermont, but died too young to get any other position; René Rapin (1621–1687), professor of rhetoric at the Collège de Clermont, wrote extensively in prose and in verse;

and in the history of Cartesianism, some of his treatises will be mentioned as well.[77] To them, I have added an Oratorian (Jean-Baptiste de La Grange), because his book was the first attack on the new Cartesian school in this period, and a member of the high clergy (Pierre-Daniel Huet), because of the interest of his *Censura philosophiae cartesianae.*[78] But I exclude the secular Jean Duhamel's further exchanges with Régis, simply because they would add nothing to my argument.[79]

These old philosophers responded blow by blow, and often by turning their arguments back on the new philosophers, so the best way to proceed seems to consider once again the four topics identified previously. As we will show, the old philosophers, full of verve, succeeded in putting in place effective responses to their opponents: but by doing this, they helped to ensure the supremacy of the latter. There is no paradox here: rather, they had recognized the rules proposed by the new philosophers and now played the same game.

3.4.1 The Methodology of Ontology: The Multiplication of Corpuscles and the Missing Metaphysical Supplement

The old philosophers give two distinct answers to the accusation that they multiply entities without necessity. First, they denounce the pretensions of the new philosophers, and assert that they are on the same footing as them. The new philosophers claim to have only three principles, matter, figure and motion, and thus to be more economical in beings than the old philosophers. In reality, they proceed exactly as do the old philosophers, because what explains this or that quality is not matter, figure and motion in general, but particular corpuscles with a particular figure and a

---

Antoine Rochon (1637–???), professor of philosophy in Bordeaux, *prédicateur* in Toulouse, finally quit the Jesuits for the Benedictines in 1685; it is said that Pardies polished up Rochon's *Lettre d'un philosophe à un cartésien de ses amis*. Arnauld, *Plusieurs raisons pour empêcher la censure ou la condamnation de la philosophie de Descartes*, § 1, pp. 303f., notes that this Jesuit mobilization was no coincidence: "On dit que le général des jésuites a écrit une lettre circulaire à toutes les maisons de la Société, pour obliger les jésuites d'écrire partout contre la philosophie de M. Descartes."

[77] For a general presentation of Fabri's project in natural philosophy, see Roux, "La philosophie naturelle d'Honoré Fabri." Brockliss, "Pierre Gautruche et l'enseignement de la philosophie de la nature," p. 201, notes that Fabri's *Physica* being too long for the students, it was probably aimed at teachers: among these teachers were obviously the Jesuits here studied.

[78] Jean-Baptiste de La Grange (1641–?) taught philosophy at Montbrison and Le Mans, then theology at Troyes; towards the end of his life, he was *curé* in Chartres; Pierre-Daniel Huet (1630–1721), preceptor to Dauphin with Bossuet (1670–1680), member of the Académie Française since 1674, was successively abbot of Aunay, bishop of Soissons, bishop of Avranches, abbot of Fontenay.

[79] Jean Duhamel taught philosophy at collège du Plessis, attached to the Sorbonne; his polemic with Régis began in 1692 with his *Réflexions critiques sur le système cartésien de la philosophie de M. Régis*.

particular motion. Perhaps old philosophers invented new forms and qualities each time that they had a phenomenon to explain; but, in such circumstances, new philosophers likewise invoke specific figures and specific motions without giving us any reason for this specificity, except their own pleasure.[80] In other words, the ancients accept the principle that one must not multiply beings without necessity,[81] but they assert that modern philosophers perform no better than they do when it comes to the explanation of specific phenomena; they multiply the figures and motions of their corpuscles at will, in just the same way as the old philosophers used to multiply forms and qualities. It is even said that they do worse, insofar as they admit Descartes' thesis that any corpuscle can be changed in any corpuscle: according to Huet, it is impossible in these circumstances to attribute particular effects to particular corpuscles, in the way that particular effects are attributed to particular Aristotelian forms or to particular Epicurean atoms.[82]

Of course, one might ask just how pertinent these responses are. First, the relation of the notion of form in general to peculiar forms is not the same as the relation of matter to specific corpuscles: in the first case, it is a grammatical relation of subsumption; in the second case, it is a physical relation of specification. Second, the fact that corpuscles change over time does not mean that it is impossible to establish a relation of causality between a kind of corpuscle and some specific effects: the same amount of matter does not have the same effects when organized into different corpuscular configurations. In this paper, however, I do not want to discuss the pertinence of the arguments, but to show that there was an argumentative debate and to reconstruct its lines of force.

The second answer the old philosophers offer is different, but likewise does not contest the principle that beings should not be multiplied without necessity. Namely, this second answer amounts to restricting the application of the traditional catchphrase to ontologically determined domains. The multiplication of forms should not be considered a multiplication of physical beings, simply because forms are not physical beings, but metaphysical principles. With their first answer, the old

[80] Fabri, *Physica*, Auctor lectori, § 12, n.p.: "[C]um enim corpus, modò quiescat, modò moveatur, modò illo, modò isto motu, centies quaeram, cur & unde sit ille motis; ergo motus principium non est: deinde innumeras atomos & figuras agnoscunt, quarum tamen nulla ratio est; igitur gratis, ut aiunt, & ex mera fingendi libidine, haec statuunt." See as well id., *Epistolae tres de sua hypothesi philosophica*, I § 23, resp. p. 52; id., *Œuvres de Fabri*, vol. IV, f. 208r.

[81] That they admit this principle is confirmed by particular explanations, see for example what La Grange, *Les principes contre les nouveaux philosophes*, chap. 26, § 9, p. 374, notes before giving his own explanation of elasticity: "Comme il ne faut point multiplier les Estres sans necessité, aussi ne faut-il pas avoir recours à aucune Forme Accidentelle, quand on peut expliquer les choses autrement." See as well, ibid., chap. 41, § 2, p. 518.

[82] "[V]erum si attentius spectemus principia Cartesii, quam specie simplicia, tam multiplicia effectu comperiemus. Nam continua illa tritura partium, quae quantumvis comminutae magis magisque tenuari, ac propterea novas semper figures induere possunt … feracissima seges est ad quidvis comminiscendum.… Id autem Aristoteli aut Epicuro nequaquam possis objicere: nam cum hic certas figures immutabiles, certumque figurarum numerum individuis suis corpusculis adscripserit; ille vero ad certas formas definitas retulerit rerum omnium ortum atque statum, certi inde effectus & definiti jure elici potuerunt" (Huet, *Censura philosophiae cartesianae*, chap. 8, § 3, pp. 171f.).

philosophers claim to be on the same footing as the new philosophers; here, they assert that forms give them an advantage, namely metaphysical supplements that do not serve to explain phenomena in their specificity, but that constitute their metaphysical foundation. This is well expressed by Rochon, who, after having recalled that Aristotelians also used corpuscles in their explanations, describes the opposition of physics and metaphysics as an opposition between the exterior and the interior:

> You can see, sir, that so far we are equal, and that your philosophy and ours are similar: the difference is that you stop there without going farther, and without even recognizing that there is something else in nature: whereas we believe that we would be stopping at the first surface if we did not seek to penetrate further to discover that beyond all that thus appears outside, there is yet inside something that is the principle of all these dispositions and all these effects, what we call *form*.[83]

That equals pulling the rug from under the feet of the new philosophers: if the forms of the ancients are not redundant in the explanation of phenomena, it is not because they are necessary, but simply because their function is not to explain phenomena, but to designate their essence.

3.4.2 *The Way of Physics: One Should Not Indulge in Hypotheses, Ignore Experiments and Use Empty Words*

Contrary to what might be expected, against the reproach of not explaining phenomena because they ignore efficient causes, as a rule, the old philosophers do not defend final causes. They rather attack the new philosophers because, devoting all their time to the invention of probable hypotheses regarding corpuscles and motions that could cause this or that phenomenon, they overlook the experiences of both common sense and the new sciences, without providing us with any new knowledge.

New philosophers intended to find explanations that could reduce phenomena to their corpuscular causes, but since these corpuscular causes are not subject to observation, they are hypothetical and probable in a context where science (*scientia*) was still equated with absolute certainty. Old philosophers insist that corpuscular causes are *only* probable, *just* hypotheses, *simple* fictions.[84] Moreover they deride with verve the different strategies of the Cartesians to pretend that they could bypass the

[83] Rochon, *Lettre d'un philosophe à un cartésien de ses amis*, §§ 59–60, pp. 142–144. See as well ibid., § 66, p. 154: "[V]ous ne ferez que repeter ce qu'enseigne Aristote. Mais souvenez-vous qu'outre cela Aristote reconnoit des Qualitez et des Formes que vous ne connoissez pas, & qu'il n'avouera jamais que l'odeur ou la chaleur soient ces vapeurs ou ces pyramides."

[84] La Grange, *Les principes contre les nouveaux philosophes*, Preface, § 27, p. 24; Pardies, *Discours de la connaissance des bêtes*, § 110, pp. 210f.; Huet, *Censura philosophiae cartesianae*, comments the Cartesian fable concerning the origin of the world, "Haec ... miracula, non disserentis Philosophi, sed somniantis; non docentis, sed optantis..." (chap. 6, § 2, p. 154) and picks up ironically the Cartesian expression *fabula mundi* (p. 158).

"only probable" character of their hypotheses.[85] Huet, for example, compares the Cartesian asserting that he can explain the probable origin of the world to a man seated at the gates of Paris who would explain to travelers that they came from one city or the other, simply because it could have been the case that they actually came from one city or the other city.[86]

It is known that in their teaching the Jesuits at least made way for new observations and experiments that were performed in the first half of the century[87]; in the polemical texts that I am here dealing with, tribute is indeed paid to the experimental philosophy that would have been that of Galileo.[88] The point to understand is that when experience is contrasted with hypothetical reasoning, this concerns both sensory experience and the scientific experiment. Indeed, what is first at stake here is to rehabilitate the sensible qualities and to rely on common sense. Everyone knows that light is spread around the world, a diamond is hard, that snow is white, that fire has heat, and here are the Cartesians telling us "what vulgar philosophy calls *sensible qualities*, are no accidents of bodies, but are rather modes of our soul, that is to say, real thoughts that we have when encountering objects that present themselves to our."[89] Who then will believe them? As systematically as Rohault stresses for each quality that the same word (e.g., "heat") could mean both the feeling (e.g., the heat that I feel) and the physical cause of this feeling (e.g., some corpuscles able to cause this heat), La Grange distinguishes the feeling from the physical cause of this feeling for a number of qualities. For him the point is not to assume that one can assimilate the feeling and its physical cause, or jump from the one to the other, but in case of heat, he explains that a series of reasonings makes it possible to attribute some heat to fire:

> [T]he Peripatetians do not draw their conclusion that fire is hot from the fact that it produces heat in the hand,… because it is not always necessary that a cause be like its effect. But they conclude that fire must necessarily be hot, because it generally heats all sorts of bodies, however different and opposed they may be, which is very well: because … when an efficient cause produces the same effect in an infinite number of different materials, the production of

[85] On the hypothetical character of corpuscular causes, see La Forge, *Remarques*, in Descartes, *L'homme*, pp. 216–218; Rohault, *Traité de physique*, I, chap. 3, § 3, vol. I, pp. 22f.; Poisson, *Commentaire*, VI, pp. 173–195; Gadroys, *Discours sur les influences des astres selon les principes de M. Descartes*, chap. 10, pp. 216–218; Lamy, *Entretiens sur les sciences*, pp. 257–261; Régis, *Système de philosophie*, Physique, I, Avertissement, vol. I, pp. 274f.; id., *Réponse à Huet*, chap. 8, § 3, pp. 304–306. For an analysis of the Cartesians' ambivalence towards hypotheses, see Clarke, *Occult Powers and Hypotheses*, pp. 131–163, 228–231, *passim*.

[86] "Nam si vel aliam caussam [sic] habere potest hic effectus, vel alium effectum habere postest caussa [sic] isthaec, hunc ex illa arcessere, divinare est…. Quam ineptus vero ille sit ridiculus qui sedens ad Lutetiae portam, profiteatur se dicturum unde viatores omnes illuc adventantes sint profecti; hunc Lugduno illum Divione dicat profectum; quia hic Lugduno, ille Divione proficisci potuit?" (Huet, *Censura philosophiae cartesianae*, chap. 8, § 3, pp. 169f.).

[87] Brockliss, "Descartes, Gassendi, and the Reception of the Mechanical Philosophy," pp. 454–456; id., "Pierre Gautruche et l'enseignement de la philosophie de la nature," pp. 190–194, 199, 209–216.

[88] Rapin calls Galileo the "Father of Modern Philosophy" (*Réflexions sur la philosophie*, § 18, p. 365) and the "Founder of Modern Philosophy" in Italy (*Réflexions sur la physique*, § 9, p. 450).

[89] Pardies, *Discours de la connaissance des bêtes*, § 7, pp. 10–13, and pp. 12f. for the quotation.

the effect cannot come from the disposition of the material, since the materials are different and the effect the same. This is why we are obliged to say then that the cause is like its effect.[90]

No ingenuity here, rather the will to reason so as to make a difference, in the sense, between what belongs properly to the body, which seems to be the origin of a particular feeling and what belongs either to the person, who actually has this experience, or to other bodies. Although this is not the place to go into detail, one can note that the purpose of Fabri's *Physica* was to establish a physics *more geometrico*, taking basic qualities (dry and wet, tense and compressed, heavy and light) as first principles endowed with such certainty that it would not be necessary to use any hypothesis whatsoever.[91]

Finally, the inability of new philosophers to take the experimental character of the new science seriously is underlined. Rochon notes for example that, once again, moderns pretend to more than they are actually able to accomplish. They pretend that they can anticipate the outcome of experiments (*prévenir les effets de la nature*), but the best that they can do is to retroactively predict this outcome when the experiment has already been performed. In this, they are similar to Cardano, who wanted to defend and illustrate his rules in astrology by drawing up the horoscopes of the dead. When he tried to extend his computations to future events concerning the living, what actually happened obliged him to resume his computations to adapt them to the actual events, once again retroactively.[92]

Considering what mechanical explanations had turned out to be in the hands of the Cartesians, the most pervasive and pertinent criticism consists in observing that they have no more informational content than their ancient counterparts. Namely, if the explanation of a given phenomenon is to say that certain corpuscles animated by certain motions produce this phenomenon, its informational content is no more determinate than the informational content of the assertion that a substantial form is what makes a given body what it is. It is Rochon whom I would like to follow in this respect. He first recalls the argument of the new philosophers that Aristotelian qualities, virtues and forms do not bring any new knowledge.[93] In reference to that argument

[90] La Grange, *Les principes contre les nouveaux philosophes*, chap. 34, § 1, pp. 452f.

[91] On Fabri's project, see Roux, "La philosophie naturelle," pp. 90f.

[92] "Il est vray que vos Messieurs font merveilles quand ils peuvent attraper une experience qu'ils ont faite cent fois pour en estre bien assurez. C'est alors qu'ils sont heureux à faire voir la beauté de leur doctrine en prevenant, disent-ils, l'experience, & en faisant voir ce qui doit suivre de leurs principes. Cela s'appelle deviner tout ce que l'on voit & predire exactement le passé. Je n'entends jamais parler de cét avantage qu'ils se donnent de prevenir ainsi les effets de la nature, que je ne me souvienne de ce qui arriva autrefois à Cardan" (Rochon, *Lettre d'un philosophe à un cartésien de ses amis*, § 85, pp. 197–198). "Prévenir les effets de la nature" is what Rohault pretended to do with his third sort of experiment, see *supra*, note 61; Cardan's horoscopes are also scoffed at by Daniel, *Voyage du monde de Descartes*, pp. 138f.

[93] "[V]ous avez oüy dire cent fois à vos Messieurs que dans la Philosophie de l'ecole on n'enseigne rien de la nature, qu'Aristote ne dit rien que ce que tout le monde sçait déja, qu'on répond à toutes les questions par une Qualité, par une Vertu, par une Forme, qui ne donnent aucune nouvelle connoissance: au lieu que Monsieur Descartes passe bien plus avant: qu'il explique les choses comme elles sont en elles-mesmes: qu'il fait entendre leur nature et qu'il rend raison de tous leurs effets" (Rochon, *Lettre d'un philosophe à un cartésien de ses amis*, § 50, pp. 128f.).

Rochon shows in the specific instance of the growing of plants that neither do the Cartesian explanations add anything new to common knowledge:

> Everyone knows that plants have fibers and pores by which the juices penetrate and then form all the parts of the plant.... If you say that in the School one says these things only in general and in a vague fashion, without explaining in particular, we say to you that you do the same. Everything you say gives no knowledge of the particular and of what is in fact in a plant. You content yourselves by saying that the pores are arranged *in a certain manner*, that they are *of a certain figure*, that the parts that *conform* to *certain openings* pass through, and that others are stopped. You try to get off with "*a certain*." But if I ask you what this certain figure is, and what the certain manner and what is this certain juice and these certain parts, you have nothing to say other than that you know no more. What more are you saying than the *per intus susceptionem* of ordinary philosophers?[94]

Like the moderns laughing at the ancients because they are unable to explain the transformation of flour into bread or the transition from a living dog to the cadaver of a dog, and pointing out that Descartes had challenged the Jesuits to find an issue on which his philosophy would not be more satisfactory than the philosophy of the School, Rochon challenges a whole assembly of Cartesians at the end of his book to explain the simplest thing, the formation of a pumpkin in a single night. His prognosis is, as one might guess, that they will be unable to do better than to parade once again with their "certain ways," "certain motions," and "certain figures."[95] Therefore, here the criticism does not bear on the mechanical explanations considered in their principle, but on what were in fact a number of them. In their desire to supplant the ancient philosophers and to show that they were able to explain absolutely every phenomenon, the Cartesians often merely asserted that a mechanical explanation of a given phenomenon should be possible, provided one assumed the adequate corpuscles and the appropriate motions. So they came to offer explanations as empty, circular or tautological as the explanations they blamed the old philosophers for. In this sense, it is no wonder to find this criticism of Rochon in Pardies,[96] La Grange,[97]

[94] Ibid., §§ 59–60, pp. 140–144, and pp. 142f. for the quotation.

[95] Ibid., § 84, pp. 194–196.

[96] "C'est une chose admirable que tous ces Philosophes qui nous reprochent perpetüellement que nous voulons les payer de mots qui ne signifient rien, & que nous leur répondons à toutes leurs demandes par une Vertu, ou par une Forme, pensent nous donner un grand éclaircissement sur ce sujet, en nous disant ce qu'ils disent à toutes les questions, que ce sont de certains atomes, de certains esprits, ou un certain feu, qui assûrément ne sont que des mots aussi vagues que le sont ceux de formes ou de vertus, & qui ne nous donnent pas plus de lumière pour voir le détail des choses, que font les qualitez occultes" (Pardies, *Discours de la connaissance des bêtes*, § 100, p. 188).

[97] "Lors qu'il s'agit d'expliquer la nature de quelque Qualite corporelle, le party des cartistes paroist assés fort pour tenir teste aux Peripateticiens: Ils ont recours à la disposition des parties, comme à un azile tres-assuré; si les corpuscules quarrés ne leur sont pas propres, ils prennent les pointus & les crochus, & dans une necessité, ils les font courir les uns aprés les autres, de sorte qu'on a toutes les peines du monde à les attraper" (La Grange, *Les principes contre les nouveaux philosophes*, chap. 3, p. 65).

Rapin[98] and Daniel.[99] It might be fitting to conclude with the condemnation of Descartes in Daniel's *Histoire de la conjuration*:

> [A]fter having led us to believe that he would explain everything in an easy and natural manner, he [Descartes] only explains physical effects by certain elements, certain assemblies of parts, certain movements and certain figures; that is hardly different from certain *entities*, certain *forms*, certain *virtues*, and certain *qualities*, and after all, by rights of seniority, the scholastic *I know not what* should win out over the Cartesian *I know not what*.[100]

3.4.3 *The Ontological Categories and the Controversy Over Animal Souls*

There were two distinct ways to counter the ontological bipartition between souls and bodies: either to display beings that are neither body nor soul, or to display beings that are intermediate between body and soul. The old philosophers took both paths, but the first was the most frequently used, especially when the old philosophers discussed animal souls. To use Father Daniel as a guide,

> the essential point of Cartesianism, its touchstone,... is the doctrine of the automata, which makes pure machines of all animals, by taking from them all feeling and all consciousness.... This single point includes or supposes all the principles and bases of the sect.... Here is the spirit and the essence, if I can speak this way, of pure Cartesianism.[101]

Indeed, the question of the soul of the animals involved issues well beyond the case of animals: first, what was psychologically at stake was to defend the idea that there is certainty in the senses distinct from the certainty of the mind and thus, in continuity with the questions dealt with in the preceding paragraph, to defend the idea of knowledge based on observations and experiments. But the question of animal

---

[98] "Et quand, pour rendre raison des choses, il [Descartes] a dit qu'elles se font par une certaine figure, par un certain mouvement, par une certaine extension, il a tout dit" (Rapin, *Réflexions sur la physique*, § 10, p. 455).

[99] "[L]es Péripatéticiens disent que toute la difficulté consiste à expliquer les choses en détail, qu'il n'y a que ce détail qui leur fait de la peine; que si vous vouliez bien leur faire comprendre cette *certaine manière*, cette *différente manière* que vous nommez si souvent et que vous n'expliquez jamais, ils seraient aussitôt à vous.... toute la science des Cartésiens en cette matiére se réduit à nous assurer que Dieu est tout-puissant et qu'il peut exécuter l'idée très-confuse, qui leur est venuë à l'esprit, d'une machine de chair & d'os, qui feroit par le moyen de ces ressorts ce que nous voyons faire aux bestes" (Daniel, *Suite du voyage du monde de Descartes*, pp. 43–46).

[100] Ibid., p. 230.

[101] Ibid., pp. 3f. The thesis that animal souls are superfluous and even, considering the bipartition of bodies and souls, impossible, is expressed by Cordemoy, *Six discours sur la distinction et l'union du corps et de l'âme*, pp. 205f., 266f.; Arnauld and Nicole, *La logique*, III, chap. 13, pp. 224–226; Poisson, *Commentaire*, V, pp. 147–168; Rohault, *Entretiens sur la philosophie*, pp. 138–152; Malebranche, *De la recherche de la vérité*, IV, chap. 11 and VI, II, chap. 7, resp. pp. 467–469 and 713f.

souls was also, and perhaps mostly, a way of defending an ontological point of view, for to confer a soul on animals is to admit that there exist more than two species of beings, or at least that there exists what Daniel calls "intermediate beings," that is to say, beings that partake of both the material body and the immaterial soul.[102]

The most elaborate argument in this respect is Pardies' *Discours de la connaissance des bêtes.*[103] The first part set out the Cartesian arguments against the existence of animal souls so clearly that Pardies was accused by some of his fellows of being a crypto-Cartesian.[104] In the second part however, Pardies presents a rigorous defense of animal souls. As a first step, Pardies proceeds psychologically: he seeks to establish the existence of a purely sensible knowledge, distinct from the intellectual or spiritual knowledge, by which we know that we know, something that implies reflexivity.[105] Thus, seeing is not only a physical process, since all the physical phenomena that accompany vision may occur in an artificial eye.[106] But that does not make seeing an intellectual process, since we can see without knowing that we see, that is, according to Pardies, with no consciousness that we see, with no attention to this perception, with no reflection about it. The most telling example taken to illustrate this purely sensible vision is the reading of a book: when we read, we see the characters, but we pay no attention to them, since we are unable, in general, to say if they were well formed, or if they were roman or italic characters.[107] The question at this point is to know what may be the ontological foundation of the sensitive knowledge that mere perception is, as distinguished from intellectual perception.

In a second step, Pardies once again asserts that, since it is distinguished from intellectual perception, sensible perception need not be associated with the mind.[108] On the other hand, figures, arrangements of parts, dispositions,

> none of this can enable us to understand how an animal could feel: we must then say that there is beyond all that some other principle, which we call *form*, and because these

[102] Daniel, *Suite du voyage du monde de Descartes*, p. 83.

[103] The argument is less tight in Grange, *Les principes contre les nouveaux philosophes*, chap. 24, pp. 339–354, and Daniel, *Suite du voyage du monde de Descartes*, pp. 83f., but identical in its principle: to establish a distinction between thought and reason, or between sensible knowledge and reasonable knowledge, which has the same effect as Pardies' distinction between mere perception and perception of oneself, namely, to assign to animal and to man two operations that are independent one from the other.

[104] On this, and for a more thorough analysis of Pardies' position, see Roux, "Pour une conception polémique du cartésianisme."

[105] Pardies, *Discours de la connaissance des bêtes*, § 78, pp. 150f.: "La connoissance spiritüelle, ou, si vous voulez, intellectüelle, est … une perception qui emporte essentiellement avec une espece de réflexion qu'elle fait invisiblement sur elle-même, en sorte que nous connoissons fort bien que nous connoissons. Mais la connaissance sensible est une simple perception d'un objet sans cette réflexion."

[106] Ibid., §§ 80–81, pp. 155–158. See in particular ibid., § 81, p. 158: "Car enfin, voir n'est pas recevoir des raions de lumiére, ni avoir une image de l'objet representée au fond de l'œil; voir, dit quelque chose de plus, puisque toutes ces representations optiques pourraient très bien se faire dans un œil artificiel."

[107] Ibid., §§ 82–83, pp. 159–161.

[108] Ibid., § 103, p. 195: "[C]es pensées qui emportent cette réflexion qu'elles font indivisiblement sur elles-mêmes, sont le seul caractére de la spiritüalité…."

operations are not beyond corporeal power, there is no need to say that this is a pure spirit, but rather that it may be a material form.[109]

Once the concept of material form is advanced, the third and last step is logically to show that this concept can be conceived, that is, that it is not contradictory; Pardies actually contents himself with showing that it is conceivable by his opponents. Indeed he proceeds *ad hominem*, noting that to explain the phenomena the new philosophers themselves need to admit the existence of motion, which is neither a bodily substance, nor a spiritual substance, but a mode of the body.[110] In the same way, the ancient philosophers would be entitled to admit substantial forms, which, "being neither bodies, nor modes, nor accidents of bodies, are nevertheless something corporeal," namely, in the case of animals, this something that makes us say of them that they are, precisely, animated.[111]

3.4.4 Another Social Twist

As we have noted, the strategy adopted by the old philosophers is often to turn the arguments of the new philosophers back on themselves. This concerns in particular the use of language: as we have seen, the old philosophers criticize the new philosophers for being all words when they state that some corpuscles endowed with certain motions are the causes of certain phenomena. Another example would be the way they accuse Descartes of playing with words when, introducing a distinction between "indefinite" and "infinite," he qualifies the world as "indefinite," and not "infinite."[112] But the explanation that the ancient philosophers give of this relation to language is not the desire to appear more knowledgeable than common people.

According to the old philosophers, the problem with the modern philosophers is rather that they are superficially only interested in public approval. Again and again, their ignorance of old philosophy is outlined and it is generally noted that they did not bother to acquire the skills necessary for the practice of philosophy.[113] According

[109] Ibid., § 105, pp. 198f.

[110] Ibid., § 107, p. 202.

[111] Ibid., § 108, p. 205.

[112] Babin, *Journal*, p. 42; Rochon, *Lettre d'un philosophe à un cartésien de ses amis*, § 14, pp. 27–29; Huet, *Censura philosophiae cartesianae*, chap. 5, § 5, pp. 149–150. For a contextualization of this distinction, see Ariew, *Descartes and the Last Scholastics*, chap. 8, pp. 155–171.

[113] See for example La Grange, *Les principes contre les nouveaux philosophes*, Preface, § 33, chap. 34, § 1, chap. 39, § 3, chap. 45, § 9, chap. 50, § 1, resp. pp. 30, 450, 503, 569 and 599. Huet, *Censura philosophiae cartesianae*, chap. 8, § 7, pp. 196–202, insists that Cartesians profess to despise erudition in general, and Latin in particular, see p. 201: "Quinetiam adeo inconsiderate imperitos se rudes produnt, ut vix alterius quam alterius quam vulgaris linguae usum concedant in scribendo, nec aliam probent Latinatem, quam simplicem, incomtam, & facilem; ne sibi scilicet, cum legent, saepius recurrendum sit ad interpretem. Jam ergo ludibrium debemus Cartesianis, quod eruditi sumus." Even if he presents philosophy as written for *honnêtes gens* (see *supra* the references given in note 39), Rapin (*La comparaison*, Avertissement, pp. 275–277) insists that it implies erudition and arduous work.

to the old philosophers, they do not seek to establish anything serious, but only to call attention to themselves in the *salons* by making fun of their opponents, even by insulting them. Rochon is here most telling:

> Your gentlemen speak easily; when in the midst of a circle, they say whatever they please regarding the doctrine of Aristotle, with no one there to contradict them. They then ridicule everything. ... One mustn't be astonished if so many people who have never read Aristotle and who have never heard of the ordinary philosophy, except in the manner it pleases these gentlemen to speak of it, think that Aristotle's philosophy is as they describe it. This manner is assuredly not very honest.[114]

Later on, Rochon notices, not without wit, a kind of pragmatic contradiction between the public of the *salons* that the new philosophers cultivate and the mechanical knowledge that they claim. This contradiction would be fully revealed if the new philosophers actually possessed the knowledge in question: of course the audience does not want to know what corpuscles actually produce this phenomenon, because such knowledge would be better suited to a locksmith or a watchmaker.[115]

Concerning the principle of authority, we can distinguish two types of strategies in the counterattack of the old philosophers. On one hand, they admit that one must follow the truth rather than Aristotle. Highlighting the historical and doctrinal distance that separates Aristotle from his later commentators, some of them call for

[114] Rochon, *Lettre d'un philosophe à un cartésien de ses amis*, § 51, pp. 129f. See as well ibid., § 58, p. 140: "[C]es Messieurs prennent tant de plaisir à ne parler jamais qu'en riant de la Philosophie vulgaire: qu'ils en prennent les lambeaux qu'ils jugent estre les plus propres pour donner à leurs auditeurs l'idée qu'ils pretendent de cette Doctrine"; La Grange, *Les principes contre les nouveaux philosophes*, Preface, § 44, p. 40: "Pour ce qui est de Gassendi & de Descartes,... n'ayant pû, ou n'ayant point osé combattre par raisons nostre Philosophie, ils se sont contentez de luy insulter"; Pardies, *Discours de la connaissance des bêtes*, § 6, p. 9: "[T]out ceci ... semble d'abord plus tenir de la galanterie d'un faiseur de Romans, que de la pensée serieuse d'un Philosophe"; Huet, *Censura philosophiae cartesianae*, chap. 8, § 3, p. 185: "Ac tam absurdis commentis ut fidem quaerant factionis hujus participes, adversarias sibi sententias *Praejudiciorum* ... infamare solent; tum praesertim cum rationibus carent ipsi argumentis."

[115] "Car enfin, Monsieur, à vous entendre parler on diroit que d'une école de Philosophie vous voudriez faire une boutique de Serrurier. ... Un honnête homme se doit-il mettre en peine de toutes ces petites particularitez. ... Quoy voudriez-vous que les Dames se fissent écolieres des Horlogers pour apprendre le nombre & l'engrainement des dents de chaque roüe, & de chaque pignon de leurs montres? faut-il donc qu'elles sçachent le biais dont sont inclinées les pallettes du balancier, ou la proportion qu'il faut donner à la diminution de la fusée ? n'est-ce pas assez qu'on sçache en general que tous ces mouvemens sont faits par la disposition par l'engagement des rouës & des ressorts, ce que vous appelerez, si vous voulez, *vertu indicative* ou *sonorifique*?" (Rochon, *Lettre d'un philosophe à un cartésien de ses amis*, § 78, pp. 182–184). See as well "que dorenavant ils [Cartists] ne s'appliquent pas tant à l'étude de l'Anatomie ..., c'est la Science des Medecins et des Chirurgiens, qu'il ne faut point leur envier. Je leur conseil encor ... de ne pas mettre tant de temps à faire des Experiences" (La Grange, *Les principes contre les nouveaux philosophes*, chap. 50, § 10, p. 611). This criticism contradicts what we explained on the relevance of experiences; notwithstanding the fact that the attitude of all the old philosophers towards the new sciences is not identical, their criticisms are sometimes more eristic than constructive.

a return to the true meaning of Aristotle, that is to say, the truth and nothing but the truth. In particular this is the case with Fabri, who intends to end the scholastic digression and to restore the thought of Aristotle to its literal and authentic meaning, and eliminate the corruptions imposed upon it by the Arabs and, to a lesser extent, the Spanish.[116] This analysis was made popular by Rapin in his *Comparaison d'Aristote et de Platon*, and, much more radically, in his *Réflexions sur la philosophie*, that are both referred to by new philosophers.[117]

On the other hand, they denounce the manner in which Descartes became an authority for the new philosophers just as Aristotle was an authority for the old philosophers. Although the term "sect" is not necessarily pejorative, as it is still commonly used for a school of philosophy, it certainly is pejorative when La Grange compares Cartesians to sects of heretics, which, having fallen in love with some extravagant opinion, then support it against all odds.[118] Descartes is called the master and doctor of the Cartesian philosophers and Régis their prince.[119] Therefore, to follow Rapin, what one believed to have won by freeing oneself from the philosophy of Aristotle is at once lost by submitting to the yoke of Descartes:

> [A]ll these fine precepts, which are given to remove us from the preoccupations of education, custom, authority, and to heal us of all popular preconceptions, are only traps for our credulity. One speaks of liberty only to impose a new yoke. This is only giving to the Moderns what one wants to take away from the Ancients, and one wants to destroy the credit of Aristotle only to establish that of Descartes.[120]

[116] Fabri, *Physica*, Auctor lectori, §§ 7–9, n.p.; id., *Epistolae tres de sua hypothesi philosophica*, I §§ 26–27 and II § 4, resp. pp. 57–58 and 70; *id.*, *Œuvres de Fabri*, vol. IV, f. 210r. These texts are quoted and commented in Roux, "La philosophie naturelle d'Honoré Fabri," pp. 91f.

[117] "Les Arabes s'étant rendus les Maîtres du monde, par leurs conquêtes, dans les siecles suivants, firent une espece de revolution dans les lettres, aussi bien que dans l'Empire. Le caractere de leur esprit, subtil, rêveur & profond, qui les attacha trop litteralement au texte d'Aristote, leur fit prendre une maniere de raisonner abstraite, qui s'écarta un peu de la solidité des Grecs & des Latins.... Outre que la Philosophie devint pointilleuse sous les Arabes, par ces precisions & par ces concepts abstraits, qu'elle introduisit dans l'école, elle devint aussi tout à fait sauvage dans ses expressions" (Rapin, *Réflexions sur la philosophie*, § 15, pp. 358–359). Ibid., § 16, pp. 359–362 says that the Spanish scholastic period was full of disputes and extravagances. On the Arabic period, see as well id., *La comparaison*, IV, chap. 6, pp. 407f., and, on the necessity of returning to a more literal understanding of Aristotle, pp. 415–417.

[118] "Je ne m'étonne plus de lire dans l'Histoire Ecclesiastique, qu'il y a eu autrefois des Sectes d'Heretiques, qui ont enseigné des Opinions qui meritent plus le nom de folies & d'extravagances, que celuy d'Erreur et d'Heresie. Quand les gens sont d'humeur à faire cabale, & qu'ils s'attachent à quelqu'un qui dogmatise, ils ne manquent pas d'entrer dans ses sentimens, quelques absurdes qu'ils puissent estre" (La Grange, *Les principes contre les nouveaux philosophes*, chap. 24, § 5, pp. 344f.). The association of "sect" with heresy is to be found as well in Daniel, *Voyage du monde de Descartes*, p. 131, but one finds by him also neutral uses of "sect," see for example ibid., p. 5.

[119] La Grange, *Les principes contre les nouveaux philosophes*, chap. 24, § 7, p. 349; Huet, *Nouveaux mémoires pour servir à l'histoire du cartésianisme*, p. 49.

[120] Rapin, *Réflexions sur la philosophie*, § 20, p. 369.

3.5 Conclusions

Both from a scientific and a philosophical point of view, the texts under examination are rather poor in quality. I noted in the introduction that scientific practice is definitely autonomous with respect to the philosophical concern for the ontological principles and the norms of natural philosophy. Even if the protagonists of this confrontation were sometimes good scientists, like Rohault and Pardies, their scientific work was not involved in this debate. Nor is this confrontation always philosophically convincing; some of these arguments, if not most of them, are rehashed and inspired more by eristic considerations than by the desire of conceptual elaboration. Moreover, as a rule I did not discuss in detail their philosophical relevance, but indulged in a somewhat artificial and schematic reconstruction of the configuration they made up. All that being said, I think that this reconstitution better allows us to understand the way in which modern and ancient philosophers opposed each other in this crucial period.

First and remarkably enough, the two sides agree on a certain number of points: both accept the methodological principle that one must not multiply beings without necessity, they oppose the use of empty words, they criticize the abuse of the principle of authority, they want to ensure respect for the established religion. On all these points, old philosophers respond in echo to new philosophers, and as I will explain shortly, they had thus already conceded a great deal, not only in rhetorical techniques, but also in philosophical principles.

This is not to say that these new philosophers and these old philosophers agree on everything. Our systematic confrontation allows us to isolate the key issue of their opposition, the question of whether or not one needs substantial forms when doing natural philosophy. This question appears in three of the arguments I have isolated:

1. For the new philosophers substantial forms lead to the unnecessary multiplication of beings since they are not necessary for causal explanations of phenomena. For the old philosophers, they are a metaphysical supplement, the function of which is not to explain physical phenomena, but to serve as their foundation.
2. For the new philosophers, the elimination of substantial forms is tied to the bipartition of beings between body and soul. For the old philosophers, one needs substantial forms, and more generally entities that can be reduced neither to body nor to soul, to account for sensible qualities and for the sensitive knowledge of animals.
3. As for religion, which I deliberately left aside, the question is to know, if it is the elimination or the conservation of substantial forms that is the most respectful of religion. Both ancients and moderns affirm that the positions of their opponents on substantial souls can lead to certain difficulties with respect to the beliefs of established religion. For example, moderns say that substantial forms are material and perishable, and thus constitute a precedent that can lead libertines to wonder if our human souls are in fact immaterial and immortal.[121] But, turning

[121] Arnauld and Nicole, *La logique*, III, chap. 19, § 2, p. 245; Rohault, *Entretiens sur la philosophie*, p. 145; Clerselier, Preface, in Rohault, *Œuvres posthumes*, n.p.

their arguments against the new philosophers in a procedure that we have often encountered in this paper, ancients say that refusing substantial forms to animals will sooner or later lead most of us to think that there is no reason to make an exception for men.[122] Pardies notes that, given the symmetry between the two arguments, such a *reductio ad libertinum* does not constitute a way to conclude in either direction:

> Some think that this opinion that denies souls for animal is dangerous and that it favors the impiety of the libertines…: For, they say, once one admits that all the operations of animals can be carried out without a soul and by the sole machine of the body, we will soon take the next step and say that all operations of men can also be done by a similar disposition of the machine of their body.… They don't perhaps reflect that one can oppose a similar reasoning and say that once you admit that everything admirable that happens with animals can happen by means of a material soul, would you not soon take the next step and say that everything that happens in men can also be done by means of a material soul? Up to then everything is equal: one has no more right than the other to reproach their feelings and to make them odious on the grounds of the consequences that could be drawn in favor of the impious.[123]

To tell the truth, one gets the impression that this is rather used as a deterrent for one's opponent than an opportunity to explore in greater depth the general concepts concerning the world and what we can know of it. Considering the symmetries existing between arguments in general, one wonders if this controversy had any winner. And in a sense, we knew the answer to this question right from the beginning. As a rule, newcomers are by definition the winners, and this is no exception: it has already been established that between 1670 and 1690 most teachers, beginning with Parisian seculars, gave up substantial forms and replaced them with mechanical explanations.[124] But a systematic study such as mine allows to reach more nuanced conclusions.

I wanted to give the old philosophers a chance to speak for themselves, and to avoid writing a triumphalist history where the new follows the old as day follows night. Their actual competences in physics are quite diverse, the two extremes being Pardies and La Grange: the first one was a brilliant scientist, who offered a pertinent criticism of Newtonian optics, the second one seems to be totally uninformed about seventeenth-century physics, and goes from one bit of nonsense to the next when dealing with the relativity of motion.[125] However, because of their long acquaintance with the scholarly practice of erudite commentary, all of the old philosophers I dealt with here developed an advanced knowledge of the works of the new philosophers, referred to precise passages and were able to discuss different theses. Thus the general

[122] Daniel, *Suite du voyage du monde de Descartes*, p. 71.

[123] Pardies, *Discours de la connaissance des bêtes*, § 49, pp. 99f.

[124] See Brockliss, *French Higher Education*, pp. 357–360; for more detailed case studies on the vanishing of Aristotelian principles in the context of teaching, see id., "Aristotle, Descartes and the New Science"; id., "Descartes, Gassendi, and the Reception of the Mechanical Philosophy"; id., "Pierre Gautruche et l'enseignement de la philosophie de la nature."

[125] La Grange, *Les principes contre les nouveaux philosophes*, chap. 9, §§ 2–6, pp. 136–144.

impression is that unlike the obscure physicians muttering away in bastardized Latin in *The Imaginary Invalid*, they managed to reply to the new philosophers. Some of them even knew how to get the mockers from high society on their side, adapting the tone of the *honnêtes gens* and manipulating with ease the polemical tools of satire and parody. But to have recourse to such techniques is in fact already to concede a good deal.

This can be easily shown by wondering what exactly the old philosophers meant to defend in the scholastic entities, and in particular with regards to substantial forms.[126] In the Aristotelian tradition saying that a natural being has a substantial form is a first approximation to answering three questions: why is this being *a* substance rather than a collection of properties? Why is it a *substance*, so that some of its properties reestablish themselves after undergoing a change? Why is it a substance *of this species* rather than of another? The issue is not only to affirm the existence of entities known as "substantial forms," but to cover reality with a complex network of distinctions, for example between natural and artificial forms, or between substantial and accidental forms. Once these distinctions were made, one could take on the discussion of serious questions: how to explain the transformation from one form to another or from privation to form? Can forms exist separated from matter? And matter separated from any form? What distinguishes a being from one species from another being of the same species? Are there substantial forms associated with all the parts of an organic being? If so, what is their relation to the substantial form of this organic being?

But in the texts we have been examining, our ancients do not make these distinctions and enter only rarely into this type of discussion.[127] To make themselves understandable to the new public of *honnêtes gens*, they do not burden themselves with these details, and sometimes seem to content themselves with the affirmation that substantial forms exist. And in so doing, they concede a lot, for the new philosophers' criticism of the old philosophy did not simply concern the existence of substantial forms, but also that of a multitude of lesser beings, of complex distinctions, and the endless discussions that accompanied them. In other words, the old philosophers came to defend substantial forms in a spirit that was no longer scholastic. But substantial forms without scholasticism were nothing, or at least not much. They were no longer philosophical tools, but hollow and empty shells, symbols of the social positions that the old philosophers wanted to defend, inasmuch as they were teachers in the schools and universities, preceptors and spiritual directors. As is often the case, only when a belief is utterly dead it becomes important to defend it.

However, one last swing of the pendulum is necessary. Namely, it is striking to note that the criticisms the old philosophers addressed against Cartesianism are

[126] On the emergence of the notion of substantial form, see Copenhaver, "Scholastic Philosophy and Renaissance Magic"; on its complexities in late scholasticism, see Des Chene, *Physiologia*, chap. 3, pp. 53–81.

[127] Rather surprisingly, the most detailed discussion on the different scholastic options concerning substantial forms is to be found in Rohault, *Entretiens sur la philosophie*, pp. 112–117.

precisely the same that will be found again during the Enlightenment: to neglect experience, to try to get away with using general words, to neglect phenomena that do not square with a predefined categorization of beings. In this sense, one can say that the ancient philosophers did not really lose their war, at least not completely: they participated in discussions of their time that determined the subsequent criticisms of Cartesianism. No doubt there is also a lesson in here, one general enough for us to conclude with. When the history of philosophy is not confined to a given work or to the works of a given author, it is often written like a play: stars lead, followed by a second-rate supporting cast, heroically taking turns on center stage, one coming to dethrone the preceding ones. Without doubt we will have to learn to write the history of this multitude of so-called extras: without always being aware of it, they may happen to define the plot.

Part II
Matter, Motion, Physics and Mathematics

Chapter 4
Matter and Form in Sixteenth-Century Spain: Some Case Studies

Victor Navarro-Brotons

4.1 Introduction

In the last few decades, scholars have rethought the history of matter theories in important ways, particularly with respect to the sources, origins and antecedents of atomism and corpuscularianism in the seventeenth century. In particular, recent commentators have shown that the supposed opposition between atomism and Aristotelianism is insufficient for understanding the historical development of matter theories and their conceptual plurality. Along with corpuscular doctrines dating back to Antiquity, such as those of Heron or Asclepiades, well treated in Lasswitz's classic work, certain aspects of the Aristotelian corpus and tradition have contributed in important ways to subsequent corpuscular doctrines.[1] Even though Aristotle opposed atomism and maintained that matter was continuous, his thought did not completely rule out corpuscular explanations.

In a recent book, *Late Medieval and Early Modern Corpuscular Matter Theories*, which offers an excellent panoramic view of the state of the art, the editors note that "the rich and disorderly results of contemporary research ... indicates just how difficult it has become to defend old essentialist distinctions."[2] In her contribution to the book, Danielle Jacquart analyses some Salernitan and Chartrian works on medicine and philosophy, dating back to the late twelfth and thirteenth centuries. She demonstrates that philosophers and physicians interested in the make-up of mixed

This work has been partially financed by grants from the Spanish Ministry of Education and Science. I am most grateful to Encarna Pastor for his help with the transcription of the Latin texts.

[1] On corpuscular theories in Antiquity, see Lasswitz, *Geschichte der Atomistik*; on Aristotelian tradition and its influence, see below, note 11.

[2] See Lüthy et al. "Introduction."

V. Navarro-Brotons (✉)
Consejo Superior de Investigaciones Científicas, Universidad de Valencia,
9 Mestral, 46110 Godella, Valencia, Spain
e-mail : victor.navarro@uv.es

D. Garber and S. Roux (eds.), *The Mechanization of Natural Philosophy*, Boston Studies in the Philosophy of Science 300, DOI 10.1007/978-94-007-4345-8_4,

bodies often relied on ancient works, which presented the structure of composite bodies as particulate, integrating such ideas into their own explanations. The result was a definition of a mixture as "the union of the mixable ingredients, which have been joined through their *minima*," as stated by Bartholomew of Salerno.[3] These texts undermine the assumption often made that Julius Caesar Scaliger was the first philosopher to adjust Aristotle's definition of mixture to the needs of corpuscular theories through his influential definition of mixture as "the movement of minimal bodies toward mutual contact, so that a union occurs." "The question thus arises," the editors ask, "whether there existed, in the medical commentary tradition, a continuous corpuscularian subcurrent that links Bartholomew of Salerno with Scaliger."[4]

In his contribution to the book, John Murdoch offers an excellent examination of the medieval and Renaissance tradition of *minima naturalia*. Murdoch criticizes some prominent historians for not taking sufficient care in distinguishing the *mixtio* and *minima naturalia*, unlike Aristotelian, Medieval and Renaissance philosophers.[5] While the available evidence does not allow us to conclude that the late medieval theory of *minima* can be considered a corpuscular theory of matter, an important exception is the Latin Geber's *Summa perfectionis*, according to William Newman the starting point of a tradition of experimental corpuscularianism in alchemy. This tradition has as one of its principal sources Aristotle's *Meteorology*, which includes a detailed corpuscular description of matter expressed in terms of *poroi* (pores) and the *onkoi* (corpuscles); another source is certain passages from the Aristotelian *De generatione et corruptione*. This tradition includes, for example, the work of Giovanni Agostino Pantheo, *Ars et theoria transmutationis* (1518), who redefined the mixture as "the union of the altered miscibles conjoined *per minima*." Pantheo states very clearly that these *minima* are indivisible and that "the element is the smallest of perceptible bodies is obvious from its definition. For element is the smallest particle of the body."[6]

This essay is part of a larger study of Spanish works in natural philosophy and science (with all the necessary caution regarding the use of the term "Spanish") in the sixteenth and seventeenth centuries.[7] The Iberian Peninsula was without doubt one of the principal loci of the so-called "Second Scholasticism," becoming the bastion of later scholasticism.[8] However, as Charles Schmitt has pointed out, this

[3] Jacquart, "*Minima* in Twelfth-Century Medical Texts from Salerno," p. 47; Lüthy et al. "Introduction," p. 19; on Bartholomew of Salerno, see also Jacquart, "Aristotelian Thought in Salerno," and Pabst, *Atomtheorien des lateinischen Mittelalters*, p. 194.

[4] Lüthy et al. "Introduction," pp. 19f. On Scaliger, see Lüthy, "An Aristotelian Watchdog."

[5] Murdoch, "The Medieval and Renaissance Tradition of Minima Naturalia".

[6] Newman, "Experimental Corpuscular Theory in Aristotelian Alchemy," p. 303, translation altered. See also Newman, *The Summa Perfectionis of Pseudo-Geber*, and Newman, *Atoms and Alchemy*. Lüthy

[7] I offered a preliminary sketch and synthesis in Navarro Brotons, "De la filosofía natural a la física moderna" and id., "La filosofía natural."

[8] But Spanish or Iberian scientific and philosophical activity can not be reduced to Scholastic thought. See Navarro Brotons and Eamon, "Spain and the Scientific Revolution."

predominance of the Aristotelian tradition did not imply doctrinal uniformity and conceptual homogeneity, but a wide spectrum of different interpretations of a large corpus of texts. Schmitt himself insisted that many texts on natural philosophy from the Renaissance period considered to be simply Aristotelian have yet to be read or analyzed. Some of them have been studied, but often with inappropriate hermeneutic and historiographic criteria not currently accepted.[9] The last few decades have witnessed a renewed interest in the works of the most prominent Peninsular authors. This interest is evident in studies going from the *Calculatores* of the end of the fifteenth century and beginning of the sixteenth, such as Alvaro Thomaz, Juan de Celaya, Pedro Margalho and Gaspar Lax and their influence on Domingo de Soto, to the studies of the main neo-Scholastics of the final decades of the sixteenth century and the beginning of the sixteenth, such as Benito Perera, Francisco de Toledo, Francisco Suárez, the Conimbrecenses, Gabriel Vázquez and Juan de Santo Tomás. All this is closely linked to the problems of continuity and change in the formation of modern thought, in particular to the study of its main protagonists, such as Galileo, Descartes, and Hobbes.[10]

In this paper I would like to contribute to this very important task by focusing on some lesser figures, not widely studied, and on some figures who have produced works that are important but which have not yet received due recognition. In the first part of this paper, I will focus on the work of the Majorcan physician Pere Bernat d'Ólesa i Rovira, who worked in Valencia and proposed a corpuscular theory of material substance. In the second part, I will briefly refer to the theories of two other Spanish physicians of the second half of the sixteenth century: Gomez Pereira and Francisco Valles. Since the three authors in question were all physicians, an obvious question is the degree to which their ideas concerning matter and, in general, their treatment of natural philosophy was dictated by the fact they were doctors and influenced by the demands of the rational practice of medicine. Although I cannot now give a comprehensive and satisfactory answer to this question, I want to note its relevance to this study; doubtless it influenced my choice of authors.

4.2 The Corpuscular Theories of the Physician d'Olesa

Pere Bernat d'Olesa i Rovira (ca. 1430–1531) was born in Palma de Mallorca. He studied art and medicine at Pisa between 1490 and 1495. In his last year, due to the arrival of the army of Charles the Sixth of France, he moved to Montpellier, and from there to the University of Lerida. In 1497 he settled in Valencia where he lived

[9] See Schmitt, *Aristotle and the Renaissance*.

[10] On Iberian *Calculatores*, see Wallace, *Prelude to Galileo*. On neo-scholastics and Descartes, see Ariew, *Descartes and the Late Scholastics*; Des Chene, *Physiologia* and *Life's Form*. On neo-scholastics and Hobbes, see Leijenhorst, *The Mechanisation of Aristotelianism*. See also Ariew and Gabbey, "The Scholastic Background."

until his death. In Valencia he must have enjoyed considerable professional prestige. He was an examiner of physicians and was appointed professor of the Faculty of Medicine, but renounced it in favor of his work as a clinician. His work, *Summa totius philosophiae et medicinae* (Valencia, 1536), was finished shortly before his death and was printed by his eldest son, Gaspar d'Olesa. It consists of a thick volume of about 400 pages, divided into three treatises, each with its own foliation. The first treatise deals with Olesa's ideas concerning natural philosophy, especially the structure of material substance and the analysis of sensations. The second part concerns the fundamental concepts of medicine, following the lead of the *Canon* of Avicenna and leaning on the ideas set out in the *Summa*'s first treatise: concepts of health and sickness, *de elementis*, *de complexionibus*, *de humoribus*, *de membris*, and *de virtutibus*. The third treatise, much larger than the previous ones, consists of a discussion of 21 propositions put forward in the first treatise, using a typical scholastic approach: *argumenta*, conclusions (with *probationes*) and *responsiones*. In addition, the treatise included a study of medicines: *de actuatione medicinarum* and *de gradibus medicinarum*, and finally an appendix on the causes of dreams, prophecies and divinations.[11]

Olesa devoted considerable attention to the question of the structure and principles of material substance and the structure and formation of compounds or mixtures. Olesa's main reference is without doubt Aristotle, which he interprets and understands mainly through the commentaries of Averroes. Apart from Aristotle and his main commentator, Olesa refers to a considerable number of authors, both in Antiquity and the Middle Ages, not only philosophers but also physicians such as Hippocrates, Galen, Avicenna, Avenzoar, Gentile da Foligno, and Arnaldus de Villanova.

4.2.1 Elements, Minima and Qualities

For Olesa, as for Aristotle, the elements are the first principles of natural things and of the mixtures. But contrary to Aristotle's opinion, the elements are ungenerable, incorruptible and are divided into minimal parts. These parts are naturally indivisible but are themselves made up of parts.[12] In mixtures, they are whole without a

[11] Biographical data on Olesa can be found in López Piñero and Garcia Sevilla, "Pere d'Oleza (Petrus d'Olese) y su obra *Summa totius Philosophiae et Medicinae*"; López Piñero, "Pere Bernat Olesa i Rovira." There is no thorough analysis of Olesa's work in the literature. López Piñero considers it "una exposicion sistemática de la filosofia natural, la biologia y la medicina desde el atomismo," but he does not make any distinction between atomism and corpuscularism.

[12] Olesa, *Summa*, Treatise I, f. 3r, col. 2: "Elementa ergo per minima naturalia divisa ingenerabilia et incorruptibilia et inalterabilia per spheras omnes disseminata sub ea tamen proportione sub qua denominatio cuiusque illorum salvari potest, neque a formis quas materie elementorum habent expoliari possint." See also ibid., f. 3v., col. 1: "Prima propositio: non est dandum naturaliter aliquod elementum purum maius minimo naturali eiusdem elementi. Secunda propositio: elementa sunt formaliter in quolibet mixto perfecto, nullo modo in formis suis remissa. Tertia propositio: minima naturalia elementorum sunt naturaliter indivisibilia, cum quo stat ipsa habere partes. Quarta propositio: elementa sunt ingenerabilia et incorruptibilia et inalterabilia naturaliter."

diminution in their substantial or accidental form. The *minima* have a figure that is an accident with respect to the form, and all the *minima* of each element have the same shape, although Olesa did not in general know what it was except in the case of fire, which, he thought, might be pyramidal because of the sharpness of the angles, or spherical because of the speed of the particles' movements.[13]

Olesa adopted an idea advanced by Galen and noted by Aristotle himself, namely that water, earth, air, and fire that we perceive are not pure substances but composed of other more basic pure principles. This allowed him to distinguish between the apparent, perceptible and sensitive *minima* and the true imperceptible and insensitive *minima*: "True *minima* of elements cannot be perceived by the senses."[14] In the same way, he affirmed that there neither exist nor is there a natural process of creating a pure element (without a mixture) greater than the natural minimum of the same element.[15]

For Olesa, elements contain qualities, not only hot, cold, wet, and dry, the first four postulated by Aristotle, but also others such as gravity, levity, lucidity, opacity, diaphaneity, color, and so on. Therefore fire is warm, dry, light, lucid, red, clear, bitter, an intense color, and productive of a sharp sound; water is cold, humid, heavy *secundum quid*, white, diaphanous, not intensely odiferous, insipid up to a degree and a productive of heavy sound; the earth is dry, heavy *simpliciter*, black, opaque, rough or unpleasant, and productive of a heavy sound. All these qualities are primary, each differing from the others. The elements possess the appropriate qualities to a very high degree, without being able to change into other elements via a change of qualities.[16]

In the Aristotelian tradition, the primary qualities (hot, cold, wet, and dry) were distinguished from the secondary qualities. The latter were divided into two types: secondary tactile qualities and specific, non-tactile, qualities. The secondary tactile qualities are derived directly from the primary qualities of the elements, all of which are tactile. But the secondary non-tactile qualities, such as smell, taste, and color come from particular combinations of the primary qualities that cannot be produced in the elements themselves, but only in compounds. Sound was usually treated as an exception, since it was thought to result solely from local motion.[17]

The derivation of secondary qualities from primary qualities was far from clear, especially in the case of specific sensory qualities (hardness for example). A particularly difficult question was the derivation of gravity and levity. In any case, the exact manner of the derivation of sensory qualities was not defined; philosophers in the tradition only asserted that secondary qualities resulted from a mixture of the

[13] Ibid., Treatise I, f. 5r.; Treatise III, f. 6r.-v. and f. 22r., on the shape of elements.

[14] Ibid., Treatise I, f. 3v., col. 2: "Elementorum minima quae vere talia sunt a nullo sensu aprehendi naturaliter posse." See also Treatise III, f. 32r. ff.

[15] Ibid., Treatise I, f. 3r. ff.; Treatise III, f. 32r. ff.

[16] Ibid., Treatise I, f. 7; Treatise III, f. 33r. ff.

[17] I follow Maier, *An der Grenze von Scholastik und Naturwissenschaft*, pp. 3ff. I have also consulted Maier, *On the Threshold of Exact Science*, an English translation of some of Maier's selected papers. See also Emerton, *The Scientific Reinterpretation of Form*.

primary ones. Some other philosophers thought that the secondary qualities of a mixture came directly from its substantial form, and therefore they held that the derivation of the qualities of the mixture from the primary qualities of its constituents was not strictly necessary. In his discussion on the question, Avicenna proposed a hybrid solution, concluding that the mixture of the primary qualities of the elements in a compound creates a disposition toward the secondary qualities, which, nevertheless, in fact proceed from the new substantial form.

Scholastic thinkers also confirmed that intermediate qualities, such as the colors red or green, could come from the extreme qualities such as black and white. But it is not clear how this idea can be reconciled with the theory that those same qualities derive from a mixture of the primary qualities. In *De sensu*, Aristotle discusses the individual specific qualities and how the intermediate quality proceeds from a particular mixture of extremes. Colors, for example, are supposed to derive from mixtures of black and white. But the connection between this theory and the theory that secondary qualities depend on primary qualities was not made completely clear. One solution to the derivation of the intermediate qualities from the extremes was to consider not a genetic or causal derivation but a natural affinity between the intermediates and the extremes (*convenientia naturalis medii cum extrema*). It involved postulating an ontological affinity between the species: the intermediate qualities contained these extremes merely in the sense of affinity (*secundum conventientiam* or *secundum virtutem*).

Olesa confronted all these problems and concluded that the secondary qualities are purely apparent, so that for him the elements retain their primary qualities in the mixture. Olesa did not accept the view that secondary qualities result from actions and passions of the primary qualities. Thus for Olesa the color green perceived in a mixture is not a quality resulting from the action of the elements, intermediate to and participating in the extremes or *per abnegationem*. Olesa attributed four true and basic colors as primary qualities of the elements in such a way that the mixtures of the elements form the apparent colors.

The color of the mixture for Olesa is just the way the mixture, with all its parts, is presented to us. He argues that if one wants to conclude that our senses are deceiving us, one has to distinguish two types of deception. If what we consider to be white appears to be green, there are two kinds of deception: the thing in question is neither white, nor does it appear to be white. Certainly, one could say that given that the mixture, for example, may contain water, which is white, our judgment was correct, although our visual faculties may be indisposed and see it as other than white. But the apparently green object is not really white, although it contains a white element; neither does it appear white since there is an insufficient quantity of white *minima* to make it appear white. So, we can say that as far as the colors of mixtures are concerned, *all* are apparent: only the qualities of the elements are genuine. The same can be said for smells, tastes, and so on. Through touch the soul makes judgments about them grounded in the mixture of corpuscles of diverse qualities.[18]

[18] Ibid., Treatise I, f. 7r. ff; Treatise III, f. 33r. ff.

4.2.2 The Problem of Mixture

Aristotle distinguished between mere alteration ("when the substratum is perceptible and persists but changes its own properties"), total change or generation and corruption ("when nothing persists in its identity as a substratum") and the intermediate phenomenon of combination, in which the components of a new substance "neither persist actually ... nor are they destroyed." Medieval scholars called combination *mixtio*, the components *miscibilia* and the compound a *mixtum*. *Mixtio* differs from mechanical mixture (e.g. of two sorts of grains) because the latter involves no change but in *mixtio* the miscibles change into a single homogeneous *mixtum*.

To solve some basic problems of mixture, Aristotle used the notions of potentiality and actuality, postulating some powers exercised by one body on another. Aristotle himself did not conceive of mixture in terms of matter and form, even though the later scholastics did. In particular, the scholastics were concerned with the relationship between the miscibles and the mixtures in terms of their substantial forms, and whether the forms of the miscibles or elements remained in the mixture, and if so, whether they experienced any strengthening or weakening (*intensio* or *remissio*). Avicenna affirmed that the forms remain intact, but their qualities are lessened and "refracted" (*remissae*, *fractae* or *castigatae*) by their mutual activity and passivity, because the substantial forms cannot be strengthened or weakened in the way in which the qualities can. For Averroes, the forms and the qualities remain, but are refracted, limited and reduced to an intermediate state. He maintained that the substantial forms did not vary, and that the elementary forms were not completely substantial, but intermediate between the substantial and the accidental:

> Out of each of miscibles, as they become mixed, there emerges a third thing in actuality, which is uniform in its properties but different in form from each of the miscibles, in that each one of the latter exists in it with a potentiality bordering on, not remote from, actuality.[19]

For Scotus the forms and qualities were destroyed in the mixture, while for Thomas Aquinas the forms were destroyed and the qualities were reduced to an intermediate level. Roger Bacon considered the mixture to be a gradual process; a series of forms, each one more specific and elevated than the previous ones that leads to and prepares the final form of the mixed.[20]

For Olesa, it is fundamental "that the miscibles are arranged according to a certain order, weight and measure."[21] He further adds that the mixture does not require the alteration of the mixibles nor that the substantial forms be refracted, as Averroes held, but that they are "so arranged in place, order, weight, and measure, that from

---

[19] *Epitome*, in Averroes' *Middle Commentary and Epitome on Aristotle's De generatione et corruptione*, p. 121, cited by Emerton, *The Scientific Reinterpretation of Form*, p. 81.

[20] Maier, *An der Grenze von Scholastik und Naturwissenschaft*, pp. 28f. and Emerton, *The Scientific Interpretation of Form*, pp. 76ff.

[21] Olesa, *Summa*, Treatise III, f. 18r. col. 2: "Est ergo mixtio apud nos miscibilium certo ordine, pondere et mensura, situs aggregatorum a causis agentibus universalibus et particularibus pro universi connexione et duratione, secundum exigentiam introducende forme substantialis unio."

them a substantial form can be introduced into matter." Furthermore, Olesa holds that elements in a mixture retain their essences:

> [T]he elements that make up the mixture are in it with its complete being, contrary to what Averroes said, and with its highest being, including even contrary qualities, contrary to the opinions of Avicenna and Aristotle; the said elements remain (in the mixture) with its ungenerated being unchanged and with its own essence in act.[22]

Olesa also thought the sublunar world has a multitude of "astral" corpuscles, that is to say, corpuscles of the same nature as that of the stars, and it is through these astral corpuscles that the connection between the earth and celestial regions is formed and the mixtures are generated. On Olesa's view, intelligences move the stars, which, in turn, cause motion in the terrestrial corpuscles of the same nature.[23] In this way, the action of the heavens participates in the generation of mixtures.

The astral corpuscles also allowed Olesa to give his own interpretation of the so-called occult qualities or "indestructible properties," as Galen called them. Olesa made it very clear that it was not a question of hidden "qualities," but properties since, for the Mallorcan physician, only the elements possess qualities and the so-called secondary qualities are only apparent. Occult properties in mixtures are occult in two ways. On the one hand, we have the proportion or quantity of the elements, deriving from their substantial forms. This is what is called *complexion*; it is not perceptible, and consequently can be called hidden. On the other hand, in mixtures, there are also astral corpuscles of different kinds from which comes another *complexio*, equally hidden. So, according to Olesa, in any mixture there are two dispositions: one that comes from the elements and another from the corpuscles of the stars. And thus there are two *complexiones* and two kinds of occult properties, that is, the elements and the corpuscles of the stars, arranged in some particular shape. For that reason there can exist innumerable hidden properties.[24]

---

[22] Ibid., Treatise III, f. 18v. col. 1: "In tali mixtione non requiritur miscibilia alterari, et consequenter generationem tali mixtioni annexam non procedit alteratio correlarium; hoc ex definitione aperte colligitur, quum ex quo miscibilia possunt a causis universalibus, et particularibus, ut paulo post dicemus, ita dispositi in situ, ordine, pondere et mensura, ut ab eis in illa materia possit introduci forma substantialis. Non est necessarium illa prius alterari, nec substantialiter refrangi ut voluit Averroes, sed solum ita disponi elementa in situ, ordine, pondere et mensura ut mixtum remaneat totum et tale quantum. Et quale forma exigit instrumentum illud esse per suis operationibus exercendis. Quale autem non acquiritur illud per alterationem, sed per proportionem subiectorum hoc est elementorum qualium.... Neque est verum quod nullum sensibile remaneat actu essentiali tale, quum elementa quae sunt in mixto in esse completo contra Averroes et in esse summo quo ad qualitates contrarias contra Avicena et Aristoteles sunt eadem elementa ingenito quae erant incorrupto sub proprio eorum actu essentiali."

[23] On astral corpuscles, see ibid., Treatise I, f. 5v. ff.; Treatise III, f. 48r.-v., and passim.

[24] Ibid., Treatise I, f. 9r.: "Dicamus ergo quod in mixtis duo sunt quae ipsum integrant..., una consurgit ex certa mensura elementorum veluti introducende forme convenit: et hec apud me est complexio quae insequens formam appellatur...." And the conclusion of this chapter *de occulta proprietate*: "Occulta proprietas de qua medici et philosophi loquuntur non est aliqua qualitas elementorum nec astrorum, sed est ipsa elementa sic mensurata et ipsa astra sub tali aut tali numero situ, proportione et mensura disposita.... In quolibet mixto sunt due mensure quarum una est

Olesa applies his theory of the structure of material substance to various physical processes similar to those described by Aristotle. As we have seen, for Olesa the mixture is composed of minimal sensibles that are also formed of insensible *minima naturalia*. On the other hand, the mixtures have pores. But since Olesa rejects the existence of the vacuum, he says that these pores are full of bodies of kinds different from that of the body in which they are found.

Although he attributes qualities to the elements, Olesa holds that the forms of these elements act by means of their figures, penetrating the pores of the mixtures. Mixed hot bodies are, everything equal, rarer than cold ones, because they have larger pores, so that they can be more easily penetrated by the elemental *minima*, not only by those with sharp angles but also by those with obtuse ones, so they are more susceptible to being acted upon.[25] Concerning combustion, Olesa holds that we must realize that fire does not beget fire but only makes manifest fire already present in a mixture, that is, it dissipates the *minima* of fire that already exist in the mixture. Fire acts by means of its subtlety, its rarity and spherical or pyramidal shape.[26]

Unlike Aristotle, who affirms in *De caelo* (305b) that "when air is generated from water, it adapts to a greater place and reciprocally," Olesa believes that water is not formed by a concentration of air but by the separation of the *minima* of the air and fire that exist in the sensible *minima* of the *minima* of water.[27] Olesa repeatedly denies the transmutation of the elements, and adopts a theory closer to atomism.

Olesa compares his ideas with those of the ancient atomists, though he does not accept that the principles of things (atoms) are infinite. On the indivisibility of atoms, Olesa comments that indivisibility can be understood in many ways: that which does not have parts, such as the intellective soul; that which is incorporeal and infrangible, such as the heavens, and that which cannot be divided through natural means, in such a way that the parts remain uncorrupted and are of the same sort as the whole. This last definition is, says Olesa, what some call indivisibility with respect to form. This is what Averroes holds, and *minima naturale* are indivisible in

elementorum et alia est corpusculorum astrorum, ex quibus due complexiones et due proprietates occulte resultant, quae tamen innumere dici possunt et hec non sunt nisi ipsa elementa et ipsa astra sic proportionata." See also Treatise III, f. 49r. ff.

[25] Ibid., Treatise III, f. 22r., col. 1"Et dicamus corpora calida quae mixta sunt ceteribus partibus rariora esset frigidis: et hoc aut quia plures poros habeant aut quia maiores aut quia ampliores. Quo sit ut facilius minima elementorum ingrediantur per illud mixtum et sic non solum illa quae habet angulos acutos sed quae obtusos et sic passibiliora dicuntur."

[26] Ibid., Treatise III, col. 1–2: "Dicamus ergo ignem (arbusta aut quaevis alia ligna aut combustibilia comburens) non generare ignem, sed preexistente detegere … Sit autem huismodi minimorum ignis in mixto latentium de ocultatio per segregationem partium heterogenearum quae sunt in combustibili: quae minima ignis occultare poterant.... Ignis habeat eam figuram quae exigitur velocitate et penetrationi et hanc puto magis esse piramidalem."

[27] Ibid., Treatise III, f. 23r., col. 2: "Et ex hiis facile potest solvi ratio Aristotelem III *De coeli* comento LXVII quum aer non inspissat ut fiat aqua neque unquam expissatur. Sed bene possunt minima aeris et ignis depurari a minimis aque, cum quibus involuta erant minima aque."

this way, according to Olesa.[28] That is to say, Olesa accepts the indivisibility of the *minima* if you understand that the *minima* are indivisible with respect to form, that is to say, as naturally indivisible (*quantum naturalis*) in relation to quantity. So Olesa excludes natural divisibility, but not mathematical divisibility. He also observes that, according to some authors, followers of Aristotle, Democritus probably did not say that atoms are indivisible with respect to quantity.[29]

To sum up the discussion to this point, Olesa held a syncretic corpuscular theory of matter derived from different sources: Empedocles and the Greek atomists, Aristotle himself and his followers—mainly Averroes, Galen and the medical tradition, though there may be other sources direct or indirect.[30] In Olesa's theory some fundamental Aristotelian notions, such as that of substantial form, were not rejected, but wind up significantly reinterpreted and approach the structural conceptions of the atomists.

4.2.3 A Corpuscular Theory of Light and Vision

Olesa's corpuscularianism, especially his account of astral corpuscles, is also evident in his discussions of light and vision, which are found in his discussions of the sensitive soul.[31] To explain light and its propagation, Olesa rejects the doctrine of multiplication of the species (agreeing with Ockham, whom he quotes), and rejects the propagation of any quality.[32] For Olesa, the fact that the medium is illuminated does not mean that there is any property called *lumen* propagated from the sun or from any of the stars, but that in the air, there is a mixture of

[28] Ibid., Treatise III, f. 6, col. 1–2: "Aliquid tamen indivisibile dici multifariam contigit: uno siquidem modo pro eo quod nullo modo habet partem nec situm, veluti sunt intelligentie et anime intellective. Alio modo per incorporali et infrangibili isto modo celum indivisibile appellari potest: alio modo pro quod naturaliter partiri nequid. Ita ut partium qualiter non corrupta remaneat totius denominationem habens : et isto modo ab aliquis indivisibile indivisibilitate forme vel nature appellant : et istud dicit apud Averroem primo *De generationem* comento LXXXV indivisibile apud naturam : et sic minimum naturale indivisibile ac impartibile dicitur."

[29] Ibid., Treatise I, f. 8 r., col. 1: "Et dico forte quia Democritus, ut dicitur ab aliquibus sectantibus Aristotelis, non posuit hec indivisibilia nisi secundum speciem et non secundum quantitatem." See also ibid., Treatise III, f. 10r. col. 1–2 : "[Aristoteles] existimem virum illud eruditissimum [Democritus] non existimasse athomos suos esse indivisibiles in quantum quantos sed in quantum naturales et sic in quantum erant minima naturalia, veluti diximus aliquos expositores Arist. illud dixisse de mente Democriti."

[30] To date our search of direct sources has been unsuccessful.

[31] Ibid., Treatise I, f. 13v. ff. See also Treatise III, f.102v. ff. on vision and light.

[32] Ibid., Treatise I, f. 14 r., col. 2: "Sed apud me tales species non sunt dande et sic nec producuntur ab intelligentia nec ab obiecto et hanc sententiam tenuit ille ingeniosissimus Ockham in secundo *Sententiarum* quamvis illic ponat ab obiecto produci aliquam qualitatem actus visionis terminativam seu productivam quam nos negamus."

corpuscles of the same type as the sun, whose motion is determined by heavenly intelligences, since they lack the particular natural movement of the elements or any of their attributes.[33] In sum, Olesa reinterprets Aristotle in a corpuscularian way.

As for the sensitive soul, Olesa says that it acts by means of animal spirits, everyone having his own particular temperament in accordance with the complexion of his organs. The animal spirits, which are processed in the brain and in the *rete mirabile*, are subtler than all the other kinds of spirit; not only are they subtle, but also clear and resplendent, smooth and clean, and have a specular nature so that they can represent images of objects in the organs. This is realized paradigmatically in the eye. The crystalline humor, the chief instrument of sight, has a specular nature, adapted by nature to represent the image of objects, so that the sensation of vision is produced in the soul. Olesa insists that the crystalline humor does not actually receive the image but merely represents it. The soul's production of the sensation is, for Olesa, not any quality actuated in the soul but it is the soul itself that sees the object and is the act of seeing.[34]

According to Olesa, some hold that in the mirror, the image represented is not a quality that exists in the mirror or in the air, but is the very mirror, made suitable for representing by its smooth and clean nature. Therefore, he argues, the humor does not receive any species coming from (or diffused from) an object, as Averroes and the Peripatetics claim. Olesa suggests an analogy with an impression in a piece of wax, in which an image is produced without adding anything to the wax but a new arrangement of its parts. On analogy, the crystalline humor and the animal spirits receive nothing when they represent images of thing; they receive neither a species produced

[33] Ibid., Treatise III, f. 43r. col. 2: "Quando ergo dicit medium est illuminatum non est intelligendum quod in medio sit subiective aliqua qualitas quae dicatur lumen, producta a sole aut astrorum aliquo, sed quod medium, id est, aer, habet secum involuta tot corpuscula solis, hoc est eiusdem speciei cum sole, quod totum videtur lucidum veluti cernes aquam claram, videt ipsam dyaphanam et albam…. Et si hec [corpuscular] ab illis moveatur per intelligentie nutum non tamen eque, nata quoque sunt moveri omni differentia possitionis, cum motu proprio elementorum cuilibet attributo careant et astrorum celi, cum quibus specie et accidentibus omnibus participant, motum insequantur."

[34] Ibid., Treatise I, 14r. col. 1: "Sunt enim spiritus animales in prima cerebri officina eo temperamento ab anima conflati … namque his spiritus peculiaria temperamento in propriis organi ab ipsa anima mediante organi complexione, eo modo quo de complexione nos loquimur: et sic his animales spiritus qui in cerebro et rethe mirabili elaborantur subtiliorem ceteris omnibus spiritibus per partium grossarum et terrearum desicationem, non solum subtiles sunt sed clari et splendidi, tersi et politi quo sit ut habeant speculi naturam et representare possunt imagines obiectorum propiorum illi organo in quo dearticulatur; et hec maxime faciunt illi qui oculo implantantur. Hoc idem faciunt oculi tunice et humor proprie christalinus qui organum in oculo ponitur et hic humor naturam speculi habens : natus est representare obiecti imaginem ni impediatur ad quem humorem sic representantem anima advertens elicit sensationem quae est visio. Et in hac representatione humor christalinus nil recipit ab obiecto ni quia humor hic natus est … et etiam elisio ab anima facta quae est visio non est aliqua qualitas in anima producta, sed est ipsa anima sic eliciens denominata videre et actu videns."

from the object (as the Peripatetics usually think) nor from the intelligence (as Averroes affirms), and nothing prohibits the soul from producing the sensation without receiving the species. Olesa also rejects Ockham's idea according to which the object produces some kind of quality that produces or defines the act of seeing. Not only does he reject this idea for sight, but for all the sensations produced through the exterior organs.[35]

So vision takes place in the following way. Given a luminous medium and appropriate conditions, both on the part of the object as well as the organ, vision is produced through the representation of the object in the crystalline humor with the help of the integrated animal spirits in the eye, without the reception of any species; that which is represented in the crystalline humor is also represented in the sensitive soul, and this is what constitutes vision.[36] The mirrors of the spirits represent the image in the front part of the eye, and in the middle and the back part of the brain, as Averroes argues, using two opposing mirrors. Consequently, the mirrors facing one another represent the image of the image.[37] Elsewhere, Olesa says that at the moment in which the facing mirrors represent the images of the objects, all the mirrors that are located in the brain's ventricles represent the images of the images, as can be easily seen in opposite mirrors (facing each other) in which the first mirror represents the image while the other represents the image of the image, and if there are no impediments, with more perfection. Thus, the sensitive soul, which contains the images of the images of the objects represented by the mirrors located in the first ventricle of the brain,

[35] Ibid., Treatise I, 14r., col. 1–2: "Nec de hoc mirari debet aliquis, ubi iam sunt qui concedunt imaginem ab speculo representatam non esse aliquam qualitatem in speculo existentem neque etiam in aere speculi immerito : sed esse speculum sic representans natum quippe sic representare : ex hoc solum quod est nature terse et polite … ergo humor non recipit speciem ab obiecto productam ut tenet Averrois et comunis peripateticorum opinio et si apud Averroim species quae sunt in medio non producantur ab obiecto, ut alii volunt, sed ab intelligentia ut in questione ostendimus esse de mente Averrois, sed apud me tales species non sunt dande et sic nec producantur ab intelligentia nec ab obiecto et hanc sententiam tenuisse ingeniosissiumum Ockham in secundo *Sententiarum,* quamvis illic ponat ab obiecto produci aliquam qualitatem actus visionis terminativam seu productivam quam nos negamus." On the analogy of wax, see ibid., 14v. col. 1: "Si forma cere esset anima, tunc ipsa sentiret figuram a sigillo impressam sine hoc quod aliquid reciperet. Cum ergo humor christalinus et spiritus representet rerum imagines sine hoc quod aliquid recipiant, nil prohibet animam elicere sensationem visivam sine alicuis speciei receptione."

[36] Ibid.,Treatise I, 14v., col. 1: "Existente medio, ut decet, luminoso et cum aliis conditionibus quae se tenent tam ex parte obiecti quam organi, quam advertentie sequitur elisio sensationis quae dicitur visio absque alicuius specie visibilis receptione in aliqua tunica aut aliqua medii parte aut organi. Sed sola representatione facta ab humore christalino principaliter et spiritibus animalibus in oculo dearticulatis, quam etiam obiecti similitudinem a dicto humore representatam anima sensitiva representat; quae representatio, quando ad ipsam advertit, est visio."

[37] Ibid.,Treatise I, 14v., col. 2: "Concedimus enim spiritum specula in eodem instanti representare imaginem obiecti in oculo anteriori parte cerebri, in media et in posteriori veluti Averroys ostendit de variis speculis contra se positis in libro *De somno et vigilia*…. Sic quoque specula contra se oposita quae ita se habent: quod post primum alia representant imaginis imaginum."

conceives the image of sound (air has the nature of a mirror with respect to sound quality, etc.), color and other qualities represented in the exterior organs. Because of this, it is called the common sense.[38]

4.3 The Absence of a Tradition

Olesa's work seems not to have been widely read in sixteenth-century Spain. The Faculty of Medicine of Valencia, founded around 1499, at the same time as the University (though it started to function only in 1502), became the most important in sixteenth-century Spain. Throughout the century, teachers there made use of the principal novelties in medical pedagogy introduced throughout Europe. From the beginning, it had chairs in surgery and basic medicine, and around the middle of the century developed into the peninsular centre of the movement to renovate anatomy, as represented by Vesalius. This renovation of anatomy was also associated with the introduction of medical humanism by the physician and professor of Greek Miguel Jeronimo Ledesma.[39] These new currents of medicine in Valencia and in the rest of Spain had little in common with the intellectual world of Olesa and with his form of expression. Even so, the question of the spread of Olesa's ideas still warrants our attention. There is no need to emphasize again how few sophisticated

[38] Ibid., Treatise II, f. 31, col. 1: "Ut diximus eodem instanti in quo specula illa sensuum exteriorum representant obiectorum imagines, in eodem specula omnia quae sunt in ventriculis omnibus cerebri representant imaginum imagines, sicut facile est viderem speculis contra se positis quae primi speculi representantis inmediate obiecti imaginem imagines illius imaginis representant. Et, si advertis perfectius semper et hoc nisi specula illa impediantur ab aliquo vapore, anima ergo sensitiva nutriens imagines imaginum obiectorum, representatis ab speculis in primo cerebri existentibus aprehendit imaginem soni et coloris et alias ab speculis organorum exteriorum representatas et sensus comunis nomen subiit." See also, Treatise III, f. 105, col. 1–2. It is interesting to note that though Olesa, following Averroes, mentions the retina in the process of transmission of the images to the common sense, neither Olesa nor Averroes consider that the image is formed in the retina, nor that this might be the principal photosensitive organ of vision. As Lindberg, *Theories of Vision from Al-Kindi to Kepler*, pp. 54–57, has explained, neither Averroes nor any author before Kepler said more than Galen said, namely that "its principal and greatest usefulness (of the retina), that for the sake of which it was brought down from above, is to perceive the alteration of the crystalline humor and in addition to convey and transmit nutriment to the vitrious humor" (Galen, *On the Usefulness of the Parts of the Body*, cited by Lindberg, *Theories of Vision from Al-Kindi to Kepler*, pp. 55f.). Galen and his medieval followers could speak of retinal sensitivity without regarding the retina as the principal instrument of vision. The better elaborated and more consistent theory of vision, Alhazen's theory, considered the crystalline to be the principal organ, showing at the same time a basic indecision over the relative functions of the various sensitive elements in the visual pathway. The change from the crystalline to the retina was a radical change in the optics and physiology of vision. For that change it was necessary to distinguish the optical phase (the eye) from the physiological phase (the brain), and to use this distinction to overcome the obstacle of the inversion of the image in the retina.

[39] López Piñero, "The Faculty of Medicine in Valencia."

analytical studies of the printed and manuscripts works of natural philosophy of the Renaissance there are compared with what was written and published in the period.

4.3.1 The Hypothesis of Menéndez Pelayo

The earliest reference to Olesa was uncovered by Marcelino Menéndez Pelayo. In his study on the *Antoniana Margarita* (1554) of the Spanish physician Gomez Pereira (published first in 1876), he writes: "Before him [Gomez Pereira], Olesa had written in a way very suggestive of atomism in his *Summa totius philosophiae et medicinae*, in which Olesa follows Democritus, and defends his opinions about natural principles, atoms and the incorruptibility of the elements."[40] With this Menéndez y Pelayo established Olesa as the point of departure of a genuinely Spanish atomist tradition, later represented by the same Gomez Pereira and by Francisco Valles. Through certain figures from the first half of the seventeenth century, this tradition was linked with the Peninsular followers of the atomism of Gassendi and Maignan in the final decades of the seventeenth century (Cardoso, Corachan, Tosca, etc.). It was the Portuguese-Jewish physician Isaac Cardoso who first brought Olesa's work to the attention of Menéndez Pelayo: he affirms that the first modern author who re-established ancient atomism was precisely the Mallorcan physician, Olesa.[41] Cardoso also considered Gomez Pereira to be a follower of Democritus.[42] Menéndez Pelayo admitted that while Cardoso may have read Gomez Pereira, he himself had not read Olesa's work. Nevertheless, Menéndez Pelayo seems to have been influenced by Cardoso's opinions, since on a fair reading of Gomez Pereira's own text, one cannot really classify him as an atomist or a follower of Democritus. But even if he is not an example of an atomist or corpuscularian, the natural philosophy of Gomez Pereira is not without interest, especially as an example of the diversity of Aristotelianism in the sixteenth century and the appropriation and elaboration of Aristotelian tradition by physicians to establish their theories and practices.

4.3.2 The Salamacan Physician Gomez Pereira

Gomez Pereira (1500–1558), trained in the University of Salamanca, owed his fame in the history of Spanish thought to his defense of the claim that animals lack

[40] Menéndez Pelayo, "La Antoniana Margarita de Gómez Pereira," pp. 331ff.

[41] Cardoso, *Philosophia Libera*, Venice, 1673, p. 10.

[42] On Cardoso, see Yerushalmi, *Dalla Corte al Ghetto*. On atomistic or corpuscular ideas in seventeenth-century Spain and beginnings of the eighteenth century, see Navarro Brotons, "Descartes y la introducción de la ciencia moderna en España"; id., "De la filosofía natural a la física"; Navarro Brotons, "El moviment *novator*."

sense (*bruta sensu carent*) in the *Antoniana Margarita*.[43] Gomez Pereira's natural philosophy was basically Aristotelian, with some noticeable influence from nominalist thought; his physiology and anatomy correspond to the medieval Galenism that he had learned in Salamanca. Gomez Pereira had a very clear exposition of atomist ideas, but also left no doubt that his ideas about the structure of material substance and the first principles of things differ from atomism. For Gomez Pereira, the first principles are the four elements, from whose aggregation come all substances.[44] Gomez Pereira rejected the Aristotelian idea of prime matter and did not accept the scholastic identification of prime matter and pure potentiality. Furthermore, he denied that the elements are composed of matter and form and affirmed that "elements are completely corrupted through the action of those things that induce dispositions contrary to their conservation, and are again engendered in the place where others were corrupted without the existence of any (prime) matter."[45] In sum, for Gomez Pereira, there is only one prime matter, the aggregate of the four elements; none of them taken separately constitutes prime matter, since no mixture is composed of only one element. On the other hand, the matter of a mixture is heterogeneous, as its *minima* are composed of heterogeneous parts that correspond to the four elements. The mixtures are more perfect than simple substances since they include elements and the form of the mixture.[46] For the eduction of the forms of the mixture, the qualities of the elements must be "refracted," and engender an accident called the temperament or complexion of the mixture, which follows the generation of the form and its eduction, with all the necessary dispositions; without this, the mixture could neither exist nor operate.[47] Furthermore, for the generation of the form, there must be a superior or heavenly cause.[48]

Though he is generally not a corpuscularian, on one occasion, however, Gomez Pereira did use the term "corpuscular." When dealing with generation, he sometimes calls the generative spirit (*spiritum genitivum*) "steam," but at other times, he refers to it as "corpuscles."[49] However, he denied that the driving force or impulse

[43] On Gomez Pereira, see Bernia Pardo, *La diferencia entre el animal y el hombre en la Antoniana Margarita*; Sanhueza, *La pensée biologique de Descartes*; López Piñero, "Gómez Pereira."

[44] See Gómez Pereira, *Antoniana Margarita*, pp. 106ff., on the principles of all the things.

[45] Gómez Pereira, *Antoniana Margarita*, p. 111, col. 1–2: "Elementa in totum corrumpuntur per actionem eorum, quae inducunt contrarias dispositiones suae conservationi, et de novo gignuntur in ibi, ubi alia corrupta fuere citra ullius materiae existentiam." See also ibid., p. 124, col. 2: "Elementa scilicet mutuo se corrumpere, ac adinvicem ex se gigni, non ullius materiae primae existentia machinata, quae novae formae, priore corrupta, nuberet."

[46] Ibid., pp. 346–348. On the question of prime matter, see Bernia Pardo, *La diferencia entre el animal y el hombre*, pp. 431ff.; Sanhueza, *La pensée biologique de Descartes*, pp. 128–131.

[47] Ibid., p. 172, col. 1, on the eduction of the forms of the mixture.

[48] Bernia Pardo, *La diferencia entre el animal y el hombre*, pp. 447ff.

[49] Gómez Pereira, *Antoniana Margarita*, p. 48, col. 2: "Tandem si corpusculum illud, appellatum spiritus, sine ullo sensu tot partes animales mire effingit...."

communicated by objects is transmitted to animals by means of "corporeal particles," and he rejected any similarity to the theories of Democritus. For Gomez Pereira what moves animals are accidents emanated from the objects, also called species.[50] Despite what Menéndez Pelayo says, it is difficult to regard Gomez Pereira as a genuine atomist or corpuscularian.

4.3.3 The Salamacan Physician Francisco Valles

The third representative of the Spanish, atomist tradition of the sixteenth century, according to Menéndez Pelayo, is Francisco Valles (1524–1592), one of the most outstanding Spanish doctors of the period. As Craig Martin has recently pointed out: "The paucity of twentieth-century scholarship on Valles does not reflect his status and influence during the sixteenth and seventeenth centuries."[51] Valles was professor of medicine at the University of Alcala from 1557 until 1572, and from this date, the Royal Physician General of all kingdoms and domain of Castille. Valles was one of the foremost representatives of the European medical humanism. He was responsible for providing direct translations of Galen and Hippocrates from the Greek, and while still considering Galen to be the main doctrinal benchmark, transformed Hippocrates into the main model of learning and above all, of medical practice. Valles also assimilated the new anatomical morphology symbolized by Vesalius, mainly via Pedro Jaime Esteve, a student of Vesalius, teacher of anatomy in Valencia and Alcalá.[52]

Apart from his works on medicine and his *Tratado de las aguas destiladas* (*Treatise on Distilled Water*), Valles wrote four works of natural philosophy. Three of them are commentaries on the eight books of Aristotle's *Physics*, and on the fourth book of his *Meteorologica*, while the fourth, entitled *De iis, quae scripta sunt physice in libris sacris, sive de sacra philosophia, liber singularis* (Turin, 1588), is focused on demonstrating that the Scriptures include in an unequivocal form the true representation of the Universe, coinciding to a large extent with Aristotelian natural philosophy. In addition, there is the *Controversarium medicarum et philosophicarum*, of which ten editions were published between 1556 and 1625 (in Alcala, Frankfurt, Lyon, Venice and Hanau). While these works of Valles and his ideas concerning natural philosophy have not been yet sufficiently studied, we can assert with confidence that the claim that Valles was a follower of atomism is totally erroneous. But even so, Valles was not an uncritical follower of Aristotle: he disagreed with important aspects

[50] Bernia Pardo, *La diferencia entre el animal y le hombre*, pp. 270ff.

[51] Martin, "Francisco Valles and the Renaissance Reinterpretation of Aristotle's *Meteorologica IV* as a Medical Text."

[52] López Piñero, "Francisco Valles"; López Piñero and Calero, *Las Controversias (1556) de Francisco Valles y la medicina renacentista*.

of Aristotle's doctrines in some cases, and proposed his own special interpretation of them in others. In his commentaries on Aristotle's physics, Valles discussed at length the question of *minima* and *maxima*, and the question as to whether or not the forms of the elements remain in mixtures. Likewise, in the *Controversarium medicarum*, Valles discussed the elements and the way in which they are present in the mixture, rejecting the atomist doctrines and accepting the Aristotelian doctrine of the four elements and the four qualities, qualities whose introduction he attributes to Hippocrates. Concerning the question as to how the substances are formed from the elements, and whether the elements are maintained in their own form in mixtures, Valles described the various opinions formulated from Antiquity, including those of Averroes, and Avicenna. Agreeing with Aristotle, Valles held that the forms of elements do not remain in the mixture in act, but only in potentiality.[53]

In the *Sacra Philosophia*, Valles again concerned himself with the elements, especially in relation to cosmological questions. For Valles, the air we breathe is not an element, nor is it in a pure state even when in its proper place. Furthermore, fire, unlike the other elements, does not move towards any place except by accident. Fire can be everywhere to fertilize and bring life to things. Neither did Valles accept the notion of ether as a fifth essence. Valles considered that the heavens are not composed of the four sublunar elements, but celestial elements; he called them elements by analogy, because for Valles, the heavens are incorruptible in a natural form.[54]

Valles, like Gomez Pereira, rejected the idea of prime matter and considered that the principle of individuation is not prime matter but magnitude. Furthermore, he pointed out that this idea is consistent with the teachings of Plato and Aristotle.[55]

Of special interest is also the discussion by Valles of gravity and levity. For Valles, weight or levity are not absolutes but relative and an expression of the natural order of things: things with more density are below things that are more tenuous. In explaining it, he referred to Archimedes and the balance and developed a relational conception of weight. Heavier or lighter weight is related to greater or lesser density, and dense or tenuous things are ordered in a way that the former are always below the latter, which they displace. Valles explained that if we suspend different substances in a balance, the weight depends on how the balance is submerged; just as Archimedes demonstrated and anyone can see, water displaces as much weight as the weight of an equal volume of water. And as denser things with less volume can weigh the same as larger things with less density, they displace less water. As a consequence, objects that weigh the same in air do not weigh the same in water. Likewise, an object that is ordered correctly in relation to its neighbors does not move from the position where it rests. If, on the other hand, it is badly ordered, it moves up or down to re-establish the natural order.[56]

---

[53] Valles, *Controversiarum medicarum ac philosopharum, Caput primum: De modo quo elementa prima sunt in compositis substantiis*, f. 1v-2r.

[54] Valles, *De sacra philosophia*, p. 235, on fire as element; pp. 346ff., on the elements.

[55] Ibid., pp. 534ff.

[56] Ibid., pp. 366–369.

4.4 Conclusion

As far as we know, Olesa was the only Spanish physician in the sixteenth century to make use of the notion of the *minima naturalia* in connection with the structure and principles of material substance and the structure and formation of compounds or mixtures. Olesa conceived of the mixture in a way similar to that of Bartholomew of Salerno in the Middle Ages, or Pantheo and Scaliger in the Renaissance. Even if we cannot consider him to be an atomist, strictly speaking, his ideas are in some respects close or similar to some eclectic philosophers of the seventeenth century with atomist tendencies. Olesa's writings, it seems, did not attract the attention of contemporary Spanish scholastic philosophers, who normally discussed questions of *minima naturalia* in contexts quite distinct from the problem of mixture, which is where they come up in Olesa's thought.[57] Take, for instance, the case of the Valencian Diego Mas, professor of philosophy from 1581 in the University of Valencia. Mas was the author of the *Commentariorum in Universam Philosophiam Aristotelis* (1581). This work is a good example of the Aristotelian post-humanist and post-Nominalist natural philosophy. In his commentary, Mas takes up questions related to *minima* and *maxima* and quotes a great number of opinions taken from ancient authors, medieval sixteenth-century authors, including those of Soto, Valles and Benito Perera. But there is no mention of Olesa.[58] Despite all the work that has been done on the question, the diffusion of Olesa's *Summa* and his ideas, as well as more general questions relating to conceptions of the structure and principles of material substance and the structure and formation of compounds and mixtures in sixteenth-century Spanish printed texts and manuscripts requires further research.

[57] This is pointed out by Murdoch, "From the Medieval to the Renaissance Aristotle," who takes into account a good number of Spanish authors.

[58] On Diego Mas, see Navarro Brotons, "La filosofía natural a la física moderna."

Chapter 5
The Composition of Space, Time and Matter According to Isaac Newton and John Keill

Carla Rita Palmerino

5.1 Introduction

In his *Mathematical Lectures*, read at the University of Cambridge between 1664 and 1666, Isaac Barrow, Lucasian professor of mathematics, criticized "those who would have magnitude constituted of a finite number of indivisibles," an opinion which he considered repugnant to the laws of mathematics. Barrow argued in favor of the infinite divisibility of all extended quantities, including material bodies:

> There is no part in any kind of magnitude, which is absolutely the least. Whatsoever is divided into parts, is divided into parts which are again divisible .... Though we are not able to comprehend how this indefinite division can be performed, yet we ought not therefore to doubt, but it may be performed, because we do perceive it to follow from the nature of matter, a thing most manifestly known to us.[1]

The young Newton, who was to become Barrow's follower as Lucasian professor of mathematics, held a view diametrically opposed to that of his master. In the *Quaestiones quaedam philosophicae*, a set of notes redacted between 1664 and 1665 and contained in the *Trinity College Notebook*, he asserted that all extended magnitudes were composed out of a finite number of extended, but partless minima.

Research for this article was made possible through the financial support of the project *Visualizing the Invisible. Representations of Matter and Motion since the Renaissance*, sponsored by the Netherlands Organization for Scientific Research (NWO). I wish to thank the editors for their valuable comments on earlier drafts of this article.

[1] Barrow, *The Usefulness of Mathematical Learning explained and demonstrated*, pp. 151, 162.

C.R. Palmerino (✉)
Center for the History of Philosophy and Science, Radboud University Nijmegen, The Netherlands
e-mail: c.palmerino@ftr.ru.nl

D. Garber and S. Roux (eds.), *The Mechanization of Natural Philosophy*, Boston Studies in the Philosophy of Science 300, DOI 10.1007/978-94-007-4345-8_5,

What Newton and Barrow had in common was the fact that they applied their conclusions to "any kind of magnitude." This choice does not need to surprise us, as it is in agreement with mainstream seventeenth-century matter theories. As I have shown elsewhere, most early modern natural philosophers believed in the isomorphism of all physical magnitudes, a belief that often led them to regard the acceleration and deceleration of bodies and the rarefaction and condensation of matter as phenomena deserving analogous treatment.[2]

If one looks at Newton's later writings, however, one sees that the belief in the existence of material atoms no longer implies a belief in spatial and temporal atoms. This clearly appears in the last *Query* to the *Optics* (1704), where Newton differentiates between the properties of physical bodies, which he describes as being composed of "hard particles," and those of space, which he asserts to be "divisible *ad infinitum.*"[3] Nevertheless, as I shall try to show in this article, isomorphist theories did leave a trace in Newton's mature writings, as appears from the fact that he continued to uphold a link between the phenomena of acceleration and of rarefaction. This might help explain why one of Newton's closest pupils, John Keill, reasserted the isomorphism of space, time and matter, which he claimed to be all divisible *ad infinitum.*

In the following pages I shall first analyze the atomism of the *Quaestiones* against the background of seventeenth-century isomorphist theories. I shall subsequently explore the evolution of Newton's ideas regarding the composition of physical magnitudes, and more in particular concerning the relation between the structure of space and that of matter. Finally I shall analyze the arguments which John Keill used, in his *Introduction to Natural Philosophy* (1720), to demonstrate that space, time and matter were all divisible *ad infinitum.*

5.2 The Isomorphism of Space, Time and Matter in Early Modern Natural Philosophy

Andrew Janiak has been the first scholar to draw attention to the fact that Newton, from his days as an undergraduate up to his latest writings on natural philosophy, established a link between the issue of the divisibility of space and that of matter.[4] In his analysis, Janiak focused on Newton's use of the distinction between mathematical and physical divisibility, which provided a clue to understanding why he abandoned the *aprioristic* commitment to atomism of his early days in favor of the experiment-based approach of the *Optics* and the *Principia.*

---

[2] Palmerino, "The Isomorphism of Space, Time and Matter in Seventeenth-Century Natural Philosophy."

[3] Newton, *Opticks*, pp. 389, 403.

[4] Janiak, "Space, Atoms and Mathematical Divisibility in Newton."

While agreeing with the general lines of Janiak's account, I wish to add a new element to it, namely the consideration of the relation that Newton established between the structure of matter and that of space and time, and more in particular between the explanation of the rarefaction and the condensation of matter and that of the acceleration and deceleration of bodies.

This link was established, for the first time, in the above-mentioned *Quaestiones quaedam philosophicae.* The *Quaestiones* begin with an essay "of the first matter," in which Newton argues that matter can neither be composed of mathematical points nor be "a simple entity before division indistinct," but must consist of extended indivisibles. A few pages later, Newton explains, in the essay "of motion," that if matter is composed of minima, the same must be true for space and time. Therefore, motion must consist of indivisible units, in which a "least space" is traversed in "a least time." This is why a physical minimum can never be said "to be moving," but only "to have moved."

> That it may be known how motion is swifter or slower consider 1 That there is a least distance, a least progression in motion & a least degree of time.... 2 These leasts have no parts for that implies that they are yet divisible neither *prius* nor *posterius*. Not least distance since it is passed over in an indivisible part of time & there cannot be a different time ascribed to the entrance of a thing into that part of space & the leaving of it. Not the least degree of motion because too that is performed in an indivisible part of time and is no sooner begun than done. Not the least moment of time because first and last imply several parts of time. 3 The least degree of motion is equal to the least distance and time.[5]

In the *Quaestiones* Newton explicitly acknowledges his indebtedness to Henry More and Walter Charleton. From More he borrows the term "indiscerpible," a neologism used in the *Immortality of the Soul* to describe something which is divisible intellectually, but not physically. For, as we read in preface of More's work,

> it cannot but be confessed therefore, that matter consists of indiscerpible particles, and that Physically & really it is not divisible *in infinitum*, though the parts that constitute an indiscerpible particle are real, but divisible only intellectually; it being the very essence of whatsoever is to have parts or extension in some measure or other. For, to take away all Extension, is to reduce a thing only to a mathematical point, which is pure Negation or Non entity.[6]

The least particles of matter are thus extended, for "a magnitude cannot arise out of mere non-magnitudes," but they are utterly indiscerpible, by reason of their "extreme littleness."[7] The fact that in the passage just quoted More describes the particles of matter as being "physically" indivisible might seem to suggest that he wants to introduce a distinction between the composition of matter and that of geometrical extension. This is however not the case. Geometrical magnitudes are, in his view, also composed of indiscerpibles. It would in fact be a great absurdity to think that a geometrical line "will consist of *points* mathematically so called, that is

---

[5] Newton, *Certain Philosophical Questions*, pp. 352–354.

[6] More, *The Immortality of the Soul, Preface*.

[7] Ibid., book I, chap. 6, p. 31.

purely indivisible," for "multiply nothing ten thousand millions of times into nothing, the product will still be nothing."[8]

A similar position is expressed by the young Newton, who, as Andrew Janiak rightly observed, follows More in trying to achieve "an *a priori* understanding of the divisibility of matter."[9]

> And that Matter may be so small as to be indiscerpible the excellent Dr Moore in his booke of the soules imortality hath proved beyond all controversie yet I shall use one argument to shew that it cannot be divisible *in infinitum* & that is this: Nothing can be divided into more parts than it can possibly be constituted of. But matter (i.e. finite) cannot be constituted of infinite parts.[10]

That the same discourse applies to space and time becomes clear from the entry "of motion" where Newton writes that "there are so many parts in a line as there can stand mathematical points in a row without touching (that is, falling into) one another in it, and so many degrees of motion along that line as there can be stops and stays, and there are so many least parts of time in an hour as there can be *tò nûn*'s."[11]

Another work that certainly left a trace in the *Quaestiones* was Walter Charleton's *Physiologia Epicuro-Gassendo-Charltoniana.* Although it is true, as Janiak says, that "there is only one sentence in which Charleton distinguishes mathematical from physical divisibility," there is little doubt that the account of the composition of continuous magnitudes contained in the *Physiologia* shaped Newton's understanding of the relation between space, time and matter.

In the *Physiologia,* Charleton not only asserted the composition of matter out of extended *insectilia*—on the account that if division could proceed to infinity matter would be annihilated—but also argued that no physical continuum could be infinitely divisible.[12] In this respect it is important to point out that Charleton distinguished between an invisible space, namely the incorporeal part of the universe, which he described as "being identical to the void" and hence infinitely divisible, and the "visible space," which is more "consentaneous to reason," as it "allows one to speculate the Catholique Principles Motions and Mutations."[13] This space, like all other physical continua, is not infinitely divisible, but is instead composed of *insectilia*. The question "whether it be convenient to transfer Geometrical demonstrations to Physical or sensible quantity" was therefore answered by Charleton in the negative:

> And this (in a word) seems to be the true and only cause why Mathematicians constantly suppose every continuum to consist of infinite parts: not that they can, or ought to understand it to be really so; but that they may conserve to themselves a liberty of insensible Latitude, by subdividing each division of parts into so many as they please; for, they well know, that the physiologist is in the right when he admits no infinity, but only an innumerability of parts in natural continuum.[14]

[8] Ibid.

[9] Janiak, "Space, Atoms and Mathematical Divisibility in Newton," p. 206.

[10] Newton, *Certain Philosophical Questions*, p. 341.

[11] Ibid., p. 353.

[12] Charleton, *Physiologia*, pp. 88, 90–93.

[13] Ibid., p. 84.

[14] Ibid., p. 97.

In the *Quaestiones,* Newton not only endorsed Charleton's belief in the existence of atoms of space, time and motion, but also borrowed his treatment of rarefaction and condensation. Like Charleton, Newton observed that rarefaction could be explained either "from vacuities interspersed or from several proportions that quantity has to its substance," and like Charleton he opted for the first hypothesis, which implied that "the body will be no such continuum as to be without distinct parts, since it will be everywhere divided by interspersed inanities."[15]

As we shall see in the following section, Newton soon grew dissatisfied with the indivisibilist theory formulated in the *Quaestiones* and abandoned his belief in the existence of spatial and temporal atoms. In his mature writings, however, he occasionally applied the same explanatory patterns to the study of motion and to that of the rarefaction and condensation of matter, which is a clear indication that isomorphist theories continued to exert an influence on his thought. This holds all the more true for his pupil John Keill, who, as mentioned in the introduction, explicitly reasserted the isomorphism of all physical magnitudes. Before analyzing the evolution of Newton's and Keill's ideas concerning the composition of space, time and matter, it might therefore be useful to reconstruct briefly the most influential seventeenth-century theories on this subject.

I shall take as a starting point the *Labyrinthus sive de compositione continui,* an anti-atomist work published in 1631 by the Leuven professor Libert Froidmont (Latinized Fromondus). After having exposed the traditional mathematical arguments against atomism, Fromondus set out to prove that atomist theories were unsatisfactory also from a physical point of view. In this context he mentioned two natural phenomena that in his view were difficult to explain on the basis of atomism, namely the rarefaction and condensation of matter and the acceleration and deceleration of motion. For those who believed in the composition of the continuum out of extended indivisibles, the only way to account for the first phenomenon was to admit the presence of interstitial voids within rarefied bodies and the mutual interpenetration of particles in condensed bodies.[16] And in order to explain how different objects could traverse the same space at different speeds, they had to postulate that "in every slow motion, some pauses and retardations occur in which the mobile rests, but which in a faster movement are filled."[17]

Fromondus' argument was based on the assumption that, if physical bodies were composed of a finite number of extended particles, the same composition would also have to be found in the space in which those bodies moved as well as in the time during which they were moving. This meant, in turn, that rarefaction and acceleration should be explained in analogous terms: those who believed that an alternation of atoms and voids was concealed within rarefied matter would have to assume that an alternation of motion and rest was concealed within slow motion. Both hypotheses seemed to Fromondus to stand in contradiction to sensory experience and also to be unacceptable from a conceptual as well as a theological standpoint.

[15] Newton, *Certain Philosophical Questions*, p. 339. For Charleton's view, see *Physiologia*, pp. 251–254.

[16] See Fromondus, *Labyrinthus*, pp. 76–97.

[17] "In omni motu tardo pausas et morulas quasdam interiiciunt quibus mobile quiescat, quae in motu celeriori complentur," ibid., p. 62.

In the *Labyrinthus* Fromondus distinguished between three different notions of "indivisible," none of which he considered tenable. He in fact addressed his arguments not only against those who regarded continuous magnitudes as being composed of extended atoms, but also against those who postulated the existence of minima of changing size, or of unextended point-atoms.

In the years following the publication of the *Labyrinthus,* the three theories criticized by Fromondus were to find new adherents. The first one was defended, among others, by the Spanish Jesuit Roderigo de Arriaga, author of a very influential *Cursus philosophicus,* published one year after the *Labyrinthus*. It is enough to look at the index of Arriaga's book to see that a close connection is indeed established between the two problematic phenomena mentioned by Fromondus. For in the *Disputatio XVI. De continui compositione* the discussion of the problem of rarefaction and condensation in *Sectio X* is immediately followed by that of the question *De velocitate et tarditate motus* of *Sectio XI.* While in section 10 Arriaga ascribes the expansion or contraction of a material body to the introduction or expulsion of particles of air (and not of void), his treatment of motion exactly corresponds to that targeted in the *Labyrinthus.* In section XI, Arriaga holds that space and time are composed of physical indivisibles and that all motions of which the speed is inferior to one minimum of space per minimum of time (that is, the velocity of the first moveable) have to be conceived as being intrinsically discontinuous.[18]

A similar account is found in Pierre Gassendi's *Syntagma philosophicum* (1658), where the rarefaction of matter is explained as the result of the interposition of void spaces within the particles of bodies and the slowness of motion as the result of the interposition of particles of rest. In this respect, it is interesting to point out that it was Gassendi who inspired Charleton's distinction between the infinite divisibility of mathematical space and the finite divisibility of physical space. For although he describes the imaginary space as infinite, continuous and indifferent to its content, when he speaks about the "mundane space," that is to say the space occupied by the world, he observes that

> neither this infinity of parts in the continuum nor mathematical indivisibility exist in nature, but are merely a hypothesis of the mathematicians, and that therefore in physics one should not argue on the basis of things that are not known to nature.[19]

The second theory criticized by Fromondus met with the support of Jean-Chrysostome Magnen and Honoré Fabri, both of whom postulated the existence of minima of changing size.

In the third disputation of his *Democritus reviviscens* (1646), Magnen maintains that space, time and motion are divisible *ad infinitum* extrinsically, or mathematically,

[18] Arriaga, *Cursus philosophicus*, pp. 428–432.

[19] "Declaratum certe est quoque iam ante & infinitatem illam partium in continuo, & insectilitatem Mathematicam in rerum natura non esse, sed Mathematicorum hypothesin esse, atque idcirco non oportere argumentari in Physica ex iis, quae natura non novit," Gassendi, *Opera omnia*, I, p. 341b.

but not physically (*extrinsece et mathematice, non autem physice*). If considered in their physical reality, space, time and motion must in fact all be composed of atoms, for

> what is true for the real space is true for the real motion, but in the real space there are atoms, hence there will be atoms also in motion, but we cannot think of atoms of motion without conceiving them as being "all at once." Hence there exist a motion "all at once," which is however successive in its connotation, that is to say extrinsic according to mathematical space. [20]

It is interesting to see that in these lines the atomic structure of motion is inferred from the atomic structure of physical space, and the atomic structure of physical space is in turn made to depend on the mere fact that in physical space "there are atoms." In his *Democritus reviviscens* Magnen admits that the acceleration and deceleration of motion and the rarefaction and condensation of matter are extremely difficult to deal with for an atomist. The last *disputatio* of his book is almost entirely devoted to explaining these two problematic phenomena. In the first chapter, *De motu locali atomorum,* Magnen argues that the various speeds of macroscopic bodies are the result of the multiplication of a minimum atomic speed, rather than of the discontinuation of a maximum speed. Each atom of the falling body acquires a new atom of impetus in each successive atom of time, and the multiplication of the atoms of impetus leads to a multiplication of speed. Interestingly, Magnen observes that the expression "uniformly accelerated motion" is strictly speaking not correct, for "there is always a space in which the motion is equal and uniform, as follows from the nature of atoms."[21]

Like acceleration, the rarefaction of matter is also interpreted by Magnen as the result of a multiplication of sorts. In chapter 2, *De rarefactione et condensatione iuxta Democritum,* Magnen explains that "a single atom, without rarefaction, inflation, or reproduction, can naturally occupy a bigger and bigger place *ad infinitum*," simply by changing its shape. For as is well known, solids of equal volume but different shapes also have different external surfaces.[22]

Fabri's theories of matter and motion, like Magnen's, rest on the assumption that physical minima can have different sizes. In the ninth book of his *Metaphysica demonstrativa,* which was published in 1648 by his pupil Pierre Mousnier, Fabri argues that "there are physical instants, because there is action through which a

---

[20] "Ut se habet spatium reale, ita & motus realis, sed in spatio reali sunt atomi, ergo et in motu erunt atomi, sed atomi motus non possunt fingi animo, quin intelligatur totae simul, ergo datur motus totus simul, qui tamen erit successivus in connotatione seu extrinsece penes spatium mathematicum," Magnen, *Democritus reviviscens*, p. 234.

[21] "Respondeo ad tertium ad sensum dari posse motum unifromiter difformem non autem stricte loquendo, quia est semper aliquod spatium per quod motus est aequalis, & uniformis idq; necessario atomorum naturam sequitur," ibid., p. 236.

[22] "Atomus est omni figurae capax, ergo occupare potest maiorem, & maiorem locum in infinitum: cum enim figurae regulares in isoperimetris sint magis collectae minoremque locum occupent, sequitur quod quo irregularior erit figura eo maiorem occupabit locum, at non potest dari, ita irregularis, quin magis irregularis esse possit, ergo etiam maioris loci capax," ibid., p. 247.

thing is."[23] In other words, time must be composed of physical instants, for nothing can exist or move in a mathematical instant. In Fabri's view, the hypothesis that space and time are composed of mathematical indivisibles is unacceptable, irrespective of whether one assumes that these indivisibles are extensionless and infinite or extended and finite. For in both cases "there could be no quicker or slower motion."[24] Given that "time, motion and space are composed in the same way," it would in fact be impossible for a point of the body in motion to acquire more or less than one indivisible of space in each successive instant of time.[25] The only way in which one could save the possibility of slower and faster speeds would be to postulate that motions are interrupted by a variable number of pauses. But this solution, which was favored, as we have seen, by the Spanish Jesuit Arriaga, is dismissed in the *Metaphysica* as untenable.[26]

Fabri's own explanation of the variety of speeds of physical bodies is based on the following principle: although a moving body can only pass through one *locus adaequatus* in each physical instant of time, successive instants can each have a different duration. To those who wonder how it is possible for an instant to be smaller than another instant, Fabri answers as follows:

> I also admit that there can be nothing smaller than a mathematical instant; but things are different in the case of the physical instant, which is potentially divisible.[27]

In Fabri's eyes, the hypothesis of the actual indivisibility but potential divisibility of physical points "makes it easy to explain all phenomena related to quantity: first, the speed and slowness of motion ...; second, rarefaction and condensation, compression and dilatation; for every point can have a bigger or smaller extension."[28] It is therefore no surprise to find that in the *Metaphysica,* Fabri takes issue not only with Arriaga's theory of motion, but also with his explanation of the rarefaction and condensation of matter:

> If air were composed of mathematical points, it would be impossible to explain how it rarefies or how it is condensed, or compressed, thereafter. For neither is it possible for a mathematical point to be bigger or smaller; nor is Arriaga right in explaining condensation by means of an expulsion of corpuscles, and rarefaction by means of an intrusion, for this contradicts empirical evidence.[29]

---

[23] "dantur instantia Physica; quia datur actio, per quam res est," Fabri, *Metaphysica demonstrativa*, p. 371. For Fabri's theory of matter and motion, see Palmerino, "Two Aristotelian Responses to Galilei's Science of Motion."

[24] Fabri, *Metaphysica demonstrativa*, p. 371.

[25] Ibid., p. 413.

[26] Ibid., p. 375.

[27] "nam equidem fateor instanti mathematico nihil esse posse minus; secus vero instanti physico, quod est divisibile potentia, ut dicemus alias," Fabri, *Tractatus physicus*, p. 110.

[28] "Facile iuxta hanc hypothesim, omnia quae pertinent ad quantitatem explicantur; Primo motus velocitas et tarditas.... Secundo rarefactio, condensatio, compressio, dilatatio; quia quodlibet punctum potest habere, modo maiorem, modo minorem extensionem," Fabri, *Metaphysica demonstrativa*, p. 414.

[29] "Si aër constat ex punctis mathematicis, non potest explicari, quomodo rarescat, vel densetur, vel comprimatur, contra post. Nec enim punctum mathematicum potest esse maius, vel minus: nec est quod

Fabri thinks that the only possible explanation of the phenomenon of rarefaction and condensation is to suppose that material bodies are composed of physical points of different shapes that under certain conditions become bigger or smaller. Fabri specifies that this possibility only applies to humid bodies, for dry bodies are not subject to rarefaction and condensation.[30] As mentioned before, in the *Labyrinthus* Fromondus had taken into account the possibility of composing the continuum out of minima of changing size, but had rejected it on the grounds that if indivisibles could become bigger or smaller, they would no longer be indivisible.[31]

The third theory discussed by Fromondus was to find an adherent in Galileo Galilei, who in the *Two New Sciences,* his last work published in 1638, abandoned the Democritean atomism defended in his previous writings in favor of a mathematical atomism. In the *Two New Sciences,* Galileo made use of a famous paradox of motion, the so called *Rota Aristotelis,* to argue that all physical bodies are composed of an infinite number of non-extended atoms, some of which are filled with matter and some of which are void. In explaining the advantages of this matter theory, Galileo underlines the fact that it "facilitates the comprehension of condensation and rarefaction," for it permits an explanation that "circumvents [both] the voids and the [inter]penetration of bodies."[32] As I have extensively argued elsewhere, Galileo's decision to abandon physical atomism in favor of a mathematical atomism may have been prompted by the need to harmonize his theory of matter with his science of motion, which was based on the assumption that falling bodies acquire a new degree of speed in each successive instant of time. "And since in any finite time, however small, there are infinitely many instants, there are enough to correspond to the infinitely many degrees of diminished speed."[33]

In presenting his matter theory, Galileo displayed confidence in the fact that his non-extended atoms would look acceptable to Aristotelians:

> But by employing the method I propose ... I believe that they [i.e., the learned Peripatetics] should be satisfied, and should allow this composition of the continuum out of absolutely indivisible atoms. Especially since this is a road that is perhaps more direct than any other in extricating ourselves from many intricate *labyrinths*.[34]

These lines may hide a reference to the *Labyrinthus,* where Fromondus had argued that "the more subtle among those who constructed the continuum out of

---

Arriaga explicet condensationem per extrusionem corpusculorum, & rarefactionem per intrusionem, quippe hoc manifestae experientiae repugnat," ibid., p. 397. Arriaga's explanation of rarefaction and condensation is criticized also in ibid., p. 424.

[30] Ibid., p. 395.

[31] Fromondus, *Labyrinthus*, pp. 57f.

[32] Ibid., p. 55.

[33] Galilei, *Two New Sciences*, 157. For a discussion of the relation between Galileo's theory of matter and his theory of motion, see Palmerino, "Una nuova scienza della materia per la *scienza nova* del moto. La discussione dei paradossi dell'infinito nella prima giornata dei *Discorsi* galileiani"; Ead., "Galileo's and Gassendi's Solutions to the *Rota Aristotelis* Paradox. A Bridge between Matter and Motion Theories."

[34] Galilei, *Two New Sciences*, p. 54 (= Galilei, *Opere*, VIII, p. 93).

atoms, composed it out of infinite rather than finite [atoms]."[35] Fromondus had however discarded the hypothesis that physical magnitudes were composed of non-extended points, on the account that where there is no extension, there cannot be any matter.[36]

Having rejected the existence of both extended and unextended indivisibles, Fromondus came to the conclusion that continuous magnitudes had to be composed of ever divisible parts. The author of the *Labyrinthus* was however not willing to endorse Aristotle's claim that these parts were finite in act and infinite in potency. Rather, he argued that space, time, and matter were divisible into an actual infinity of *partes proportionales,* proceeding towards ever smaller magnitudes as 1, ½, ¼, and so on.[37] Why Fromondus opted for this solution becomes clear in the last chapter of the *Labyrinthus,* which is devoted to the two problematic phenomena of the rarefaction and condensation of matter and of the acceleration and deceleration of motion. Here Fromondus argues that, although in the natural course of things bodies do not rarefy *ad infinitum,* the composition of the continuum out of an infinite number of proportional parts allows for the possibility that God, in his omnipotence, may produce such an endless rarefaction. In order to support his claim, Fromondus appeals to another, more familiar phenomenon:

> God can endlessly rarefy matter, either without form or with a form preternaturally conserved. This can be shown by means of the example of the slowness of motion: for slowness, being a laxness of parts in successive continua, is very similar to rarity in permanent continua.
> ... And just as from the infinite slowness of motion it can be shown that rarefaction can proceed endlessly, so from the speed of motion, which can also be increased *in infinitum* (for couldn't God create celestial spheres that were bigger and bigger *ad infinitum* and yet able to accomplish a rotation in 24 hours?), we can show that also condensation, if related to divine power, can be without end.[38]

Like his atomist opponents, Fromondus thus regarded deceleration as a key to understanding rarefaction and acceleration as a key to understanding condensation.

---

[35] "Subtiliores, inter eos qui continuum ex atomis struxerunt, ex infinitis potius quam finitis composuisse," Fromondus, *Labyrinthus*, p. 9. For the possible influence of Fromondus on Galileo, see Redondi, "Atomi, indivisibili e dogma," pp. 555–557.

[36] Fromondus, *Labyrinthus*, pp. 97–99 (Caput XXXI, *Frustra quidam conati inter Aristotelem & Epicurum medij incedere, negando ullas esse in continuo partes, aut asserendo infinitas, sed indivisibiles*).

[37] "Si spatij quod pertransitur magnitudo habet partes proportionales infinitas, quarum una prior est, altera posterior, igitur corpus quod sine replicatione per tale spatium movetur, debet infinitas partes transire, unam post alteram, partesque in eo motu successivae erunt etima infinitae: nam unicuique parti spatij permanentis, sua pars motus successivi respondet," ibid., p. 137.

[38] "Deus tamen materiam illam sine forma, aut cum ipsa contra naturam conservata, potest sine fine rarefacere. Quod etiam exemplo tarditatis in motu possumus declarare: tarditats enim in successivis est quaedam partium laxitas, simillima raritati in permanentibus.... Veluti autem ex *tarditate* motus infinita *rarefactionem* posse sine fine procedere ostenditur; ita ex *velocitate* motus, quae etiam in infinitum increscere potest (cur enim Deus caelestes sphaeras ampliores & ampliores sine fine creare nequeat, quae omnes 24 horarum spatio revolvantur?) ostendere possumus, *condensationem*, si ad virtutem divinam comparemus, nullum habere finem," ibid., pp. 191–193.

As we shall see, the theories discussed in this section were subjected to critical scrutiny by Newton and Keill, both of whom linked their speculations about the structure of material bodies to the analysis of the relation between material and spatial extension.

5.3 The Evolution of Newton's Views on the Composition of Space, Time and Matter

Although they subscribed to quite different theories of matter and motion, the authors mentioned in the previous section shared the assumption that time, space, and matter had the same structure. This assumption, as we have seen, led them to regard the acceleration and deceleration of heavy bodies and the rarefaction and condensation of matter as phenomena deserving analogous treatment.

The belief in the existence of atoms of space, time and motion, was endorsed also by the young Newton, who in the passages of the *Quaestiones* analyzed above explained the rarefaction and condensation of matter and the acceleration and deceleration of motion in analogous terms. It is interesting to see, in this respect, that the passage about rarefaction, which was part of the entry "of the first matter," was deleted by Newton together with the discussion about indivisibility and least distance in the entry "of atoms" and with the whole, still unfinished, entry "of motion." The editors of the *Notebook* list different reasons why Newton may have grown dissatisfied with his own accounts: in the first case, he must have realized that it was illegitimate to assume that atomism had "superior explanatory power (...) in accounting for phenomena such rarefaction and condensation," while the "ontological first principles of atomism" were "still in need of further justification."[39] As for the other entries, Newton possibly understood that his notion of a conceptually indivisible atom was incoherent and that minima of space and time made it difficult to conceive how motion could be swift or slow.[40]

I am, by contrast, convinced that the young Newton regarded the swiftness and slowness of motion and the rarefaction and condensation of matter as two related phenomena and that therefore he could not delete the entry on motion without also rethinking those on first matter and atoms. It is hence not to be excluded that Newton started reconsidering his whole theory concerning the composition of physical magnitudes when he realized that the minima of space and time made it impossible to account for the different speeds of moving bodies.

Although in his mature writings Newton continued to speak about "particles of time" (*particula temporis*), this expression, as I.B. Cohen has noticed, does not refer to "finite atoms of time in the sense of tiny particles of matter. Rather, for Newton,

[39] Newton, *Certain Philosophical Questions*, p. 40.

[40] Ibid., pp. 46f., 78f., 87.

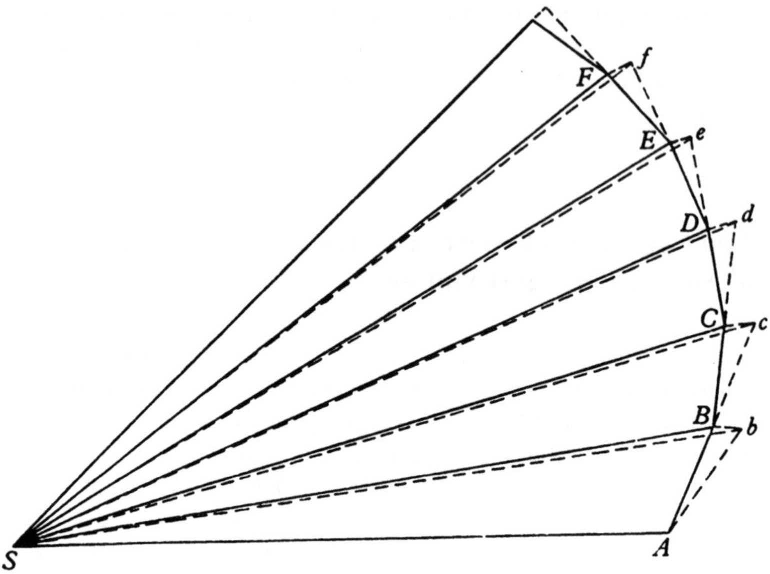

Fig. 5.1 The trajectory of a body subjected to the action of a centripetal force

time is finitely continuous and only infinitesimally discrete."[41] Take, for example, the manuscript *De motu corporum in gyrum,* which was redacted in the autumn of 1684. After describing the trajectory of a body subjected to the action of a central force S as the sum of the rectilinear lines FE, ED, DC, CB, BA (see Fig. 5.1) traversed in successive particles of time, Newton assumes that there is "an infinite number of these triangles, each of them infinitely small, so that each triangle corresponds to a single instant of time, then with the centripetal force acting uninterruptedly the proposition will stand."[42]

Michel Blay has convincingly argued that the main aim of Newton's *Principia* was "to achieve mathematical rigor in the transition from the discontinuous to the continuous," which meant to "proceed from conceptualizing action in terms of impulses or impacts to mathematizing a continuous, uninterrupted action like that of gravity."[43] As we have just seen, this transition was accomplished by assuming, first, that time was divided into equal intervals, and by imagining, subsequently, these intervals to be diminished *ad infinitum.*

[41] Cohen, "Newton's Concept of Force and Mass, with Notes on the Laws of Motion," p. 74. Newton speaks about "particles of time," which he describes as "infinitely small" or "minimally small" in *The* Principia. *Mathematical Principles of Natural Philosophy*, pp. 316, 652f., 673f.

[42] Newton, *Unpublished Scientific Papers*, p. 272.

[43] Blay, "Force, Continuity, and the Mathematization of Motion at the End of the Seventeenth Century," pp. 225f.

At a physical level, Newton's theory implied the rejection of Descartes' view of gravity as a discontinuously acting force. Interestingly enough, Newton's first attempts to devise a physical explanation of gravity alternative to Descartes' led him to establish once again a connection between his theory of matter and his theory of motion, and more in particular between the phenomenon of acceleration and that of rarefaction. In his *Second Paper on Light and Colours* read at the Royal Society in 1675, Newton put forward the hypothesis that the gravitating attraction of the Earth was caused by "the continual condensation of an aethereal spirit."[44] What Newton meant by this becomes clear in a letter to Boyle of 1678, where a number of physical phenomena, such as the diffraction of light, the solution of metals, the "changes of colours, which are often made by the mixtures of several liquors," the melting of substances and their fermentation, and also the collision between two bodies and the action of gravity are all described as the result of the action of aether, which runs "through all intermediate degrees of density." The physical explanation of gravity put forward in the letter rests on the supposition that "aether consists of parts differing from one another in subtlety by infinite degrees," so that from the top of the air to the center of the earth,

> the aether is insensibly finer and finer. Imagine now any body suspended in the air, or lying on the earth, and the aether being by hypothesis grosser in the pores, which are in the upper parts of the body, than in those which are in its lower parts, and that grosser aether being less apt to be lodged in those pores, than the finer aether below, which cannot be without the bodies descending to make room above for it to go out into.[45]

It is therefore as if Newton regarded the continuity of the motion of fall as a consequence of the "gradual subtlety" of the parts of aether pervading all gross bodies.

A more explicit link between theory of matter and theory of motion, or, to be more precise, between variations in density and variations in speed, is established in the *De gravitatione et aequipondio fluidorum,* a manuscript probably redacted a few years before the *Principia.* Among the definitions, which Newton posits as the foundation of the science of weight and of the equilibrium of bodies in fluids, we find the following ones:

Definition 11. The intension of any of the above-mentioned powers is the degree of its quality.
Definition 12. Its extension is the amount of space or time in which it operates.
Definition 13. Its absolute quantity is the product of its intension and its extension....
Definition 14. Velocity is the intension of motion, slowness is remission.
Definition 15. Bodies are denser when their inertia is more intense, and rarer when it is more remiss.[46]

---

[44] Newton, *Papers and Letters on Natural Philosophy*, p. 251.

[45] Ibid., p. 253.

[46] Newton, *Unpublished Scientific Papers*, pp. 149–50.

According to these definitions, then, one motion is faster than another if its speed is more intense and one body is denser than another if its inertia is more intense, where 'inertia' by definition 8 is "the force within a body, lest its state should be easily changed by an external exciting force."[47] In his explanation of Definition 13, Newton clarifies that the variations in the intension and in the extension of motion are totally independent of one another:

> Motion is either more intense or more remiss, as the space traversed in the same time is greater or less, for which reason a body is usually said to move more swiftly or slowly. Again, motion is more or less in extension as the body moved is greater or less, or as it is acting in a larger or smaller body. [48]

Things are different in the case of the inertia of material bodies, which due to the addition of interstitial pores can acquire greater extension without acquiring new inertia, so that the latter becomes "more remiss":

> If with Descartes or Epicurus, we suppose rarefaction and condensation to be accomplished in the manner of relaxed or compressed sponges, that is, by the dilatation and contraction of pores which are either filled with, or empty of, some very subtle matter, then we ought to estimate the size of the whole body from the quantity of both its parts and its pores as in Definition 15; so that one may consider inertia to be remitted by the increase of the pores and intensified by their diminution, as though the pores, which offer no inertial resistance to change, and whose mixtures with the truly corporeal parts give rise to all the various degrees of inertia, bear some ratio to the parts.
>
> But in order that you may conceive of this composite body as a uniform one, suppose its parts to be infinitely divided and dispersed everywhere throughout the pores, so that in the whole composite body there is not the least particle of extension without an absolutely perfect mixture of infinitely divided parts and pores. Certainly such reasoning is suitable for contemplation by mathematicians; or if you prefer the manner of the peripatetics: things seem to be captured differently in physics.[49]

In the lines just quoted, Newton distinguishes between two ways of conceiving the composition of matter. One can regard material bodies either as being composed of extended atoms or else as being divisible *ad infinitum.* Although he considers the first hypothesis to be more plausible from a physical point of view, Newton bases his definitions on the second hypothesis, which as he states is "suitable for contemplation by mathematicians."[50]

It is hence clear that the similarity between variations in speed and variations in density, which is asserted in *De gravitatione*, only holds true under the hypothetical assumption that material bodies are infinitely divided. This hypothesis is however a fruitful one, as it enables Newton to elucidate, in the concluding pages of his manuscript, the difference between hard and fluid bodies. Newton speaks about "absolutely fluid" or "absolutely hard" bodies, even though these are not found in nature, "for one cannot ratiocinate mathematically concerning ones partially so." A hard body,

[47] Ibid., p. 148.
[48] Ibid., pp. 149f.
[49] Ibid., p. 150.
[50] Ibid., p. 151.

being made of particles that do not move relatively to each other, but are "so bound together, as it were by glue," can be regarded as "a single undivided and uniform body which preserves its shape most resolutely." A fluid body, instead, is "of such kind that it has no small portion or particle which is not likewise fluid," which means that one can consider it as being "uniformly divided at all points."[51]

It is interesting to see that the explanation of hardness and fluidity found in *De gravitatione* radically differs from that offered by Charleton in the *Physiologia,* a work which, as we have seen, exerted a strong influence on the young Newton. While *De gravitatione* describes fluid bodies as being infinitely divided, the *Physiologia* explicitly asserts that the dissolution of bodies cannot "proceed to *infinity,* but must consist in some *definite term,* or *extreme,*" for "every real magnitude is incapable of interminable division." Moreover, Charleton resorts to the difference between hard and soft bodies to prove that the first constituents of bodies are solid particles:

> If we assume the principles of all things to be exquisitely hard, or solid; then do we admit the production of not only hard, but also of soft bodies to be possible, because softness may arise to a concretion of hard principles, from the intermistion of inanity: but if we assume soft principles, then do we exclude all possibility of the production of hard bodies, that solidity, which is the fundament of hardness, being substracted: Therefore is the concession of *Atoms* necessary.[52]

A similar reasoning is found in the passage of the entry "Of the first matter" which Newton deleted from his *Quaestiones quaedam philosophicae.* Here one reads that hardness arises from "plenitude of matter" so that bodies acquire a harder nature when there are few interspersed *vacua*, and a softer nature when there are more of them. In the *Quaestiones,* the supposition of a vacuum goes hand in hand with the rejection of the infinite divisibility of matter. As McGuire and Tamny clearly explain, the young Newton thinks that either "a vacuum exists, or else there is an infinity of smaller and smaller parts. But the latter assumption begs the question of how an homogeneous matter can actually get divided."[53]

In the *De gravitatione,* as we have seen, Newton introduces the hypothesis of the infinite divisibility of matter precisely to explain how bodies can rarefy and become fluid. Can one describe this change of mind as a shift from one form of isomorphism to another? A careful comparison between the two texts does not allow for such a conclusion. As we have seen, the view that space, time and matter must all be composed of extended minima lay at the very basis of the aprioristic atomism of the *Quaestiones.* In *De gravitatione* the hypothesis that material bodies are divisible *ad infinitum* in the same manner as geometric bodies is instead presented as mere heuristic tool. Newton recognizes in fact that bodily nature is a free creation of God and that it is therefore impossible to determine *a priori* the actual composition of matter.

There is also another, more important reason that prevents Newton, in *De gravitatione,* from endorsing a full-fledged "isomorphism." His explicit anti-Cartesian

---

[51] Ibid., pp. 151f.

[52] Charleton, *Physiologia*, p. 89.

[53] Newton, *Certain Philosophical Questions*, p. 45.

agenda leads him to distinguish between the properties of space, which is "eternal, infinite, uncreated and uniform," and those of physical bodies, which are "opposite in every respect."[54] One crucial difference between extension and matter lies precisely in their divisibility: "Spaces themselves are not actually divisible," for their parts, contrary to those of physical bodies, cannot be separated.[55]

Newton possibly derived the idea of the indivisibility and immobility of space from the *Enchiridion metaphysicum,* a work by Henry More first published in 1671. While in the *Immortality of the Soul* (1659) More stressed the fact that not only matter, but also geometrical extension must be composed of extended parts—"for, to take away all Extension, is to reduce a thing only to a mathematical point, which is pure Negation or Non entity"—, in the *Enchiridion* he emphasizes the differences, rather than the similarities, between space and matter. More states that "infinite extension, distinct from matter" has "no physical parts" from which it is combined and into which it can be "truly and physically divided." This means, in other words, that the whole space is "simple and indiscerpible."[56]

In More's *Enchiridion* and in Newton's *De gravitatione,* extension and matter are considered isomorphic with respect to their mathematical structure, but not to their "physical composition." Being eternal and uncreated space is not "composed" of parts and is therefore also not divisible into physical parts. This is precisely what Newton explains in his draft "Tempus et Locus," a manuscript redacted in the early 1690s and published by McGuire:

> But neither does Place argue the divisibility of a thing or the multitude of its parts, and on that account imperfection, since space itself has no parts which can be separated from one another, or be moved among themselves, or be distinguished from one another by any inherent marks. Space is not compounded of aggregated parts since there is no least in it, no small [or] great or greatest, nor are there more parts in the [totality] of space than there are in any place which the very least body of all occupies. In [each] of its points it is like itself and uniform nor does it [truly have part]s other than mathematical points, that is everywhere [infinite in number] and nothing in magnitude. For it is a single being, [most simple] and most perfect [in its] kind.[57]

The issue of the indivisibility of space plays a crucial role also in the *Leibniz-Clarke Correspondence*. To Leibniz, who in his third paper argues that "since space consists of parts, it is not a thing which can belong to God," Clarke replies that this difficulty only arises "from the figurative abuse of the word parts."[58] Given that "infinite space is one, absolutely and essentially indivisible (…) to suppose it parted, is a contradiction in terms." This answer does not satisfy Leibniz, who observes that

---

[54] Newton, *Unpublished Scientific Papers*, p. 145.

[55] Ibid., p. 137.

[56] More, *Manual of Metaphysics: A Translation of the* Enchiridium Metaphysicum (1679), p. 58. For an analysis of the influence of More's *Enchiridium* on Newton's *De gravitatione*, see Slowik, "Newton's Metaphysics of Space"; id. "Newton, the Parts of Space, and the Holism of Spatial Ontology".

[57] McGuire, "Newton on Place, Time and God," p. 117.

[58] Alexander (ed.), *The Leibniz-Clarke Correspondence*, pp. 25, 31f.

"to say that infinite space has no parts, is to say that it does not consist of finite spaces."[59] Clarke's answer is worth quoting in full:

> Infinities are composed of finites, in no other sense, than as finites are composed of infinitesimals. In what sense space has or has not parts has been explained before .... Parts, in the corporeal sense of the word, are separable, compounded, ununited, independent on, and moveable from, each other: but infinite space, though it may by us be partially apprehended, that is, may in our imagination be conceived as composed of parts; yet those parts (improperly so called) being essentially indiscerpible and immovable from each other, and not partable without an express contradiction in terms, ... space consequently is in itself essentially one, and absolutely indivisible.[60]

As observed by Janiak, the point Clarke wants to make here is that space is composed of finite parts only "in a mathematical sense, for it is only through a mathematical process of division that an object is divided into infinitesimals."[61] To this one might add that the reference to infinitesimals serves Clarke to clarify the relation between "division" and "composition": the fact that lines can be mathematically divided into infinitesimals, does not mean that they are really composed of them; in the same way, the fact that space can be mentally parted does not mean that it is not "composed" of parts. Given that physical division implies the resolution of a magnitude into its first components, only those magnitudes that are composed of ontologically independent elements are "divisible" in the physical sense of the term.

The issue of the physical divisibility of matter is explicitly addressed by Newton in the *Principia,* and more precisely in the Third Rule, where he claims that "those qualities of bodies that cannot be intended and remitted and that belong to all bodies on which experiments can be made should be taken as qualities of all bodies universally."[62] While it is certain that the "extension, hardness, impenetrability, mobility and force of inertia of each of the whole arise from the extension, hardness, impenetrability, mobility and force of inertia of the parts," it is still unclear whether divisibility can also be regarded as a universal property of matter:

> Further, from phenomena we know that the divided, contiguous parts of bodies can be separated from one another, and from mathematics it is certain that the undivided parts can be distinguished into smaller parts by our reason. But it is uncertain whether those parts which have been distinguished in this way and not yet divided can actually be divided and separated from one another by the forces of nature. But if it were established by even a single experiment that in the breaking of a hard and solid body, any undivided particle underwent division, we should conclude by the force of this third rule not only that divided parts are separable but also that undivided parts can be divided indefinitely.[63]

---

[59] Ibid., p. 38.

[60] Ibid., p. 48. Alexander wrongly replaced the term "indiscerpible," used by Clarke in his original letter, with "indiscernible." See Janiak, "Space, Atoms and Mathematical Divisibility in Newton," p. 225.

[61] Ibid.

[62] Newton, *The* Principia. *Mathematical Principles of Natural Philosophy*, p. 795.

[63] Ibid., pp. 795–796. I follow here the corrected translation found in Janiak, "Space, Time and Mathematical Divisibility in Newton," p. 215.

The question of whether the division of matter stops at the level of atoms or can instead proceed *ad infinitum* is left open by Newton. As Janiak observes in his book *Newton as Philosopher,* in the lines just quoted the word "parts" does not need to stand for "atoms," but can refer "agnostically to the insensible constituents of bodies." If matter were infinitely divisible "the parts in question would bear mass, for mass is a continuum quantity and therefore has no smallest unity."[64] It is interesting to quote, in this context, a sentence that Newton added, in the second edition of the *Principia,* to the third corollary of Book III, Prop. VI: "If the quantity of matter in a given space can, by any rarefaction, be diminished, what should hinder a diminution to infinity?"[65] The way in which Newton phrases his hypothesis is highly significant. Rather than making a positive statement, he stresses the absence of reasons for excluding the infinite divisibility of matter. Only experiments can decide how things stand in reality. For as Newton explains in the 31st *Query* to the *Optics,* it is God who in an act of free will decides about the ultimate composition of bodies:

> All these things being considered, it seems probable to me, that God in the beginning formed matter in solid, massy, hard, impenetrable, moveable particles, of such sizes and figures, and with such other properties, and in such proportion to space, as most conducted to the end for which he formed them .... And since space is divisible *in infinitum,* and matter is not necessarily in all places, it may be also allow'd that God is able to create particles of matters of several sizes and figures, and in several proportions to space, and perhaps of different densities and forces, and thereby to vary the laws of nature, and make worlds of several sorts in several parts of the universe.[66]

Andrew Janiak concludes his article on "Space, Atoms and Mathematical Divisibility in Newton," by indicating a fruitful line of research: The shift "from the *a priori* speculation of the *Quaestiones* to the somewhat more constrained speculation of the *Queries*" is "well worth future investigation."[67]

It is my hypothesis that the aprioristic commitment to atomism required the belief in the isomorphism of space, time and matter, and that therefore Newton could not abandon the latter without also giving up the former. As I have shown in the previous section, most seventeenth-century atomists distinguished between a mathematical space, which they regarded as infinitely divisible, and a physical space, to which they ascribed the same structure as matter. The same thing holds true for the young Newton, who in the *Quaestiones* ascribed the property of indiscerpibility not only to the minima of matter, but also to those of space and time, convinced as he was that no magnitude could be divisible *ad infinitum,* for "an infinite number of mathematicall points sink into one being added together."[68] Newton must have soon realized, however, that spatio-temporal atomism posited serious problems for his mathematical treatment of motion. Moreover, in elaborating his critique of Cartesian physics, he started regarding

[64] Janiak, *Newton as Philosopher*, p. 109.

[65] Newton, *Unpublished Scientific Papers*, p. 312.

[66] Newton, *Opticks*, pp. 403f.

[67] Janiak, "Space, Atoms and Mathematical Divisibility," pp. 226f.

[68] Newton, *Certain Philosophical Questions*, p. 336.

space as ontologically independent from matter. Being uncreated and absolutely uniform, space was infinitely divisible in a mathematical sense, but absolutely indivisible in a physical sense.

This does not mean, however, that in Newton's mature writings the speculations concerning the ultimate composition of material bodies are disconnected from those concerning the structure of space and time. As we have seen, in the 1670s and 1680s Newton repeatedly established a link between his theory of matter and his theory of motion, by relating the infinite degrees of acceleration to the infinite degrees of the aether's rarefaction. And although this hypothesis is not formulated in Newton's published works, both the *Principia* and the *Optics* envisage the possibility that matter may be infinitely divided. In other words, the fact that space is divisible *ad infinitum* does not imply that the same *must* be true for matter, but only that it *can* be true for matter.

As we shall see in the following section, the isomorphism of space, time and matter was explicitly reasserted by Newton's pupil John Keill, who in his *Introduction to Natural Philosophy* presented the infinite divisibility of material bodies as a necessary consequence of the infinite divisibility of space and time.

5.4 The Isomorphism of Space, Time and Matter According to John Keill

In 1700 John Keill, who is best known for his role in the Newton-Leibniz dispute on the priority of the invention of the infinitesimal calculus, held for a first time in Oxford a course on natural philosophy, which was first published in Latin as *Introductio ad veram physicam* (1701), and then translated in English under the title *An Introduction to Natural Philosophy* (1720).

As is made clear in the preface, the main goal of the *Introduction* is to criticize the natural philosophy of Descartes and his followers, who "pretended to explain all things mechanically by matter and motion" and yet introduced a philosophy "which was as contrary to the true Laws of Mechanics as possible."[69] After devoting the first lecture to the exposition *Of the Method of Philosophizing,* Keill puts forward his theory concerning the composition of physical magnitudes.[70] In lecture II, *Of the Solidity and Extension of Bodies,* Keill follows Newton in criticizing the Cartesian equation of matter and extension:

> The solidity of bodies ... is common to every species of bodies.... By this property, Body is distinguished from another kind of Extension, which we conceive to be penetrable, which we call Space, and wherein we behold all Bodies to be placed and moved.... Each of them

[69] Keill, *An Introduction to Natural Philosophy*, p. viii.

[70] For an analysis of Keill's theory of the composition of the continuum, see Thijssen, "David Hume and John Keill and the Structure of the Continua." Thijssen focuses on lectures 2 and 3 of Keill's *Introduction.*

> [=space & body] seems to be endued with Attributes, not only distinct and proper to themselves, but so contrary that it is impossible to imagine they could be inherent in the same Subject. For we conceive Body as solid, or impenetrable, divisible and capable of motion, whose parts may be easily disunited, separated and removed from one another.... On the contrary ... space ... is immovably fixed, capable of no action, form or quality; whose parts it is impossible to separate from each other, by any force however great.... We are ready to demonstrate against the *Cartesians* that in truth there is given a space distinct from Body; or, in other words, that Space and Body are not the same things.[71]

As we have seen, the need to distinguish between the properties of extension and those of bodies led Newton to deny, in the *De gravitatione,* the isomorphism of space, time and matter first asserted in the *Quaestiones.* The case is different for Keill, who in Lecture III, *Of the Divisibility of Magnitude,* sets out to demonstrate that space, time and matter are all divisible *ad infinitum.* Newton's pupil knows, however, that a *caveat* must be made: By "divisibility" he does not mean "an actual Separation of Parts from one another which supposes motion, which indeed the nature of space does not admit," but "only the resolution of any magnitude into its parts, or their distinction and assignment."[72] Given that "extension appertains and necessarily adheres to all Species of Magnitude, ... as well to space as to body," all extended magnitudes must behave in the same way with respect to their divisibility.

> Having now settled these principles, we return to our purpose; which was to demonstrate that all extension, whether corporeal or incorporeal, was divisible *in infinitum* or had an infinite number of parts; which we shall endeavour to prove by many invincible arguments.[73]

The "invincible arguments" put forward by Keill are the traditional scholastic objections against atomism of the kind one encounters in Fromondus' *Labyrinthus.* Keill argues that atomism is incompatible with the principles of Euclidean geometry and that it makes it impossible to account for the incommensurability between the side and the diagonal of a square or for the one-to-one correspondence between the points of two concentric circumferences.[74] Keill is however aware of the fact that one possible way to eschew these objections is to resort to the difference between physical and mathematical extension:

> Being compelled by the force of demonstration, they readily allow a Mathematical Body may be divisible *in infinitum*; but they deny that a Physical Body can be always resolved into still farther divisible parts. But what I would know is a Mathematical Body, but something extended into a triple dimension? Does not divisibility belong to a Mathematical Body, by reason it is extended? But a Physical Body is extended after the same manner: wherefore since divisibility depends on the nature and essence of extension itself, and owes to its origin, it is necessary that it must agree to all extensions, whether Physical or Mathematical. For, to use a Logical Expression, whatever is predicated of any *Genus,* is predicated of all the *Species* contained under that *Genus.*[75]

---

[71] Keill, *An Introduction to Natural Philosophy*, pp. 13–15.

[72] Ibid., p. 20.

[73] Ibid., p. 26.

[74] Ibid., pp. 26–30.

[75] Ibid., pp. 30f.

It is clear that Keill cannot subscribe to the solution of those seventeenth-century atomists, like Gassendi, Charleton, and Magnen, who distinguished between a mathematical space, which they regarded as being divisible *ad infinitum,* and a physical space, which they considered to be isomorphic with material bodies. All magnitudes falling under the genus of "extension" must enjoy the same composition; and given that mathematical bodies must be divisible *ad infinitum,* the same must be true also for physical bodies.

It is interesting to see that Keill, like the Aristotelian Fromondus, tries to fight atomism not only on mathematical, but also on physical grounds:

> But because the Philosophers, against whom we dispute, are not acquainted enough with geometrical demonstrations, and therefore do not easily perceive their evidence; before we end this Lecture, we shall produce one physical argument taken from motion, for the infinite divisibility of quantity: namely, if quantity consisted of indivisibles, it would follow, that all motion would be equally swift, nor would a slow snail pass over a less space in the same time than the swift-footed Achilles.[76]

As we have mentioned in the previous section, it is not to be excluded that Newton abandoned the atomism of the *Quaestiones,* when he was faced with the problem mentioned here by Keill. This would explain why he abruptly interrupted the entry "On Motion" and concomitantly deleted those "On Atoms" and "First Matter." We have also seen that Newton's way out of the *impasse* was to introduce a distinction between the mathematical and the physical divisibility of matter: The fact that physical bodies are divisible *ad infinitum* in thought does not prevent God from creating material particles of whatever sizes and figures. Although Keill does not deny that different material substances are composed of particles of different forms and dimension, he insists on the fact that the infinite divisibility of space necessarily entails the infinite divisibility of matter:

> There is yet another distinction amongst the Philosophers, not unlike the former [= the distinction between mathematical and physical body]; whereby they own that every body is mathematically divisible *ad infinitum,* but they deny that it is physically so. If these words have any meaning, it is certainly this: They acknowledge that a Body is mathematically, that is, really and demonstratively, divisible *in infinitum,* but they deny it to be physically, or according to their false *hypothesis*: and so they have a distinction, against which nothing can be replied.[77]

Having stated the principle that all magnitudes are divisible *ad infinitum,* Keill devotes the entire Lecture IV to an analysis of "the objections usually brought against the divisibility of matter."[78] It is interesting to see that the "cavils produced by the atomical Philosophers" are mostly answered with the help of geometry. To those who claim, with Epicurus, that if quantity was divisible *ad infinitum* it would contain an infinite number of parts, Keill replies that this would only be the case if the parts were equal. It is not the number of parts which counts, but their "number

[76] Ibid., pp. 31f.

[77] Ibid., p. 31.

[78] Ibid., pp. 32–45.

and magnitude conjointly," so that "if the magnitude of parts is diminished in the same ratio as their number is increased, the whole made up of all parts would remain the same."[79] Given that in Keill's hypothesis space and matter are divisible into infinitely small parts, "it is necessary that their number should be infinitely great, before they exceed any given quantity." A confirmation of this principle can be found in geometry: The example of Torricelli's hyperbolic solid shows that an infinitely long space can be equal to a finite space.[80]

In response to the second argument of the atomists, according to which "if all quantity is divisible *in infinitum,* any the least magnitude will be equal to the greatest, since the least has as many parts as the greatest," Keill observes that it is "not repugnant to the nature of Infinity, that one Infinite may exceed, be multiplied, or divided by another Infinite."[81]

The third argument analyzed by Keill is fetched from the Divine Omnipotence:

> God, they say, can resolve any quantity into its infinitesimal Parts, and separate those parts from one another: but if so, then may be given the ultimate part and the divisibility would be exhausted, therefore quantity is not divisible *in infinitum.* I answer, without doubt God is able to do whatever is possible, or what is not repugnant to his immutable nature: but since we have already demonstrated that there cannot be given any Particle of matter, however small, which may not be still divided into other infinite Particles; it is thence manifest, that God cannot so divide matter, as that there shall be given its ultimate indivisible.[82]

It is interesting to compare these lines with the main objection against atomism put forward by René Descartes, whose matter theory represents the main polemical target of the *Introduction.* In the *Principia,* Descartes reiterated the conviction already expressed in a letter to Gibieuf dated 19 January 1642, that "there cannot be atoms or parts of matter which are indivisible of their own nature … for there is nothing which we can divide in thought and which we do not thereby recognize to be divisible."[83] While Descartes put the emphasis on what God *could* do, namely divide *in actu* what was divisible in potency, Keill stresses what He *cannot* do, namely exhaust the inexhaustible. The parts in which a continuum can be resolved are by definition extended and being extended they must be further divisible:

> So, if there are given infinitely small quantities, these will be some quantities infinitely less than these: and again, there may be others infinitely less than the last, and so always on

[79] Ibid., p. 35.

[80] Ibid., pp. 33–37.

[81] Ibid., pp. 38f.

[82] Ibid., p. 39.

[83] Descartes, *Œuvres*, vol. VIII, p. 51. In the letter to Gibieuf, Descartes wrote: "Ainsi, nous pouvons dire qu'il implique contradiction, qu'il y ait des atomes ou des parties de matière qui aient l'extension et toutefois soient indivisibles, à cause qu'on ne peut avoir l'idée d'une chose étendue qu'on puisse avoir aussi celle de sa moitié, ou de son tiers, ni, par conséquent, sans qu'on la conçoive divisible en 2 ou en 3. Car, de cela seul que je considère les deux moitiés d'une partie de matière, tant petite qu'elle puisse être, comme deux substances complètes, & *quarum ideae non redduntur a me inadequatae per abstractionem intellectus*, je conclus certainement qu'elles sont réellement divisibles," ibid., vol. III, p. 477. For Descartes' rejection of indivisibles, see Roux, "Descartes atomiste?"

> *in infinitum*. First then, we shall thus prove, that there are given quantities that are infinitely less than infinitely small quantities.[84]

The first proof provided by Keill consists in showing that although in a circle one can draw a chord smaller than any given line, it will always be possible to find a line that is smaller than the given chord. A second example, which he declares to have borrowed from the Scholium of the first Section of Newton's *Principia,* consists in arguing that "any given circular or parabolical angle of contact, will be infinitely greater than the angle of contact at the vertex of the cubical parabola."[85] Moreover "there may be innumerable angles infinitely greater than the circular angle of contact, which yet shall be infinitely less than any rectilinear angle; and so you may proceed *in infinitum,* nor does Nature know any limits."[86]

The fact that Keill's proofs in favor of the infinite divisibility of physical bodies are all derived from geometry is a clear sign of his belief in the isomorphism of space and matter. Moreover, it reveals an attempt to answer the objection of those who argue that God must be able to exhaust the division of matter, for the distinction between "potency" and "act" only applies to finite beings. According to Newton's pupil, God could not resolve a quantity into its last parts without violating the laws of geometry.

In lecture V, which carries the title *Of the Subtility of Matter,* Keill devotes his attention to "those minute particles into which matter is actually divided, or of which it is compounded."[87] His goal, however, is neither to describe the composition of matter nor to account for the different properties of physical bodies, but rather to show "the smallness of the particles by calculation." By making an appeal to the authority of Rohault, Halley, and especially of Boyle, he argues that "a cubick inch of gold may be divided into 47,619,047 parts" and that a cubic inch of copper contains 2,111,400,000,000 parts.[88] But these particles are of course much bigger than those composing the effluvia of odorous bodies (the density of which diminishes in the duplicate ratio of the distance from the center of the body), which in turn are bigger than the animalcules, which in turn exceed the size of blood globules. It is not difficult to guess which conclusion Keill wants to draw from his calculations:

> For it may be gathered from thence, that ten thousand two hundred and sixty six of the highest mountains in the whole earth do not contain as many grains of sand, as one grain of sand can of the blood-globules of these animalcules. It is no wonder, if you here stand amazed, and being struck with no prodigious a thing, should call in question the infinite divisibility of matter, although it is supported by uncontrollable demonstrations.... This subtility of Nature is wonderful beyond measure; but there are other particles of matter still more subtle than these, to which if the above mentioned globules were compared, they would not only appear as mountains, but as vast earths. I mean the particles of light.[89]

[84] Keill, *An Introduction to Natural Philosophy*, p. 41.

[85] Ibid., pp. 43f.

[86] Ibid., p. 45.

[87] Ibid., p. 46.

[88] Ibid., p. 48.

[89] Ibid., p. 59.

Just as there is no limit to the subtlety of matter, so also its rarefaction can proceed *ad infinitum*. By using once again a geometrical argument, Keill tries to demonstrate that "any quantity of matter, however small, can be diffused through any space, however large."[90]

So far Keill has insisted on the similarity between space and matter, but has not touched upon the nature of time. In lecture VI, *Of Motion, Place and Time,* he shows that the ancient paradoxes of motion, such as Zeno's and Diodorus Cronus', can be solved by assuming that space and time are composed of parts decreasing in geometrical proportion *ad infinitum*. In the first section of this article we have seen that Galileo had explained the continuity of motion by assuming that a body touched upon a point of space in each successive instant of time. Keill denies instead that infinite points can make up a line, on the assumption that it would be absurd to grant motion through a point.

> What we have said of Points, the same may be accommodated to Instants or Moments of Time, by showing that as all Magnitudes, so likewise time is divisible *in infinitum;* so that there is no particle of time that can be properly be called an Instant or Point of Time: as no Part of a Line coincides with a Geometrical Point, and as infinite Points do not compose a Line, but a Point, so likewise infinite Moments or Points of Time, are equal to no Time. An interval indeed of Time betwixt different Moments, may be equal to a given Time, but the Moments themselves will be equal to no Time; for Time is not compounded of Moments, but of Parts, which are also Times, nor is Motion performed in an instant, but in Time.[91]

Having demonstrated that physical quantities are not composed of indivisibles, but that they are all divisible into ever divisible parts *ad infinitum,* Keill devotes the remaining part of his *Introduction* to deducing the laws of nature according to the principles of Newton's natural philosophy. Lecture XI, which carries precisely the title "Of the Laws of Nature," contains a passage which seems to confirm that Galileo's isomorphism theory was a source of inspiration for Keill.

In Theorem XVII Keill sets out to prove the so-called Double distance rule, which states that the space traversed in a given time by a body in naturally accelerated motion is half the space which would be traversed in the same time by a body in uniform rectilinear motion having a speed equal to the final speed of the accelerated motion. Interestingly enough, in his demonstration Keill does not make use of the infinitesimal calculus, but follows the method of proof employed by Galileo in the *Dialogue* and in the *Two New Sciences.* By means of Fig. 5.2, in which the line AB represents the time of fall and the line BC the final speed of the falling body, he demonstrates that the space passed over by the body in uniform rectilinear motion is to the space traversed by the body in uniform motion as the triangle ABC is to the rectangle ABCD. A comparison between the figure used in the *Introduction* and that found in the *Dialogue* (Fig. 5.3) reveals however, not only the similarities, but also an essential difference between Galileo's and Keill's proofs.

Galileo's demonstration was built on the assumption that the falling body passes through an infinite number of degrees of speed, which are "achieved during the infinite instants that there are in the time DA corresponding to the infinite points on

[90] Ibid., p. 64.

[91] Ibid., p. 73.

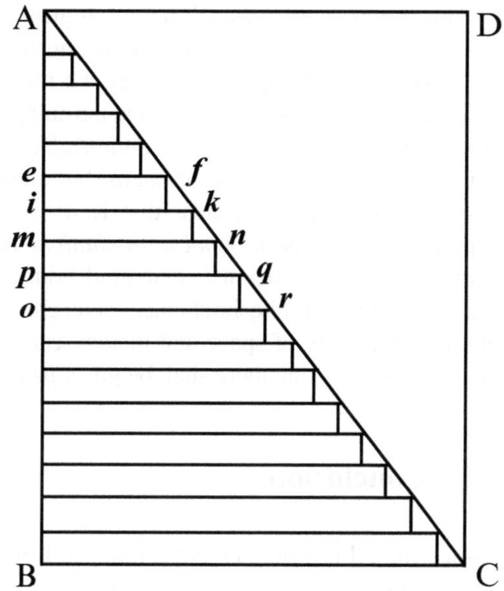

Fig. 5.2 The acceleration of falling bodies according to Keill

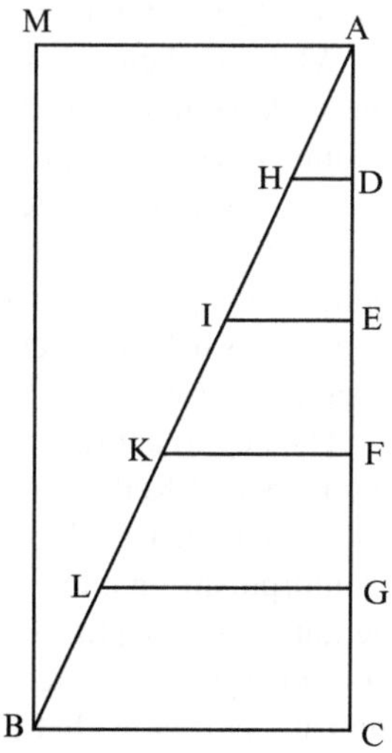

Fig. 5.3 The acceleration of falling bodies according to Galileo

the line DA."[92] Keill who, as we have just seen, explicitly denies that a line is composed of points and that time is composed of instants, divides the line AB into the small segments *ei, im, mp, po*, which he calls respectively the "infinitely small particle of time *ei*," "the particle of time *im*" and the "point of time *mp*."[93]

[92] Galilei, *Dialogue*, p. 229 (= Galilei, *Opere*, VII, p. 248).

[93] Keill, *An Introduction to Natural Philosophy*, p. 144.

This remarkable lack of terminological coherence is symptomatic of Keill's difficulty in coming to terms with continuous variation. But the passage just analyzed is also important as it reveals the main aim of Keill's demonstration: in reaction to Galileo, who had asserted the composition of all physical magnitudes out of non-extended atoms (*atomi non quanti*), Newton's pupil claims that space, time and matter cannot be composed of points, but must be divisible into ever smaller parts *ad infinitum*. It is therefore clear that Keill would not be willing to subscribe to the hypothesis, which Newton's *De gravitatione* describes as being "suitable for contemplation by mathematicians," that fluid bodies are infinitely divided.[94] According to Keill, Newton's infinitesimal calculus teaches precisely the opposite, namely that the infinite divisibility of space and matter "cannot be exhausted." Not even God can so divide matter "as that there shall be given its ultimate indivisible."[95]

5.5 Conclusion

Although in this article I have contrasted Keill's and Newton's views concerning the composition of matter, the reader of the *Introduction to Natural Philosophy* would look in vain for a passage in which the pupil explicitly disagrees with his master. What is more, Keill seems to suggest that all he writes concerning the divisibility of space, time and matter is in accordance with Newton's teaching.

Wherein, then, lies the difference between the two theories? In my view it lies in the understanding of the relation between physical and mathematical divisibility. While Newton interprets the infinite divisibility of space as an indication that God is free to ascribe whatever size and figure to the particles of material bodies, Keill uses it to argue that matter *must* be infinitely divisible. Even though the particles composing physical bodies are by definition extended, for "extension appertains and necessarily adheres to all Species of Magnitude," there is no particle, "however small, which may not be still divided into other infinite Particles."[96]

The key to appreciating the discrepancy between these two views is provided by a sentence of Keill's *Introduction*, which states that the "infinite divisibility of matter can be demonstrated from geometry."[97] The explicit endorsement of the space-matter isomorphism enables Keill, like the young Newton and many other seventeenth-century natural philosophers, to take an aprioristic approach to the issue of the composition of matter. By contrast, in the *Principia* and in the *Optics* Newton repeatedly claims that the ultimate structure of physical bodies can only be ascertained by experimental means, as the infinite divisibility of space does not necessarily imply that matter is also infinitely divisible.

---

[94] Newton, *Unpublished Scientific Papers*, p. 151.

[95] Keill, *An Introduction to Natural Philosophy*, p. 39.

[96] Ibid.

[97] Ibid., p. 21.

Chapter 6
Beeckman, Descartes and Physico-Mathematics

Frédéric de Buzon

The phrase, "there are very few physico-mathematicians,"[1] written by Isaac Beeckman in his *Loci communes*[2] on the occasion of his encounter with Descartes in November 1618 is well-known. The language appears to be new,[3] and is not found in Beeckman before this date. He comments on Descartes in this way:

> This person from the Poitou, has spent time with many Jesuits and other clever and learned men. He says that he has never met anyone, except for me, who makes use of this way of investigation, which delights me, and joins physics to mathematics in an exact way. And as for me as well, I have never spoken with anyone apart from him about this kind of investigation.[4]

But the compliment is odd. Beeckman had meditated on this subject for about a decade and a half; from the very first remark in his *Journal* (probably from 1608 to1610), he wondered why all of the arts are not subordinated to one another, why there is not "a general science or art of all mathematics, and again, of mathematics and physics, and again of physics and ethics, and again of physics and alchemy, etc."[5]

I would like to thank Daniel Garber, the translator of this chapter, and Sophie Roux deeply for their invaluable remarks.

[1] "Physico-mathematici paucissimi." *Journal*, vol. I, p. 244.

[2] This is the title Isaac Beeckman gave at the beginning of his manuscript.

[3] See, though, a contemporary book by Philipp Müller, *De cometa anni 1618 commentatis physico-mathematica specialis et generalis*.

[4] "Hic Picto cum multis Jesuitis alijsque studiosis virisque doctis versatus est. Dicit tamen se nunquam neminem reperisse, praeter me, qui hoc modo, quo ego gaudeo, studendi utatur accurateque cum Mathematicâ Physicam jungat. Neque etiam ego, praeter illum, nemini locutus sum hujusmodi studij," *Journal*, vol. I, p. 244.

[5] Ibid., vol. I, p. 5.

F. de Buzon (✉)
Faculté de Philosophie, Université de Strasbourg, Strasbourg, France
e-mail: fdebuzon@unistra.fr

D. Garber and S. Roux (eds.), *The Mechanization of Natural Philosophy*, Boston Studies in the Philosophy of Science 300, DOI 10.1007/978-94-007-4345-8_6,

But obviously, Descartes had had much less experience with these kinds of questions. This compliment shows the constant care with which Beeckman drafted his reading notes, experiments, and reflections over 30 years. He sometimes judges other authors on their way of harmonizing mathematics and physics, and in a more particular way, on the ways in which they agree with the small number of philosophical theses that he considers his own and to which he returns again and again. With regard to Bacon and Stevin, he writes that the first did not try hard enough to join *mathesis* to physics (he believed, for example, that the cause of the interval of an octave was obscure), while the second was too devoted to mathematics and dealt too rarely with physics. Thus, the phrase "this way of investigation (*hoc modo studendi*)" in the quotation, is what is most important. In fact, it is not just a question of unifying mathematics and physics in general, but the specific way in which it is done. In making his judgment about Descartes, Beeckman enters him into a very select list of authors, those who are most important for him in the renovation of science, and gives him the particular distinction of being a kind of *alter ego*. But is Beeckman right to assume, or to presuppose, that he and Descartes have a common style?

As we know, only certain important but limited problems (music, hydrostatics, free fall, etc.) were discussed during the meetings of autumn 1618. The texts that relate to these meetings have often been commented on, and in a way, they undermine the validity of Beeckman's assertion, which is, perhaps, somewhat premature. Quite to the contrary, commentators generally contrast, often in a very schematic way, Descartes the supposedly abstract mathematician with Beeckman the physicist.[6] If we take the question of music as our starting place, we can show that insofar as they have something in common in their appreciation of the genesis of harmonic intervals, the schematic opposition in question disappears: Beeckman appreciates Descartes as someone who brings an experimental way of distinguishing the major third from the fourth, and as someone who constructs consonances from the continuous division of a string segment, in agreement with Beeckman's own theory of vibration.

Nevertheless, it would be useful to redefine the relations between Beeckman and Descartes concerning physics in general and concerning the new concept of physico-mathematics. In this essay I would like to see how the most general principles of physics were received by these two thinkers. For a long time indeed, since the first discovery and publication of the *Loci communes*, commentators have recognized principles of the conservation of motion in Beeckman, principles of the conservation of the speed of a simple body, but also something much more problematic, principles of the conservation of direction and the communication of motion.[7] This has sometimes transformed older suspicions of borrowings or dishonesty on part of

[6] See Koyré, *Études galiléennes*, p. 121.

[7] On the interest in the rediscovery of the manuscript of the *Journal*, see Koyré, *Études galiléennes*, pp. 108f., n. 2.

Descartes[8] into genuine accusations of plagiarism. In this study I would like to discuss three problems, which have an intimate relation to the formulation of Descartes' laws of nature in *Le monde* and in the later *Principia*. This will show that the question of Descartes' supposed appropriation of Beeckman's rules is rather complicated, and that what is in question concerning the relations of these two figures is precisely the very concept of physico-mathematics.

6.1 Beeckman

When Descartes first met Beeckman, the latter had already been quite deeply involved in thinking about the different aspects of the persistence of motion in isolated bodies and the global conservation of motion in the world. The publication of extracts from Beeckman in the Adam-Tannery edition of Descartes' writings,[9] then in the *Correspondance* of Mersenne,[10] and finally in the complete edition of the *Journal* allows us to assemble the materials necessary to evaluate a discussion that has gone on for quite some time. On the one hand for Cornelis de Waard, Descartes owes Beeckman the essential elements of his discoveries.[11] On the other hand, Alexandre Koyrtries to emphasize their fundamental differences regarding the question of the conservation of direction or determination of motion in one direction.[12]

In his *Journal*, Beeckman had clearly touched on the three principal aspects of the laws in *Le monde*, namely (a) the persistence of motion in a vacuum, (b) direction, and (c) the global conservation of motion in impact. But is that to say that Descartes follows Beeckman in all of them? Let's take up the three points in question by noting first of all that one does not find in Beeckman a systematic exposition of the notion of a law. Furthermore, one must add to Beeckman's physical edifice at least one supplementary principle, that of isoperimetric figures. Evidence for this can be found in the presentation Beeckman made of his theses to Gassendi, which

[8] Thus Leibniz, *Remarques sur* l'Abrégé de la vie de M. Descartes, in *Die philosophischen Schriften*, vol. IV, p. 316: "1630. It seems that one wrongs M. Isaac Beeckman in treating him harshly solely on the basis of the reports found in the letters of M. des Cartes. I learned that one shouldn't put your trust in them to the handicap of others, since M. des Cartes puts a strange twist on things when he is offended with someone."

[9] Descartes, *Œuvres*, vol. X, pp. 15–169.

[10] Mersenne, *Correspondance*, vol. I, pp. 632–644 (Rules of impact).

[11] De Waard's positions are repeated in the only synthetic treatment of Beeckman available, see van Berkel, *Isaac Beeckman (1588–1637) en de Mechanisering van het Wereldbeeld*, with English summary pp. 317–319, and van Berkel, "Beeckman, Descartes et la philosophie physico-mathématique."

[12] Koyré, *Études galiléennes*, pp. 108f., n. 2.

contain (a) but not (b) and (c), while a physical principle indicating that large bodies have a small surface, but small bodies have a large surface,[13] is presented immediately after (a) and immediately derives from the more general principle of isoperimetric figures. In sum, for us to see Beeckman only as a predecessor of Descartes presupposes that we take the Cartesian presentation of the three laws as the norm. Of course we should not over-systematize what Beeckman did not present in a carefully worked-out form, but one could also show that in Beeckman, there are not three but four laws. This will allow us to better see the ways in which Descartes did not follow him.

6.1.1 Persistence of Motion

Beeckman's notes on the conservation of the speed of bodies moving in a vacuum are found from the time when he first began to keep his *Journal*. A remark from July 1612 describes the persistence of the circular motion of the heavens through the sole continuation of an initial motion.[14] Once put into motion a body will not come to rest unless it is hindered in its motion by an external cause; it has no reason spontaneously to change its state either of rest or of motion. From this moment on Beeckman thus considers motion as a stable state and not as a tendency toward rest, a *quietatio*.[15] The general epistemological background to this proposition is the principle of parsimony, which he often recalls; he cites at least twice the formula "*male fit per plura quod bene fit per pauciora*."[16] The perseverance of motion renders both celestial intelligences and a "continual command of God" useless.[17]

A year later he wrote, "Once moved, they [i.e. bodies] never rest unless they are hindered,"[18] followed immediately by some scattered remarks on the motion of the sun and on stones thrown in a vacuum. In this way, Beeckman progresses towards another principle in noting that, for bodies acting in a plenum, the capacity to overcome the resistance of the medium is a function of the size of the body. On the other hand,

---

[13] "Tum quam utile sit axioma rebus physicis indagandis: corpora magna habere superficiem parvam, parva vero magnam." *Journal*, vol. III, p. 123.

[14] "Coelum semel motum semper movetur." Ibid., vol. I, p. 10.

[15] This text from Toletus illustrates the scholastic notion of a *quietatio*: "Motus ex se tendit in quietem termini ad quem, in quo stat et quiescit res; unde motus quietatio quaedam, id est, via in quietem dici potest, quae partim est cum ipso motu." Toletus, *Commentaria*, bk. V, chap. 6, text 54, f. 163r.

[16] *Journal*, vol. I, p. 51 à propos of the notes on music, and *Lectio*, in ibid., vol. IV, p. 122 regarding the architecture of the world.

[17] "Censendum videtur coelum nec ab intelligentiis moveri, nec continuo Dei nutu, sed sua et situs natura semel motum, nunquam per se posse quiescere. Quod ergo fieri potest per pauca, male dicitur fieri per plura." Ibid., vol. I, p. 10.

[18] "Mota semel nunquam quiescunt nisi impediuntur." Ibid., vol. I, p. 24.

he appears to establish the persistence of motion by the commonplace metaphysical argument of the first mover.[19] Many notes in the *Journal* mention this principle, which Beeckman always claims as his own discovery. In its canonic and definitive form, he presents it in this way: "What is once moved always moves unless it is hindered."[20] With the help of this principle, he criticizes authors, such as Galileo, who failed to understand why a runner cannot stop instantly.[21] Beeckman thus reverses the traditional problem of knowing why bodies persevere in motion. In 1629, after having recalled the principle, he confides to Mersenne:

> Nothing more certain has come to mind than this reason, and in twenty years I have read nothing, heard nothing, or meditated on nothing which could provoke in me the least suspicion of error.[22]

6.1.2 Persistence of the Form of a Motion

Even though the conservation of motion is thus consistently affirmed, the persistence of direction is articulated in a problematic and generally allusive way. Beeckman considers the question of the vortex and associates with it the supposedly rectilinear motion of a stone which is thrown off of a wheel moving around its axle. This, he claims, is directed "not circularly, but in rectilinear motion to the place to which it was tending at the moment in which it was released."[23] Beeckman, though, advances no justification of the claim. But, in passages exactly contemporary with his meeting with Descartes, Beeckman attempts a formulation of his great principle, as an accompaniment to a further explanation of the form of the trajectory of such a stone:

> That which once moves in a vacuum always moves, either in a straight line, or in a circular path, both around its center, such as the diurnal motion of the Earth, and around a center, such as annual motion. Since the smallest part of a circumference is a curve, and the periphery of a curve behaves in the same way as the curve, there is no reason why the annual circular motion of the Earth should abandon this curved line, and proceed in a straight path since the straight line is no more natural than the circular, and it is in nature and extension equal to

[19] Beeckman proposes that once something is put into motion, it never ceases to move on its own, and from this fact it follows that the world does not need the continuous effort of God to move perpetually (ibid., vol. I, p. 10). But nevertheless, he needs a universal cause of motion, which he finds in God, at once the creator of bodies and of motions (thus, ibid., vol. I, p. 131).

[20] "Quod semel movetur, semper movetur nisi impeditur." See ibid., vol. I, pp. 10, 24f., 44, 61, 167, 253, etc.

[21] Ibid., vol. I, p. 157.

[22] Beeckman to Mersenne, June 1629, in Mersenne, *Correspondance*, vol. II, p. 233.

[23] "[N]on circulariter, sed in rectum ad locum, ad quem eo momento quo solvebatur, spectabat." *Journal*, vol. I, p. 167.

> the circular since any part of the circumference behaves in the same way in relation to the entire curve as any part of the straight line behaves in relation to the entire straight line.[24]

Hence Beeckman regards the conservation of the form of motion as something over and above the conservation of the quantity of motion.

One can see here something of a hesitation. On the one hand, our experience with the stone and the vortex gives results that are hardly convincing with respect to the supposed conservation of a circular motion. On the other hand, the theoretical reflection on curves appeals to a property whose application is strange: the straight line and the circle are the only plane lines all of whose parts can be superimposed on one another, from which Beeckman infers their naturalness. These are two shapes that can be continued indefinitely without changing their geometrical nature. One can see that the principle of reason, used first to guarantee the stability of the quantity of motion, is now applied to its geometrical figure. Beeckman would thus like to be able to attach the double rotation of the Earth directly to the persistence of a motion with a particular form. This acquired circular motion only happens in a vacuum: hence Beeckman wonders why bodies in the air never move circularly.[25]

We must keep two points in mind at this stage. (1) Whatever the degree of generality of the problem, Beeckman always conceives it in the context of cosmology, and tries to associate celestial phenomena with terrestrial phenomena. (2) The addition of direction to a principle which only concerns the quantity of translation as initially formulated, is conceived by Beeckman as the addition of a geometrical form having its own properties and a certain capacity for causality; the latter has a theological dimension, since these forms are the object of a divine choice. The decisive argument is that, like rectilinear motion, circular motion can persist to infinity. But this argument only raises a strictly mathematical consideration, without considering the physical means by which it is realized.

6.1.3 Conservation in the Exchange of Motion

To the two persistences, persistence in motion and persistence in direction, Beeckman adds reflections that relate to the conservation of motion in impact. These notes also turn up for the first time exactly when Descartes is in Breda in November and December 1618.[26] They are later followed up by some other related remarks, jotted

[24] "*Id, quod semel movetur, in vacuo semper movetur*, sive secundum lineam rectam, seu circularem, tam super centro suo, qualis est motus diurnus Terrae<quam circa centrum, qualis est motus>annuus. Cum enim quaelibet minima pars circumferentiae sit curva, atque eodem modo curva atque tota peripheria, nulla ratio est cur motus circularis Terrae annuus desereret hanc lineam curvam et ad rectam procederet, nam recta non magis naturalis et aequalis naturae et extensionis est quam circularis, quia pars circumferentiae se eo modo habet ad totam, quo pars rectae ad rectam totam." Ibid., vol. I, p. 253.

[25] Ibid., vol. I, pp. 253f.

[26] Ibid., vol. I, pp. 265–267; Mersenne, *Correspondance*, vol. II, pp. 633–635.

down in the period of January–March 1619.[27] A year later, the same issue is broached again, without the results being substantially modified.[28] Finally, Beeckman returns to the problem in 1629,[29] and then one final time in 1634.[30] He envisions from the beginning a collection of rules, which permit us to predict the result of the collision of two bodies as a function of their size and their speed.

Beeckman does not formulate a general principle concerning the exchange of motions, but a rule, formulated in 1620, renders the collection of these statements coherent. This rule comes down to the proposition that the motions of bodies in collision with one another *are subtracted* from one another.

In notes from the winter of 1618, Beeckman envisages a first case where a body in motion collides with another at rest, and communicates to it a motion in accordance with a law described through a particular case, which he generalizes and then illustrates with a mechanical analogy.[31] The particular case is that of bodies equal in size, which after collision results in an equal motion in the two bodies, whose speed is equal to half of that of the body that was originally in motion. The conceptual tool Beeckman uses to measure the size of the bodies remains relatively vague: Beeckman calls it *corporeitas* and seems to connect the equality of the bodies to the fact that they contain the same number of (elementary) parts. He thus constructs a rule for exchanging motion such that if the body at rest is double that in motion, the speed of aggregate of the two bodies after the collision will be one third. In brief, the resulting speed will be obtained by redistributing the quantity of motion in proportion to the relative size of the bodies. This principle is explained through an analogy: "A double weight, carried by an equal force, goes two times more slowly than the first weight." The analogy is hardly satisfying, but it reveals the idea of a general mechanism of proportion, which unites the size of bodies and the communication of speeds.

Once this case has been set out, Beeckman takes up a second in which two equal bodies approach one another with an equal speed. Curiously (at any rate, to Descartes' mind), the result proposed is that the two bodies come immediately to rest: *directe quiescent*. When the two speeds are different, the system consisting of the two bodies moves with a speed equal to the difference of the two initial speeds. That comes down to saying that there is no global conservation of motion in the universe, since the collision of equal bodies annihilates the motion of those that participate in the collision.

Beeckman immediately draws this conclusion from the loss of motion: the note following remarks that "motion never increases in a vacuum, but decreases. Why isn't there universal rest?" Beeckman seeks a solution in cosmology: "This is why God alone can conserve motion by once moving the greatest bodies with the least

[27] Beeckman, *Journal*, vol. I, pp. 271f.; Mersenne, *Correspondance*, vol. II, pp. 635f.

[28] Beeckman, *Journal*, vol. II, pp. 45–54; Mersenne, *Correspondance*, vol. II, pp. 636–640.

[29] Beeckman, *Journal*, vol. III, pp. 128–131; Mersenne, *Correspondance*, vol. II, pp. 640–642.

[30] Beeckman, *Journal*, vol. III, p. 369; Mersenne, *Correspondance*, vol. II, p. 642.

[31] Beeckman, *Journal*, vol. I, pp. 265–267.

speed[32] which then, in turn, perpetually arouse and enliven the others, which tend toward rest."[33] Even though Beeckman can be considered one of the physicists opposing the conception of local motion as *quietatio*, that same doctrine can be found at global level of the universe. There it is saved from universal rest only by the disproportion of the celestial bodies, which are supposed to compensate for the smaller bodies' continual loss of motion. Thus, through failing to draw a sufficient distinction between the problem of building a general physics and the problem of the behavior of bodies at the astronomical level, a relatively modern conception of motion can accommodate within itself the remnants of ancient cosmology.

If one compares Beeckman's principles with their Cartesian equivalents, it is obvious that his most general physical laws do not ground a permanent conservation principle. Nevertheless, Beeckman was aware of this major difficulty and sought to resolve it by elaborating the astronomical suggestion to which we have called attention.

6.1.4 Isoperimetric Figures

I would now like to call attention to a final principle that Beeckman develops in a *Lectio* held on 3 June 1627 on the occasion of his nomination to the rectorate at the Latin School of Dordrecht, one that he called "*theorema de figuris isoperimetris*."[34] This principle gives us the solution to the aporia previously noted. For bodies with the same perimeter, "the smallest bodies have a greater surface and are thus subject to a greater degree of hindrance from the air in relation to their volume."[35] Other formulations are found in the same period. Thus, in a letter to Mersenne from 1 October 1629,[36] he proposes that light and small bodies have a large surface in relation to their *corporeitas*, their volume. This general principle explains the persistence of motion or rest of large bodies, but also of those that are mathematically more regular.[37] Thus the definitive formulation of the principle of isoperimetric figures, as expressed in the *Lectio* of 1627, is purely mathematical. It distinguishes two cases, one for dissimilar (i.e. irregular) figures, and the other one for similar (i.e. regular) figures. In the first case he says that "the most regular of isoperimetric

[32] I read here "minima celeritate," as in Mersenne, *Correspondance*, vol. II, p. 635 and p. 123, and not "maxima celeritate," as in the edition of de Waard (Beeckman, *Journal*, vol. I, p. 267); see Mersenne, *Correspondance*, vol. VIII, p. 422 n. 4, where de Waard notes this correction in his edition of Beeckman's *Journal*.

[33] Beeckman, *Journal*, vol. I, pp. 266f., also cited in Mersenne, *Correspondance*, vol. II, p. 633, with the collection of impact rules proposed by Beeckman. See Beeckman, *Journal*, vol. I, p. 196.

[34] The *Lectio* is found in ibid., IV 122–26; the cited phrase is on p. 125. Gemelli, *Isaac Beeckman, atomista e lettore critico di Lucrezio*, p. 26 emphasizes the importance of this principal.

[35] Beeckman, *Journal*, vol. III, p. 49, April-May 1628.

[36] Mersenne, *Correspondance*, vol. II, p. 281.

[37] Ibid., and note of 30 April 1618, see Beeckman, *Journal*, vol. I, pp. 170f.

figures have the greatest volume."[38] In the second he says that "the larger of equally regular figures with respect to volume has a smaller surface."[39] The first case is applied to two examples, the spherical shape of the world, as was determined by the Architect of the Universe, and the circular shape of the city of Carthage, as delineated by Dido. Beeckman takes advantage of the occasion to distinguish place and space: the place is not the surface containing the body since the same surface can contain bodies of different sizes. The other case is illustrated by a greater number of examples, fortified cities, again, and then the lengthier motion of heavy bodies, the faster advance of larger boats, the speed of arrows, the colossus of Rhodes, the anchors of boats, and so forth. The astronomical example, as noted earlier, completes the argumentation relating to the general conservation of motion in the universe, even if the collisions of small bodies tend toward a general rest.

One can see here that considerations of mathematical form are determinants for Beeckman, concerning both the direction and the behavior of a body in motion in a medium. The general approach can be summarized as follows. A general principle is first constituted, that of the conservation of motion once it is imprinted on a body. Beeckman immediately sees that the difficulty is to maintain the validity of the principle despite appearances and experiments. Concerning direction, the principle is made more precise through remarks about the form of motion, that is, the trajectory of the body in motion. But the form of the motion is not justified through its genesis, for circular motion no more than for rectilinear. Finally Beeckman begins to reflect on the transmission of motion, but only to make explicit the way in which the general conservation principle is hindered by collision with other bodies. It is here, in the last analysis, that the principle of isoperimetric figures enters.

Thus in a rather broad sense one can certainly find in Beeckman remarks which correspond closely to the Cartesian laws: he proposes, in fact, a sketch of the three conservation principles that Descartes will treat later. Because of this, one cannot simply neglect this historical element in understanding the Cartesian texts. But Beeckman sets out something more, and in a kind of organization that is at root very different. Presenting these sources should certainly not be regarded as an argument that Descartes was a plagiarist. Rather, it permits us to understand the grounds on which Descartes found some of Beeckman's results acceptable and to understand the grounds on which he rejected others. Outside of the obvious differences in content, and the fact that Beeckman's principles are never really systematized, the conceptual framework of the Cartesian laws is in reality different. After considering the formation of the notion of a law of nature in Descartes, it is to this is that we shall turn.

[38] "Figurarum isoperimetrarum ordinatissima est capacissima." Ibid., vol. IV, p. 122.

[39] "Figurarum aeque ordinatarum major minorem, respectu capacitates, habet superficiem." Ibid., vol. IV, p. 123. This principle itself was formulated much earlier, in a note from April 1614: "Want de corpulentie ofte swaerheyt drejnckt een dynck terneder ende de superficies, die teghen de locht kompt, verhindert int vallen. Unde sequitur globum ejusdem materiae, majoris tamen quantitatis, celerius cadere globulo minoris quantitatis: ratio enim superficiei minoris globi ad corpulentiam ejusdem globi majorem habet rationem quam superficies majoris globi ad ipsum corpus majoris globi." Ibid., vol. I, p. 31.

6.2 Descartes

What permits Descartes to break with the fundamentals of the medieval tradition of laws of nature[40] is the reduction of the open and indefinite collection of empirical regularities to the principles of motion alone. This conception is itself only rendered possible by the reduction of all material changes to local motion alone, and by the adaption of a uniform concept of matter, such as is defined in *Le monde*. It is only in this way that *mathesis* can connect to physics.

It is illuminating to return to the context of the first attested usage of the expression "law of nature" in Descartes, which is found in a letter dated from 1631.[41] The laws of nature are evoked in the discussion of a question posed by Mersenne related to the free-fall of bodies in a vacuum and the proportion between the time and distance fallen, a question which recalls precisely the older work Descartes carried out with Beeckman[42] and which was supposed to find a definitive response in the treatise on physics then in preparation. The solution he offers Mersenne is grounded on a principle of the conservation of motion considered as an *assumption*:

> As to what you ask me about on what foundation I base the calculation of the time that a weight attached to a cord [i.e. a pendulum] of 2, 4, 8 and 16 feet would take to fall, although I should put it in my physics, I don't want to make you wait until then and I will try to explain it. First, I assume that motion once imprinted on some body remains there perpetually, unless it is removed by some cause, that is to say, what in a vacuum once begins to move, always moves with a uniform speed.[43]

To this assumption Descartes adds another, that of a *gravitas,* which at every instant adds a new force for descending and which constrains the body constantly to increase its speed, and thus alters the uniform motion which a body left to itself would have. Descartes then situates his proposal in the context of an abstract or purely mathematical physics, which he identifies with the estimation of the displacement of an object in the vacuum. Now, it is exactly against these abstract assumptions that in the autumn of 1631 Descartes appeals to the laws of nature, using that notion for the first time:

> In the third place, you asked me how a stone moves *in vacuo*. ... I don't assume just the vacuum, but also the force that makes the stone acted upon always move uniformly. This is obviously in contradiction to the *laws of nature*: for all natural powers act more or less according as the subject is more or less disposed to receive their action, and it is certain that a stone isn't equally disposed to receive a new motion or an increase in speed when it is

---

[40] On the medieval antecedents of the notion of a law of nature, see Crombie, "Infinite Power and the Laws of Nature: A Medieval Speculation."

[41] Descartes, *Œuvres*, vol. I, p. 230.

[42] See Descartes to Mersenne, 13 November 1629, Descartes, *Œuvres*, vol. I, p. 72. The context in Beeckman is recalled in the following letter, 18 September 1629, in ibid., pp. 82–105.

[43] Descartes to Mersenne, 13 November 1629, in ibid., pp. 71f. The letter is in French, except for the last phrase, which is in Latin.

> already moving rather fast and when it is moving rather slowly. But I think that I now can determine the proportion by which the speed of a stone which descends increases not *in vacuo*, but *in hoc vero aëre*.[44]

I will overlook the fact that the question is never actually treated in the surviving part of *Le monde* and that there is little chance that it could have figured there, insofar as the question that interests Descartes there is less the calculation of the speed by which a body falls than it is the physical cause of weight.[45] But, for our purposes, we must notice that even though they are not defined precisely, the *laws of nature* are set in opposition to the collection of abstract mathematical assumptions made previously, the assumption of the conservation of motion in a vacuum and the assumption of *gravitas*. The principle according to which a body once moved always moves unless it encounters an obstacle, whose paternity Descartes rightly attributes to Beeckman,[46] does not yet have the explicit status of law: on the contrary, the indefinite collection of laws of nature concerning the capacity bodies have for receiving new accelerations in accordance with their speed concerns the concrete conditions of motion, while Beeckman's principle, even if admitted, remains an ideal assumption. Furthermore, the differential capacity of a body to receive degrees of speed in accordance with its own speed is not a true law in Descartes' later physics, but results from a combination of the two first laws of *Le monde*.

Thus, starting from *Le monde*, the real novelty is that what were earlier abstract assumptions now become laws valid for nature itself, and, as a consequence, there is no longer any place for a conflict between that which is recognized as an abstract rule and that which is produced in reality in the physical world. There is a distinction between the acceptable rules or assumptions (in the present case, the principle of conservation) and the assumptions that are not. From then on, Descartes definitively calls laws of nature the rules by which the changes in the parts of matter happen, matter being conserved by God in the way in which it had been created.[47] The central point here, as is also recalled in the *Principia philosophiae* (II 23), is that matter is exclusively conceived in terms of its capacity to be divided and to be mobile, and that all of the diversity of its forms thus depends on motion.

6.2.1 Persistence of Motion

We should not be surprised that the first known instance of the problem of the conservation of motion in Descartes makes explicit reference to Beeckman; thus, at least in private, Descartes recognized his debt. The *Cogitationes privatae* indicate that "Quod enim in vacuo movetur semper moveri existimabat,"[48] just as the Cartesian fragment with regard to the same problem of the fall of bodies, recopied

[44] Descartes to Mersenne, October or November 1631, in ibid., pp. 230f.

[45] *Le monde*, chap. 11, in Descartes, *Œuvres*, vol. XI, pp. 72–80.

[46] Descartes, *Œuvres*, vol. II, p. 91.

[47] Descartes, *Œuvres*, vol. XI, p. 37.

[48] Descartes, *Œuvres*, vol. X, p. 219.

in Beeckman's *Journal*, notes: "in vacuo quod semel motum est semper movetur."[49] Writing to Mersenne on 18 December 1629, Descartes evokes Beeckman's positions and his great principle. He notes that Beeckman assumes that that which is once moved tends to continue its motion of itself and in a vacuum, just as he, Descartes does.[50] He even remarks a little before: "We should have remembered that we assume that what was once moved always moves in a vacuum, and I shall try to demonstrate it in my treatise."[51]

The treatise envisioned is almost certainly what will become *Le monde*. One can see that one of Descartes' aims in that text is the *demonstration* of Beeckman's proposition, evoked as an admitted fact, but whose theoretical status is only that of an assumption. We thus need to *ground* a law already known, and not discover it in an absolute sense.[52] This is Descartes' first law in *Le monde*:

> The first [rule] is: that each part of matter, taken individually, always continues to be in the same state as long as collision with other bodies doesn't force it to change. That is, if it has a certain size, it will never become smaller unless other bodies divide it; if it is round or square, it will never change this figure, without other bodies forcing it to do so; if it is at rest in some place, it will never leave, as long as others don't drive it away; and if it has once started to move, it will always continue with a constant force until others stop it or slow it down.[53]

In this way *Le monde* rethinks an earlier inheritance: the first law is a repeat of Beeckman's principle, except that it is generalized to every state of a body (figure, size, etc.), before it is applied specifically to the problem of motion alone. The principle is thus grounded on the stability of a part of matter, and thus of a corporeal substance, taken apart from all interaction with another part of matter.

6.2.2 Communication of Motion

The second law of *Le monde* is directly opposed to Beeckman's theses:

> I assume for a second rule that when a body pushes another, it cannot give it any motion unless it loses at the same time as much of its own.[54]

While Beeckman allows motion to be lost in certain cases, and while speed and size compensate for one another in other cases, Descartes presents a universal principle of the compensation of motions and the sizes of bodies. The law as

---

[49] Ibid., p. 78.

[50] Descartes to Mersenne, 18 December 1629: "Supponit, ut ego, id quod semel moveri coepit, pergere sua sponte, nisi ab aliqua vi externa imediatur, ac proinde in vacuo semper moveri, in aere vero ab aeris resistentia pautalitm impediri." Descartes, *Œuvres*, vol. I, p. 91; Mersenne, *Correspondance*, vol. II, p. 341. See also *Parnassus*, in Descartes, *Œuvres*, vol. X, p. 219.

[51] Descartes, *Œuvres*, vol. II, pp. 90f.

[52] An analogous approach à propos of the theory of refraction is suggested in Costabel, "La réfraction de la lumière."

[53] Descartes, *Œuvres*, vol. XI, p. 38. While at the beginning of chap. 7 Descartes talks explicitly of "laws [*lois*]" (ibid., p. 36), when he comes to give them he calls them "rules" [*regles*]

[54] Descartes, *Œuvres*, vol. XI, p. 41.

presented in *Le monde* is not at all quantified; the first attempts at quantification, rather similar to those of Beeckman, only appear in the correspondence from 1639 to 1641.[55] But the central point is that the two laws in *Le monde* appear as two consequences of a common principle, that of the conservation of motion. The conceptual gain over Beeckman is twofold. On the one hand, the first law is drawn from a more general principle, and on the other, this same principle can account for apparent exceptions. The aporia in which Beeckman found himself is thus overcome. However, many similarities remain. In particular, the problems of direction and magnitude of the motion are treated separately, and considered as a second domain in which divine immutability can act.

Unlike Beeckman, in *Le monde* Descartes says nothing in detail about the rules of collision, limiting himself to mentioning the possibility of formulating them elsewhere in more detail.[56] Why this silence? In large part, it seems, because, as the earlier expositions show, their formulation assumes combining rules determining impact where there is a merging of motions, with those that concern reflection, where there is not. Now, since Discourse 1 and 2 of the *Dioptrique* treat the reflection of bodies by considering the body in motion as a material point and by assuming a surface of reflection that is absolutely immovable, Descartes has not yet figured out how to combine the principles in an orderly way.[57]

6.2.3 Persistence and Direction

The third law in *Le monde* is the following:

> I add for the third [rule]: that when a body moves, even though its motion is most often made in a curved line and can never move in way that is not in some way circular …, nevertheless each of its parts, taken individually always tends to continue in a straight line. And thus their action, that is, the inclination that, they have to move, is different from their motion.[58]

This law appears to contradict Beeckman's hypotheses, which admits the conservation of both rectilinear and circular motion, so as to allow him to treat the problem

[55] Descartes to De Beaune, 30 April 1639 (Descartes, *Œuvres*, vol. II, p. 543; Mersenne, *Correspondance*, vol. VIII, p. 421), Descartes to Mersenne, 25 December 1639 (Descartes, *Œuvres*, vol. II, p. 626; Mersenne, *Correspondance*, vol. VIII, p. 696), 28 December 1640 (Descartes, *Œuvres*, vol. III, p. 205; Mersenne, *Correspondance*, vol. X, p. 173), 17 November 1641 (Descartes, *Œuvres*, vol. III, p. 451; Mersenne, *Correspondance*, vol. X, p. 382).

[56] "Je pourrais mettre encore ici plusieurs règles pour déterminer, en particulier, quand et comment et de combien le mouvement de chaque corps peut être détourné, et augmenté ou diminué, par la rencontre des autres; ce qui comprend sommairement tous les effets de la Nature." Descartes, *Œuvres*, vol. XI, p. 47.

[57] In fact, the greatest difficulty with the rules set forth in the *Principia* concerns again the apportioning of the case of the reflection of one body off of another and the communication of motion. The hesitations and corrections abound, but without yielding satisfactory results, as is well known. Thus, the articulation of the rules of impact (*Principia* II 46–52) is considerably altered in the French translation, especially in accordance with the supplementary principle proposed in the letter to Clerselier of 17 February 1645, in Descartes, *Œuvres*, vol. IV, p. 187.

[58] Descartes, *Œuvres*, vol. XI, p. 44.

of astronomical bodies in motion. For Descartes here, the persistence of rectilinear motion is connected with the unique simplicity of straight motion, which should be conserved without regard to duration or without regard to the comparison between two instants. Now, at a time *t*, a body in motion has a unique direction, with relation to which every other direction is oblique. Because of this, in the third law we are not dealing with a law of the conservation of a form (rectilinear or circular) but with the conservation of a direction, of a tendency towards ..., which always has as an *effect* the form of a path that is exactly straight. Descartes establishes his central point many times, in different registers:

(a) In *Regula* VI, the straight line is said to be absolute, while the oblique line is said to be relative.[59] In fact, since obliquity is the changing of direction, the concept of obliquity is a composite made up of the basic direction, considered as a simple nature, together with another thing, a change. Since we must conceive of the straight line in order to conceive of the oblique, the latter is composed and thus more complex.
(b) In general, Descartes considers the curve as a polygon which has an infinity of sides, and thus as a line which constantly changes its direction.[60]
(c) The example of the sling provides an experimental confirmation. But the particular case of light is even more obvious, since the tendency or action to move which constitutes this phenomenon propagates itself exactly in a straight line, in accordance with the analogy of the wine vat,[61] without the intervention of a formal cause.

The central point is that the straight line is from then on something that can be conserved without change, while a curve, circular or otherwise, appears as a continual change. The geometrical or formal aspects of the problem are hidden behind the mechanical constraints which engender the curves.

Indeed, even though the question, for Descartes as for Beeckman, was to understand that which is unchanging in the dimension of motion, the two authors give a totally different response. For Beeckman, a geometrical property (homogeneity of the part and the whole) that establishes the naturalness of a form is enough to establish that a motion can persist if nothing hinders it. If an external cause is necessary for change, then it is necessary to know beforehand what is naturally stable: the circle appears to have this property, and permits Beeckman to understand the stability of the astronomical system.[62] Beeckman conceives of the forms described by bodies in motion in this way. In this way he conceives of the physical object as

[59] Descartes, *Œuvres*, vol. X, pp. 381f.

[60] Descartes to Mersenne, 23 August 1638, Descartes, *Œuvres*, vol. II, p. 309. See also Costabel, "La courbure et son apparition chez Descartes," and Cassirer, "Descartes' Kritik der mathematischen und naturwissenschaftlichen Erkenntnis," which is the preface to *Leibniz' System in seinen wissenschaftlichen Grundlagen*, which can be found in Cassirer, *Gesammelte Werke*, vol. I, pp. 29f.

[61] *Dioptrique*, in Descartes, *Œuvres*, vol. VI, pp. 86–88.

[62] It is obvious that this method is to be compared with that of Kepler.

determined by the mathematical form, mathematically speaking, or, in metaphysical language, he considers the formal cause as preexisting, or at least, as simultaneous with the efficient cause. Now, for Descartes, both ideally and in reality, the geometrical form of a body's trajectory is always secondary insofar as it is the trace of a motion of a point or of a line: motion thus becomes, in fact, the first object of Cartesian science, both in principle and in fact.

> [T]he nature of the motion which I intend to speak of here is so easy to understand that the geometers themselves who, among all men, are the most educated in conceiving distinctly the things that they have considered, have considered it more simple and more intelligible than all their surfaces and their lines; this is manifested in the fact they have explained the line by the motion of a point and the surface by that of a line.[63]

These remarks thus lead us to invert the way in which the opposition between Descartes and Beeckman is usually understood. The more mathematical and more Platonist of the two is not the one usually thought to be so. Quite contrary to our initial expectations, for Beeckman the mathematical form has a causal power, which in the end totally disappears in the Cartesian elaborations of the laws of motion. Theology plays a supporting role in both accounts, to be sure. But Descartes' God directs nature through three laws of motion which he imprints identically in all bodies as conditions of being, and in minds as conditions of knowledge, while Beeckman's God, a bit like that of Kepler or later, like that of Leibniz, chooses the best forms for bodies and thus regulates motions. Furthermore, Beeckman's world in 1618 only maintains itself in motion by an astronomical artifice. Beeckman seeks a solution to a problem in general physics by way of a local solution, however vast the locale might be. We need a particular decree from a God who organizes the heavens and optimizes the figures of bodies. On the other hand, for Descartes, since the laws are secondary causes of the formation and global conservation of the world, they should permit us to prove the sufficiency of the efficient cause. There is no better way of establishing this point except by gathering the collection of the forms of the effects that derive from motions.

6.3 Physico-Mathematics

What, then, of physico-mathematics in Beeckman and Descartes? Without doubt, these two authors both appeal to the new conception of phenomena, in opposition to the scholastic explanation through forms and qualities: everything in fact is reducible to the motions of the parts of bodies. But like the entire tradition before him, Beeckman maintains a radical difference between physics and mathematics. He advances, for example, that squaring the circle is possible in the physical world but not mathematically, since mathematical space is divisible to infinity while

[63] Descartes, *Œuvres*, vol. XI, p. 39.

physical space is not.[64] He excludes motion from mathematics, which considers its objects as outside of time and at rest. Body and motion only appear with physics. Thus, he writes:

> We must assume body and motion in physics. It appears that we must assume that there is body and motion in physics. For who could understand the cause of resistance, and where the first motion comes from?[65]

And again:

> *The reason for motion.* At a mathematical instant, the stone in motion is in a place and thus doesn't move. But in a physical instant, it moves.[66]

In other words, physics deals with bodies in motion, while mathematics determines forms for bodies considered as immobile. From this point of view, Beeckman is largely the heir of Pierre de la Ramée, who proposed that these disciplines are irreducible, rejecting even the concept of *mathesis mixta*;[67] and even further still, Beeckman is the heir of the Aristotelian separation of the disciplines.

After the *Regulae* and *Le monde*, for Descartes, in general, there is no difference in principle between *mathesis pura* and *mathesis mixta*: all mathematics is such through the consideration of order and measure, whatever the object to which the concepts apply. Furthermore, for Descartes the very concept of motion belongs to pure mathematics. The object of physics and that of mathematics is the same object, differing only in its modality, the one considered as actual and the other as possible.[68] In his response to Joannes Ciermans of 23 March 1638,[69] Descartes notes *a propos* of the *Géométrie*:

> In that treatise I didn't explain any of those things that belong to arithmetic properly, nor did I solve any of those questions in which order is regarded together with measure, examples of which are found in Diophantus. But furthermore, I didn't even treat motion, which pure mathematics (at least that which I have most cultivated) takes for its principal object.[70]

One thus understands how even the term physico-mathematics disappears in Descartes. *Physico-mathematicus* refers to someone who unites two disciplines, while maintaining the distinction between the objects that they treat. Descartes' *mathesis*, 10 years after the encounter with Beeckman, is grounded, on the contrary, in their identification, if not, as many contemporaries have thought, in their confusion.

---

[64] Beeckman, *Journal*, vol. I, p. 26.

[65] Ibid., vol. III, p. 310. Note that the first sentence of the quotation is a marginal summary.

[66] "Lapis motus τῷ νῦν mathematico est in loco et sic non movetur. At τῷ νῦν physico movetur." Ibid., vol. III, p. 357.

[67] See Ramus, *Scholarum mathematicarum libri unus et triginta*, l. IV, p. 114; see also de Buzon, "Mathématique et dialectique: Descartes ramiste?"

[68] "The difference consists only in this, that physics considers its object not only as a true and real entity but as existing actually and as such. However, mathematics [*Mathesis*] considers it only insofar as it is possible, and which doesn't exist in actual space, but yet could exist." Descartes' Conversation with Burman, in Descartes, *Œuvres*, vol. V, p. 160.

[69] Descartes, *Œuvres*, vol. II, p. 56.

[70] Ibid., pp. 70f.

Chapter 7
Between Mathematics and Experimental Philosophy: Hydrostatics in Scotland About 1700

Antoni Malet

Many years ago J.B. Conant contrasted Pascal's and Boyle's approach to hydrostatics and pneumatics in terms of "two traditions," one mathematical, the other experimental.[1] Peter Dear has brilliantly recast Conant's suggestion by linking Pascal's (so-called) mathematical approach and Boyle's experimental approach to their contrasting theological views.[2] In a more general way, there is a broad consensus that the experimental approach was the distinguishing feature of the teaching of natural philosophy in Britain from the late seventeenth century on. In the early Enlightenment in Britain, Larry Stewart and others have shown, the utilitarian, manipulative, visual, experimentalist side of natural philosophy was favored and stressed to the point that the mathematical content almost disappeared. It was an approach in which hands-on experience and observation not only helped to overcome difficulties in concept-clarification and in mathematical arguments, but appeared as real alternatives to them.[3] Although there is much truth in those accounts, we present here evidence that a British mathematical approach to hydrostatics and pneumatics was successfully developed by John Wallis, James Gregorie (or Gregory), Newton, and others. In a sense that we will specify here, their approach is more deeply and more genuinely mathematical than Pascal's. Finally we also present evidence that such a mathematical understanding of hydrostatics and pneumatics occupied a prominent place in the teaching of natural philosophy in Scottish universities from the late seventeenth century on.

---

[1] Conant, *Harvard Case Histories*, vol. I, p. 59.

[2] Dear, "Miracles, Experiments, and the Ordinary Course of Nature."

[3] Stewart, *The Rise of Public Science*, *passim*, but see particularly chap. 4.

A. Malet (✉)
Department d'Humanitats, Universitat Pompeu Fabra, c. Raman Trias Fargas 25, 08005 Barcelona, Spain
e-mail: antoni.malet@upf.edu

D. Garber and S. Roux (eds.), *The Mechanization of Natural Philosophy*, Boston Studies in the Philosophy of Science 300, DOI 10.1007/978-94-007-4345-8_7,

7.1 Between Mathematics and Experimental Philosophy

The main and almost only source for seventeenth-century hydrostatics is Simon Stevin's (1548–1620) *De Beghinselen des Waterwichts* (*The Elements of Hydrostatics*). Originally published in Dutch in 1586, it was available in Latin and French from the early years of the seventeenth century.[4] Besides new results on the weight or pressure (he does not explicitly distinguish the two notions) that fluids exert upon inclined surfaces, Stevin provides an original mathematical demonstration of the so-called hydrostatic paradox, i.e., that the force acting upon any given surface, *S*, on which some homogeneous fluid in equilibrium rests, is independent of the volume of fluid resting upon it and depends solely on the measure of the surface and the height (or vertical distance) of the upper surface of the fluid on the surface *S*. In *Discorso intorno alle cose che stanno in su l'acqua o che in quella si muovono* (1612), Galileo offers an original physico-mathematical account of the equilibrium of fluids in syphons. Then, as it is well known, in the late 1630s and early 1640s, Berti, Torricelli and others designed and performed experiments involving columns of water and mercury. In their discussions the focus was primarily on both the nature of the space appearing in the glass tube closed top end and the cause that held the water and the mercury up to specific heights within the tubes. The experiments and Torricelli's interpretation of them circulated widely, but often in second-hand reports and only in epistolary form until the posthumous printing of Torricelli's *Lezioni Accademiche* (1715). While Pascal's interpretation of the experiments was essentially the same as Torricelli's, the French savant and others (including Perier, Petit, Mersenne, Gassendi, and Roberval) crucially enlarged Torricelli's work between 1645 and 1648 with newly designed experiments—the Puy-de-Dôme and the "vacuum within a vacuum" experiments outstanding among them. Moreover, Pascal first offered a cogent systematic explanation of the barometric experiments by deductively deriving them from the principles of hydrostatics. As Alexandre Koyré put it many years ago, Pascal's treatises "contain few really new ideas; possibly none at all," and yet his work on hydrostatics and pneumatics is brilliantly original on account of "the admirable order in which the facts, real as well as imaginary, are set forth and arranged as a function of one single idea."[5]

Barometric experiments (as they were called after 1663) were also performed in England about the same years, in the late 1640s and early 1650s. In the second half of the 1650s, after news of the invention of the air-pump by Guericke (1654) reached England, Boyle, Hooke, and others improved the pump and set out to experimentally investigate the properties of the vacuum and the "spring" of the air. Boyle first published their results in *New experiments physico-mechanical, touching the spring*

---

[4] Stevin's hydrostatics was included in the *Wisconstighe Gedachtenissen* (*Mathematical Memoirs*, 1605–1608) and in its Latin and French adaptations, *Hypomnemata mathematica* (1605–1608) and *Mémoires mathématiques* (1605–1608). It was printed again in 1634 as part of *Les œuvres mathématiques de Simon Stevin*.

[5] Koyré, "Pascal Savant," pp. 155f.

of the air and its effects (1660), followed by other tracts in 1663 and 1666.[6] All of this is very well known. Perhaps not so well known is that in 1671 John Wallis published within his *Mechanica* a hydrostatical treatise that included pretty much the main physical conclusions reached in Boyle's and Hooke's experiments but recast and articulated in mathematical argument. In effect what Wallis achieved was to turn hydrostatics and pneumatics into a mixed mathematical science in the style of Archimedes and Stevin.[7] More interesting and probably still less known is that Wallis' mathematized hydrostatics and pneumatics were transformed and improved by James Gregorie (or Gregory, 1638–1675), and probably by other Gregories, and became a widely used text for the teaching of natural philosophy in the Scottish universities probably up to the 1740s. The evidence we have—to be discussed in this paper—is a complete, neatly copied, 40-page long manuscript titled "Hydrostatica," dated 1740, that offers a full account of hydrostatics and pneumatics as they became organized in the last third of the seventeenth century. Although its authorship poses many problems (to be discussed in the Appendix), its content is ultimately to be traced back to Wallis' "De hydrostaticis."

In his important article about Pascal's hydrostatics and the way in which it differs from Boyle's, Dear claimed that Pascal's treatises on hydrostatics and the weight of the air "fall squarely into the genre of the mixed mathematical sciences: Pascal's hydrostatics is ... simply an extension of the classical mathematical science of mechanics." Dear deals with Pascal and Boyle within a deep, insightful discussion of the different value of experiments and experimental reports in the contrasting theological contexts of Catholic France and Protestant England. He convincingly elucidates why different understandings of miracles and the regularities of nature in Catholic and Protestant traditions contribute to explaining why Pascal and Boyle granted different epistemological value to facts and "matters of fact." Dear takes both of them as being representative of the "dominant forms" of natural philosophy pursued then in France and England—and I think he is largely right in doing so.[8]

On the other hand, it is far from obvious that Dear's understanding of Pascal's hydrostatics as mixed mathematical science bears close scrutiny. As we shall see now, Pascal's hydrostatics fails to show many basic features of mixed mathematics and we would oversimplify the differences between Pascal's and Boyle's natural philosophies by making them hinge on their degrees of mathematization. More generally, Dear's account somehow makes mathematization in general, and mixed-mathematics articulations of natural philosophy in particular, to be deeply in conflict not only with Boyle's experimentalism, but with mainstream English natural philosophy in Boyle's time. This is widely an accepted view about seventeenth-century English natural philosophy and might be one of the reasons why Wallis'

[6] Middleton, *The History of the Barometer*, pp. 55f., 59, 61–67. On the introduction of the word "barometer," see pp. 71f.

[7] I know of no full study of Wallis' *Mechanica*, but see the chapter devoted to it in Scott, *The Mathematical Work of John Wallis*, pp. 91–121.

[8] Dear, "Miracles, Experiments, and the Ordinary Course of Nature," p. 675.

mathematization of hydrostatics has been overlooked. However, Wallis' mechanics and the role of his hydrostatics in Scottish universities suggest that mathematics was more important for British natural philosophy from the late seventeenth century onwards than many standard accounts allow.

When Pascal's treatises (written about 1651 or shortly thereafter) on fluid equilibrium and the weight of the air were finally published in a more or less finished form in 1663, they drew criticism from Boyle and others for their poorly specified experimental procedures and for the loose, ambiguous way in which results and conclusions rested on experimental facts.[9] Dear is completely right in that Pascal's reports of his experiments, even when he used them for demonstrative purposes, stand no comparison with Boyle's excruciatingly detailed experimental reports. Experiments are understood differently and play a different role within Boyle's and Pascal's works. From beginning to end, Pascal's hydrostatics have a stronger and more clear systematic deductive structure so that theoretical arguments play a role without equivalent in Boyle's works. I would claim, however, that the most substantial difference lies elsewhere, in that Pascal at times uses experiments as crucial arguments to prove theoretically foundational results but at other times he dismisses their apodictive value. That is to say, experiments do not play a consistent role in Pascal's treatises. Let us consider in detail this fundamental ambiguity in Pascal's hydrostatics.

Pascal sometimes explicitly assumed experiments to be mere "confirmations" of important theoretical arguments. Experiments are useful, he said, mainly because they are more persuasive than theoretical arguments:

> Since in physics, experiments [*experiences*] have a greater force to persuade than reasonings do, I have no doubt that one would want to see the ones confirmed by the others.[10]

Pascal further argues in this context that experiments have little demonstrative power. Experiments may fail to show the expected effects, he says, but there is always a way to explain the failure *and* save the conclusions theoretically predicted. In discussing the experiment of taking a partially inflated balloon up to the top of a 500-toise high hill, and then down, he describes it perfunctorily—little by little it becomes swollen, just by itself, until at the hill top it appears to be fully swelled; then the contrary happens when taking it down.[11] Yet, this is a particularly important experiment, according to Pascal, because it shows one of his crucial theoretical "conclusions," that the pressure exerted by a given sample of air depends on the volume of air that it supports, so that if we took the given sample to a place where it

---

[9] Boyle criticized Pascal's approach in *Hydrostatical paradoxes* (1666). On Boyle's criticism, Pascal's experiments, and the material difficulties they pose for seventeenth-century technology, see Middleton, *History of the Barometer*, pp. 45–50; Koyré, "Pascal Savant," pp. 150–155; A.W.S. Baird, "Pascal's Idea of Nature," pp. 297–320.

[10] Pascal's treatises on hydrostatics were published posthumously as *Traités de l'équilibre des liqueurs et de la pesanteur de la masse de l'air* (Paris, 1663); quotations are from Pascal, *Œuvres complètes*, p. 245.

[11] Pascal, *Traité de la pesanteur de la masse de l'air*, p. 245. One *toise* amounts to one *fathom*, roughly two meters.

supported a different volume of air, its pressure would change accordingly.[12] For all of its importance (or perhaps because of it), Pascal addresses the hypothetical failure of the balloon showing the predicted effects. Such an event "will not destroy my conclusion," he says, because he always may argue that the hill is not high enough:

> But if one were to do the experiment, I would have this prerogative, that if it should happen that there was no difference in the swelling of the balloon on the highest mountains, this would not undermine what I have concluded, since I could say that they are not high enough to cause a sensible difference.[13]

On the other hand, Pascal's hydrostatics (*Traité de l'équilibre des liqueurs*) is demonstratively grounded on one proposition, "That fluids weigh according to their height." According to Pascal this proposition is proved by "experience" (*sic*).[14] Pascal takes pains, first, to stress the unobvious, seemingly paradoxical nature of this proposition. Then he introduces a balance mechanism that demonstrates it experimentally and "exactly (*exactement*)."[15] That is to say, Pascal does not ground his treatise on mathematical axioms or principles, and even less so on some kind of self-evident, universally experienced knowledge that everyone knew to be true: these were the kinds of axioms or principles upon which mixed mathematical sciences were grounded. In order to prove by "experiment (*l'expérience*)" that the weight of a fluid varies with its height, Pascal describes a series of differently shaped containers fixed to a wall whose equally sized open bottoms are all provided with tightly fitting stoppers that do not allow the water poured in the containers to flow down. Stoppers are linked to a balance mechanism that would measure the weight they support. Provided that the height of water in all the containers is the same, "experience shows that the same force (*force*) is needed to prevent every one of the stoppers to go down, notwithstanding the fact that the amounts of water are all different in the different containers."[16] He further proves by the same method that the force needed in all the containers equals the weight of the water contained (up to the common height) in the right cylinder whose base is the equally sized open bottom.[17] Pascal has used here an experimental fact to prove the proposition that is in effect grounding almost all subsequent propositions.

Next, Pascal provides a series of "experiences" about the so-called "hydrostatic paradox," which he explains by means of his first proposition. Then he investigates the

[12] This is conclusion number 7, in Pascal, *Traité de la pesanteur de la masse de l'air*, p. 244.

[13] Ibid., p. 245.

[14] The first chapter of the *Traité de l'équilibre des liqueurs* is devoted to experimentally demonstrate "Que les liqueurs pèsent suivant leur hauteur," pp. 236f.

[15] Pascal's proof comprises a set of vessels fixed on a wall, of widely different capacities, all of which have at their bottom a hole of equal size and shape. Once filled to a certain height, the balance allows one to measure the force necessary to retain the stoppers of the vessels' bottoms in place, which "experience" shows to be the same when the height of water is the same, no matter how much (or how little) water is contained in the different vessels, pp. 236f. On Pascal's experiments in hydrostatics and pneumatics, see Harrington, *Pascal philosophe*, pp. 47–51; Koyré, "Pascal savant."

[16] Pascal, *Traité de l'équilibre des liqueurs*, pp. 236f., quotation on p. 236.

[17] Ibid., p. 237.

"true cause" (*sic*) of such "multiplication of force." He finds it in the metaphor of the "mechanical machine," which he applies to any vessel full of water: "a vessel full of water is a mechanical machine to multiply forces." Here Pascal brings in the mechanical principle that applies to "all the ancient machines," and which he assumes to be accepted by everybody: "distance [in which any weight or "force" is moved] decreases in proportion to the increase in force"—or to put it anachronistically, the work is the same when the weights moved and the distances traveled are inversely proportional:

> [S]o that the path is to the path [inversely] as the force is to the force. This can even be held to be the *true cause* of this effect: it being clear that it is the same to move ten pounds of water an inch as it is to move a pound of water ten inches.[18]

This mechanical principle was used in grounding mechanical treatises adopting the mixed mathematics format. When this was the case, however, the principle was explicitly highlighted as one of the axioms or principles of the work and was given some geometrical disguise. Pascal, on the other hand, avoids any mathematical idiom and chooses to present it as an *obvious* general principle. Furthermore, when Pascal eventually introduced some mathematical mechanics in his treatise (to prove the equilibrium of weights resting upon fluid surfaces that are inversely proportional to the weights), he did so almost as an afterthought, in a sketchy way and as a side issue. His geometrical considerations are grounded on the well-known principle that in the natural motions of heavy bodies their center of gravity always falls.[19] Pascal's demonstration involving the center of gravity is prefaced by the warning that only geometers will be able to understand it: "Here again is a proof that can only be understood by the geometers." [20] In Pascal's treatises the deductive, theoretical structure is strong, and the truths about nature are not as tightly connected to experiments as they are in Boyle's. However, mathematics play a secondary role in Pascal's hydrostatics in the sense that it neither dictates its organization nor constitutes the main instrument of proof. It rather is a sparingly used auxiliary language.

7.2 The Mathematical Hydrostatics of Wallis, Gregorie, and Newton

As mentioned above, the University of Aberdeen preserves the only known copy of a hydrostatical manuscript titled "Hydrostatica." The neat amanuensis copy that has come to us is dated 1740, therefore suggesting that it was still used in that year.

---

[18] Ibid., emphasis added.

[19] "Voici encore une preuve qui ne pourra être entendue que par les seules géomètres.... Je prens pour principe, que jamais un corps ne se meut par son poids, sans que son centre de gravité descende." Pascal, *Traité de l'équilibre des liqueurs*, p. 238.

[20] Ibid.

Since it incorporates material from different sources, including Wallis' 1670 "De hydrostaticis" and two hydrostatical propositions from Book I of Newton's *Principia*, its authorship is questionable. We shall discuss its authorship in the Appendix and focus now on its contents.

"Hydrostatica" opens with four definitions (specific gravity, absolute gravity, relative gravity, and fluid) and two propositions from which all the remaining results are derived by mathematical argument. It is divided in two parts. The first six propositions deal with general results on the equilibrium of fluids. They are complemented with an interesting scholium devoted to the doll experiment (more about that below). The second part contains five propositions devoted to demonstrating mathematically the physical effects of the elastic force (*vis elasticam*) of the air as well as those observed in some of the most significant barometric experiments. It contains many long scholia devoted to comment on well-known experiments.

The first two propositions of the manuscript along with their long lists of corollaries and the definition of fluid (on the whole some eight pages of text out of the 39 written pages of the manuscript) come almost verbatim from Newton's *Principia*, Book II, Section 5, Propositions 19 and 20 and their corollaries. (As is well known, Proposition 19 comes from Newton's *De gravitatione* (c. 1672), and Proposition 20, not actually found in *De gravitatione*, was also likely composed about the same time.[21]) The borrowing is explicitly acknowledged on page 10 of the manuscript: "we bring over [these two propositions] from the principles of the most illustrious Newton." Proposition II is the mathematical cornerstone of the manuscript, for which Proposition I is only preparatory. Proposition II reads thus:

> If all the single parts of a spherical fluid that is homogeneous at equal distances from the center and rests on a concentric spherical bottom gravitate towards the center of the whole, then the bottom supports the weight of a cylinder whose basis is equal to the surface of the bottom, and whose height is the same as that of the incumbent fluid.

In Fig. 7.1, AEI is the external spherical surface of a fluid that is homogeneous at equal distances from the center. It rests upon a concentric spherical bottom and gravitates towards the center of the spheres. The bottom, according to Proposition II (and the result is correct according to our understanding of the laws of hydrostatics), sustains a weight which is *not* equal to the weight of the whole mass of the fluid, but equal instead to the weight of the fluid contained by the cylinder whose base equals the surface of the spherical bottom and whose height equals the fluid's. The proof divides the fluid in "innumerable" (*sic*) concentric spherical surfaces that divide the fluid in "innumerable" equally thick concentric "orbs." Then the reader is first required to assume "the force of gravity to act only on the upper surface of every orb"; and, secondly, "the actions to be equal on equal parts of the surface." The second requirement seems obvious and unnecessary, yet it

[21] Shapiro, "Light, Pressure, and Rectilinear Propagation." Shapiro carefully analyzes the conceptual nuances separating Newton's articulation of hydrostatics in *De gravitatione* and the *Principia* (pp. 276–284); about the date of composition of Proposition 20, see p. 283.

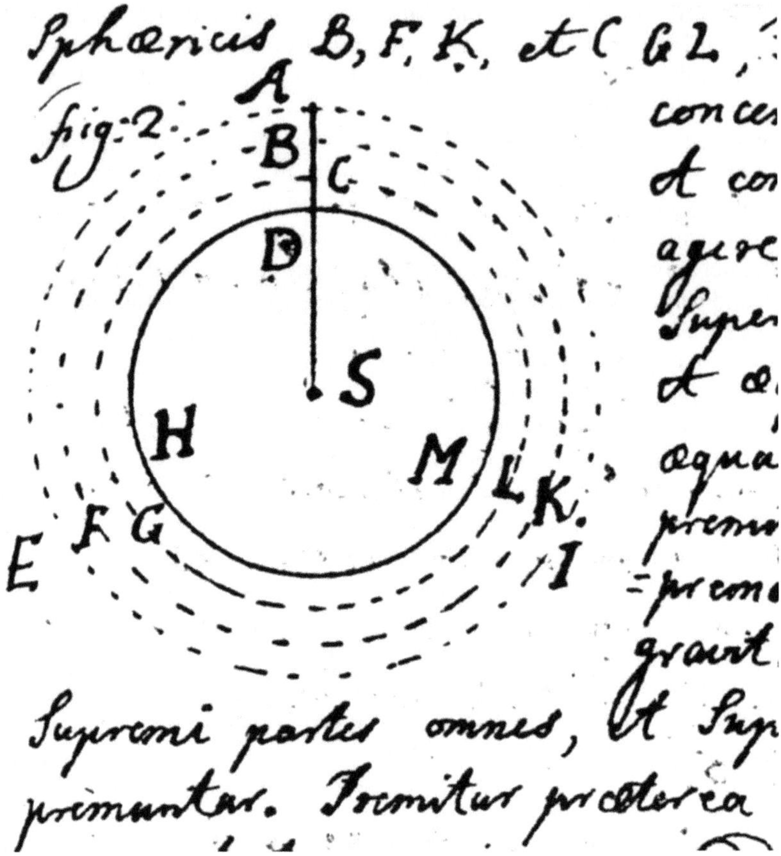

Fig. 7.1 From "Hydrostatica," p. 7

plays an implicit role in the demonstration through the notion of "pressure," not explicitly defined:

> Let DHM be the surface of the bottom [see Fig. 7.1], AEI the upper surface of the fluid, and BFK, and CGL [etc.] innumerable spherical surfaces. Let the fluid be divided into concentric [orbs] of equal thickness, and conceive the force of gravity to act only on the upper surface of every orb, and the actions to be equal on equal parts of surface. Therefore, the upper surface, AEI, is pressed by the single force of its own gravity, by which [force] all the parts of the upper orb and the upper surface [of the second orb], BFK, are pressed. Hereafter, the second surface, BFK, is pressed by its own force of gravity, which added to the previous force doubles the pressure. This pressure plus its own force of gravity, i.e. a triple pressure, will press the third surface, CGL. And similarly a 4-fold pressure will press the fourth surface, a 5-fold [pressure] the fifth [surface], and so on. Therefore, the pressure that presses upon the bottom is not as the [weight of the] solid quantity of the incumbent fluid, but equals the weight of the lowest orb multiplied by the number of orbs, i.e. the weight of the solid body that is equal to the cylinder above mentioned. Therefore, the bottom supports the weight (*pondus*) of the cylinder above mentioned. Q.E.D.[22]

[22] "Hydrostatica," pp. 6f. As specified in the Appendix below, Gregorie's hydrostatics manuscript is bound together with other mathematical treatises, each numbered separately.

The basic and repeated step in the foregoing demonstration is that the weight that pulls down any orb presses down on the next downwards orb as much as the latter is pulled down by its own weight, and so on. The argument seems to rest on a subtle play between pressure and weight that makes the proof plausible. The proof is presented as independent of the nature of gravity and its quantification. The pressures (or weight on every element of surface) here involved are equal on any spherical orb provided that the weights are as the spherical surfaces, $w_1: w_2 = s_1: s_2$. However, the proof might be said to be circular if one questions the equality between corresponding surface elements of contiguous spherical shells—and it could be questioned because it was a source of paradoxes in discussions about the indivisibles of concentric circles. In any case, the terse formulation of the argument makes hard to pinpoint any conceptual weakness.[23] The proof also includes a daring leap from "the number of orbs" to the height of the fluid, which is therefore taken to be equivalent to the collection of its infinitesimal heights. However, no care is taken to clarify mathematically the precise nature of the infinitesimally thin orbs involved.

From propositions I and II it is mathematically deduced that the upper surface of any homogeneous motionless fluid that is everywhere equally pressed from above is horizontal—or, to speak accurately, spherical and concentric with the earth, which in practice is indistinguishable from a horizontal surface. The following propositions of the manuscript (numbers III–VI) demonstrate by geometrical argument that bodies whose specific gravity is greater than (respectively, equal to) the specific gravity of the fluid in which they are submerged will subside all the way to the bottom (respectively, will remain in whatever position they are left). These propositions III to VI paraphrase propositions 1, 2, 5, 6, and 7 in John Wallis' "De hydrostaticis" (published as part of Wallis' *Mechanica* in 1670), although the arguments are not identical. Those in the 1740 manuscript attributed to Gregorie rely on propositions I and II just mentioned (which have no equivalent counterparts among Wallis'), while Wallis' arguments are grounded on his own foundations of mechanics.

The manuscript concludes its hydrostatical part by an explanation of the so-called experiment with the "doll (*nino*)," to which we shall return presently. Proposition VII follows, which proves that in the barometric experiment the column of mercury must stop its falling down at that precise height in which it presses upon its basis as much as the air presses upon the other points of the mercury surface. Proposition VIII is devoted to demonstrating the results observed in some controversial experiments (involving weighing barometric tubes with mercury in atmospheric equilibrium) proposed by Jonathan Goddard to the Royal Society in 1662.[24] Proposition IX demonstrates the equivalence of the elastic force of the air and the weight of the incumbent air. Finally all of this is applied to explain the workings of syphons and water pumps (Propositions X and XI). The Scottish manuscript of 1740 follows Wallis' text rather more closely in the second, pneumatical part than in the first, hydrostatical part.

---

[23] I am indebted to Xavier Roqué and Luis González for their insightful comments on this difficult argument.

[24] The experiments are described in Middleton, *History of the Barometer*, p. 73.

The 1740 "Hydrostatica," which opens with the explicit claim that hydrostatics is "a part of Mathematics,"[25] is a neat instance of classical mixed mathematical sciences. Not only do the propositions follow a consistent mathematical argument, but experimental evidence, although not neglected, is placed in a subordinate position—as complementary rather than fundamental. In particular, experiences or experiments are never part of a proof. In fact, the "Hydrostatica" manuscript formally separates between, on the one hand, the mathematically proved theoretical results, set forth in *propositions* and their *corollaries*, and on the other the reports of matters experimental, set forth in *scholia*. When phenomena such as the sinking of mercury in the barometric experiment, or the stabilization of mercury at one specific height, or the independence of this specific height from the inclination, shape or size of the mercury container, when such phenomena appear within a *proposition*, they are presented as mathematical conclusions, not as Boylean matters of fact.

Experimental matters are generously mentioned, both within propositions and as illustrations of celebrated natural effects, but experimental reports fall squarely within the category of what Dear has aptly called "universal statements about how things happen in the world," not of *what happened* on a particular historical date and time under specified circumstances.[26] Experimental reports in the 1740 manuscript are *not* first-hand accounts of "historical" experiments done at specific times and places by specific people. On the contrary the manuscript acknowledges that they are made out of the many "observations of many [people]" and their purpose is that of showing that "what is demonstrated from experiments wholly agree with what has been [mathematically] demonstrated here."[27] When the manuscript introduces the elastic force (*vis elastica*) of the air, it makes reference to "innumerable experiments set up in the pneumatic pump by recent authors" that agree with the existence of the elastic force (no more details here) and then derives by mathematical argument from the previous results on hydrostatics that the elastic force in an open vessel is measured by the weight (*onus*) of the incumbent air.[28] "Hydrostatica" contains long asides on other experimental phenomena. For instance, it reviews the evidence showing the presence of tiny air bubbles in the blood, which it links to experiences with a sealed flaccid bladder in a pneumatic pump[29]; or it discusses at length Mersenne's and Boyle's numbers for the proportion between the maximum compression and maximum dilation of the air.[30] Mathematics and experimental matters are well integrated in the manuscript. There is a balance of sorts between them in the sense that the importance the manuscript allocates to experimental matters is qualified by its ancillary role as illustrations, corroborations, or aids to

[25] "Hydrostatica," p. 3.

[26] Dear, "Miracles, Experiments, and the Ordinary Course of Nature," p. 667.

[27] "Hydrostatica," p. 22.

[28] Ibid., pp. 24f., Proposition IX.

[29] Ibid., pp. 27–29.

[30] Ibid., pp. 30–33.

facilitate the understanding of knowledge which truth is grounded on and guaranteed by mathematical demonstration. The role of experiments in the 1740 "Hydrostatica" manuscript will be clarified by focusing on a popular experiment widely discussed in contemporary treatises.

7.3 The Long Life of the "Doll Experiment"

The "Hydrostatica" manuscript concludes its hydrostatical part by one scholium devoted to a "well-known experiment (*notissimum experimentum*)," the so-called experiment with the "doll (*nino*)." The experiment shows a doll (see Fig. 7.2) that may "swim" in fluids, now sinking to the bottom, now remaining in suspension somewhere within the fluid. In the water-filled phial, AB, is submerged a small vessel, D, flexible but able to bear compression, perfectly closed, and containing air so that the vessel and the air within have jointly almost the specific gravity of water, but not quite. The manuscript explains that if the water in the phial is strongly compressed (by the thumb or by any other means), since the water is free from compression, the

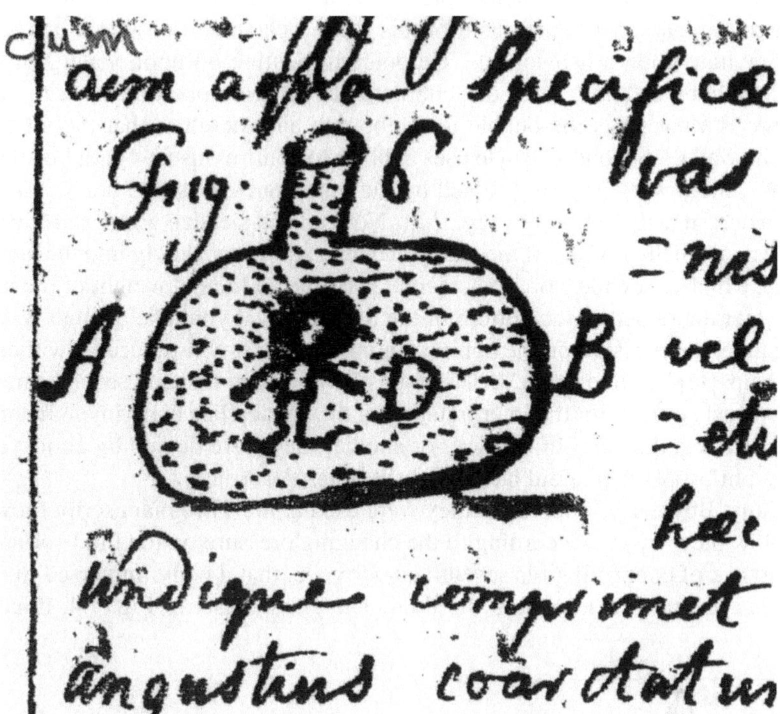

Fig. 7.2 From "Hydrostatica," p. 16

added pressure will press on all sides the vessel D, which will then occupy a smaller space than the water that weighs the same as itself. Therefore it will subside. On the other hand, when the thumb is withdrawn, the elastic force (*vi elastica*) of the air enclosed in the vessel D will restore it to its original size, and therefore D will ascend. The reader is told that the experiment and its theoretical explanation are included to deactivate it as evidence contrary to the propositions proving that bodies either sink or remain stationary within a fluid according to whether their specific gravities are heavier or equal than that of the fluid.[31]

Different versions of the doll experiment were well known in the early decades of the eighteenth century. It was included in 'sGravesande's popular *Mathematical Elements of Natural Philosophy*, but as far as we know it first appeared in print in 1666, in Robert Boyle's experiments with "bubbles" within his *Hydrostatical Paradoxes*. Boyle's "bubbles" are small round empty pieces of glass with slender necks and very narrow openings. The glass bubbles are made in such a way that they are slightly lighter than an equal bulk of water. They float, but the addition of a very small weight would sink them. If the pressure upon their opening increases, the air included in the bubble is compelled to shrink, and some water will come in. The bubble becomes heavier than an equal bulk of water, and therefore sinks.[32] Now, if the pressure is removed, the spring of the air pushes a drop of water out, the bubble becomes lighter and presently rises to the top.

Boyle uses his bubbles, which he presents as an "instrument" (*sic*) of his invention, for three purposes. First, they confirm his second paradox, i.e., that a lighter fluid may gravitate upon a heavier one. He pours turpentine oil upon water (on which surface a bubble floats) until the bubble sinks. Boyle concludes that the added pressure is what sinks the bubble to the bottom and therefore that the oil weighs upon the water.[33] Secondly, Boyle uses bubbles to confirm his first paradox, i.e., that in fluids, the lower parts are pressed by the upper ones. Cast the bubble in a long tube, where it will swim (see Fig. 7.3). Now take a slender wand and thrust the bubble beneath the surface of the water, and then add water slowly into the tube. The bubble, which first endeavored to emerge, is by the additional weight of the incumbent water depressed to the bottom of the tube. Next, if you take enough water out of the tube, the pressure on the bubble diminishes and it will proceed, "without any other help, [to] begin to swim."[34] The results Boyle called first and second paradoxes are obviously crucial for the interpretation of the Torricellian experiment in terms of mechanical equilibrium. In one way or another they were denied by almost all the natural philosophers that did not accept that interpretation.

Finally, Boyle uses bubbles as they were used in the 1740 manuscript (now they sink, now they "swim" according to the changing pressure on the fluid surface) for the purpose of confirming his seventh paradox, i.e. that a body immersed in a fluid sustains a lateral pressure from the fluid. The glass vessel in Fig. 7.4, filled with

[31] Ibid., pp. 16f.

[32] Boyle, *Hydrostatical paradoxes*, in *Works*, vol. V, pp. 221, 249.

[33] Ibid., p. 222; for the reference to them as an "instrument," see p. 220.

[34] Ibid., p. 222.

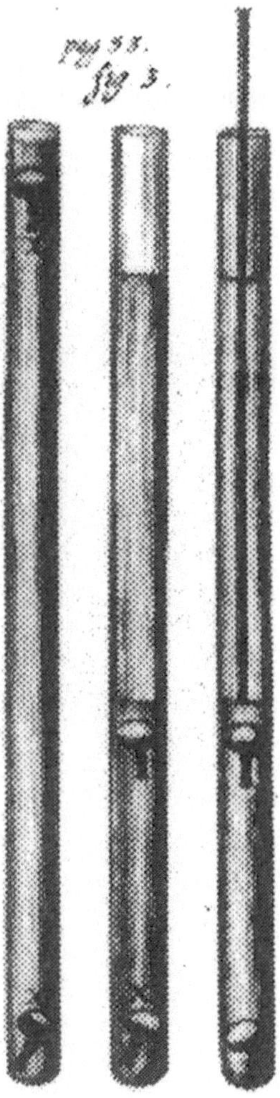

Fig. 7.3 From R. Boyle, *Hydrostatical paradoxes*, in M. Hunter, E.B. Davis, eds. *The Works of Robert Boyle* (London: Pickering & Chatto, 1999), V: 189–279, p. 212

water on which a bubble floats, is provided with a cork carefully and tightly closed with "cement." As shown in the figure, the cork is provided with a slender pipe of glass EF whose lower end reaches well below the surface of water. Through E water is added until it reaches a convenient height upon the cork, say K, in which situation the added pressure makes the bubble sink to the bottom. This experiment, says Boyle, not only teaches us that the upper parts of the water gravitate upon the lower ones (as confirmed before), but

> That in a Vessel, that is full, all the lower parts are press'd by the upper, though these lower be not directly beneath the upper, but aside of them, and perhaps at a good distance from the Line in which they directly press.[35]

[35] Ibid., p. 249.

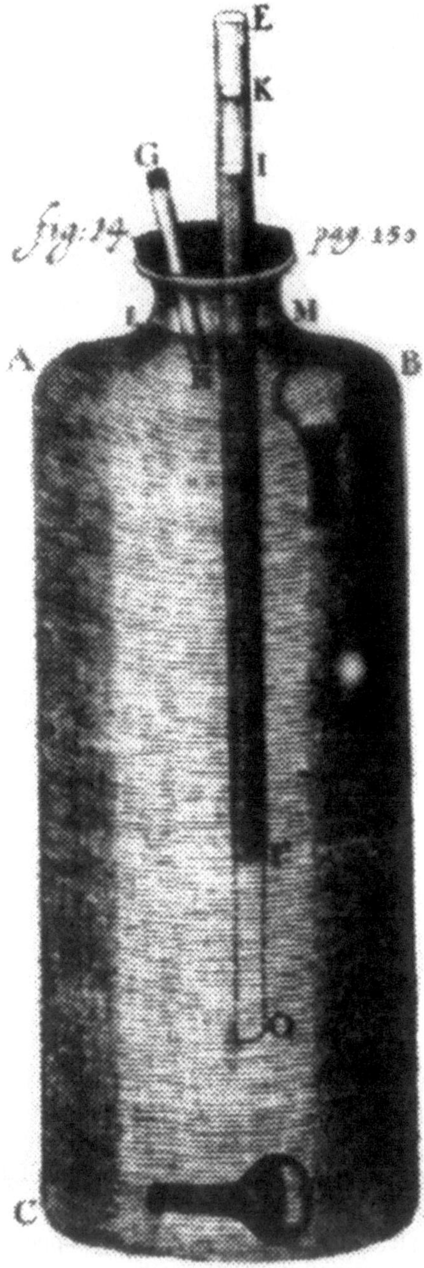

Fig. 7.4 From R. Boyle, *Hydrostatical paradoxes*, in M. Hunter, E.B. Davis, eds. *The Works of Robert Boyle* (London: Pickering & Chatto, 1999), V: 189–279, p. 242

Boyle's use of his bubbles is typical of his general approach. Each one of his paradoxes is carefully built "to disprove the received errors" and "to make the Truths the better understood." As has been often remarked, there is, so to speak, a simple, short thesis concerning a matter of fact behind each one of his experiments.

As mentioned above, another variant of the doll experiment was included in Willem Jacob 'sGravesande's influential *Mathematical Elements of Natural Philosophy, Confirmed by Experiments*.[36] In 'sGravesande's version (see Fig. 7.5), the 15-in-high glass AB is full of water and tightly covered by a bladder, yet with a small quantity of air left between the bladder and the water surface. Immersed in the water are "little Figures of Glass ... hollow, of an Inch and a half long, representing Men, ... [which] have a small Hole in one of their Feet, and are lighter than Water."[37] When the finger presses the bladder, the water is compressed and therefore "enters in the little Men thro' the Hole at their Feet, and compresses the Air in their Bodies more than it was." So the figures become heavier and descend. Taking away the finger, the air in the figures expands itself, drives out the water, and therefore they rise again. In 'sGravesande, the doll experiment is just a funny, entertaining experiment.

In one sense, the "doll" or "bubble" experiment presented in the three sources here considered is one and the same, but in another, non-trivial sense it is not. There are obvious differences in the physical set-up of the experiments in Boyle's treatise, in the 1740 "Hydrostatica" manuscript, and in 'sGravesande's work. More importantly, there are noticeable differences in both the role each experiment plays and the theses it confirms or establishes. In 'sGravesande's it belongs to a chapter titled "Several Experiments concerning the Air's Gravity, and its Spring." It offers an archetypical experimentalist approach to pneumatics in which dozens of experiments seamlessly follow one another to confirm, illustrate, make visible, and demonstrate "effects" produced by the gravity of the air, its pressure, and its "elasticity." Here the doll experiment is just one more piece among many in a big mosaic of experiences and observations only loosely connected to one another. It follows experiments some of which show "the spring of the air" to be equal to the "weight of the whole atmosphere" while others illustrate the power of atmospheric pressure to raise water to great heights. It is followed by experiments that show that "animals cannot live without air" and by others that show the effects upon animals of compressed air.[38] 'sGravesande's use of experiments is fully consistent with what Larry Stewart and others consider the typical teaching of natural philosophy of the early Enlightenment—utilitarian, experimentalist, and subordinating mathematics to manipulation and visualization to the point that the mathematical content almost disappears. In this approach, hands-on experience and observation of experimental results not only facilitate concept-clarification and the understanding of mathematical arguments, but are construed as real alternatives to them.[39]

---

[36] It was first published in Latin in 1715 (with a 3rd edition appearing in 1742), and translated into French (twice, in 1746 and 1747) and English (1720), in which language it knew its 6th edition in 1747.

[37] 'sGravesande, *Mathematical Elements of Natural Philosophy*, vol. I, p. 231.

[38] 'sGravesande, *Mathematical Elements of Natural Philosophy*, vol. I, pp. 224–233.

[39] Stewart, *The Rise of Public Science*, *passim*, but see particularly chap. 4.

Fig. 7.5 From 'sGravesande's *Mathematical Elements of Natural Philosophy* (London, 1721–1726), I, p. 234, Plate 31, Figure 2

In 'sGravesande but even more markedly in Boyle, the doll experiment is not connected with mathematical deductions, nor with properties of the vacuum established mathematically. 'sGravesande presents it essentially as an amusing experience fitting loosely into a series of experiments connected by the physical notion of the weight and spring of the air. Boyle presents different variations of the experiment and each of them serves him to prove a different particularized thesis. Notice that Boyle deduces several results from the doll experiment, but none comes from 'sGravesande's presentation—at least explicitly. By contrast, the "Hydrostatica" manuscript discusses the doll experiment only within a scholium, which formally marks it out as a remark or example not essential to the main train of reasoning. In fact, the correct interpretation of the experiment (provided by mathematically argued propositions) is used to reinforce mathematical deduction by clearing away any doubts that empirical observation may produce. As the scholium concludes, "[a]ll these things [the doll's behavior] far from contradicting the foregoing propositions are in fact [their] consequences."[40]

The "Hydrostatica" manuscript mentions experiments other than the one with the doll, always having them play similar roles. For instance, experiments corroborate results mathematically deduced in which a scale weighs a barometric tube full of mercury at different heights above the mercury's surface.[41] In another series of described experiments, the elastic force of the air is used to explain experimental effects such as the formation of bubbles in blood within a pump receiver, or the increase in volume of animals introduced in a receiver when it is emptied.[42]

The contents of 1740 "Hydrostatica" manuscript are therefore a powerful combination of mathematics and experiments. We know that it originated as part of John Wallis' *Mechanica* (1670–1671), which provided a systematic and mathematical exposition of everything known in mechanics—from statics to the computation of centers of gravity (including powerful new mathematical methods of his own), to Galileo's laws of motion, to the recent (1668) discussions about the laws of impact within the Royal Society, to hydraulics.[43] One wonders how Wallis' strongly mathematical approach to hydrostatics and pneumatics was received by Boyle. As is well known, Boyle explicitly contrasted his approach to the mathematician's. He claimed his approach to be superior because mathematical demonstrations in physical matters are built upon "suppositions or postulates" about which it is easy to be mistaken, while his approach uses sense and experience in matters that are known through the senses. Boyle also thought mathematics was not an appropriate idiom for gentlemanly discussions about matters philosophical.[44]

[40] "Hydrostatica," p. 16.

[41] Ibid., pp. 22f.; these are the Goddard's experiments mentioned above, see n. 24.

[42] Ibid., pp. 27–30.

[43] Scott, *John Wallis*, pp. 91–111; for the innovative mathematics Wallis used in it, see Maierù, *John Wallis*, pp. 257–259.

[44] Shapin, "Robert Boyle and Mathematics"; Henry, "Robert Boyle and Cosmical Qualities."

7.4 Conflicts at the Royal Society

For decades the Royal Society had a fair amount of disagreements about its philosophical priorities. These have sometimes been construed as conflicts between those primarily involved with the physical sciences versus those involved with the life sciences, but sometimes as a confrontation of experimentalists plus "naturalists" versus "mathematicians."[45] Notice that these categorizations do not coincide. In the latter Boyle (for instance) would side with the naturalists, while in the former he would oppose them. Both categorizations do probably oversimplify the conflicts. Boyle, for instance, was likely to agree with the mathematician William Molyneux's comment (in a 1686 letter to Halley, on Halley's being elected Secretary of the Society facing opposition from naturalists) against those "that were for rejecting all kinds of useful knowledge except ranking and filing of shells, insects, fishes, birds, etc. ... and reckoning chemistry, astronomy, mathematics, and mechanics, as rubs in their course after nature." But he would disagree with the interest of experimentalists such as Robert Hooke and Henry Power (1626–1668) or of the mathematically inclined Sir William Petty (1623–1687), among others, in framing Cartesian-style mechanical hypotheses.[46]

The mathematician William Neile wrote about 1669 a memorandum with "Proposalls" to reform the Royal Society's organization and the ways in which it pursued the advancement of knowledge. Neil's text has been presented as evidence of the tensions between Royal Society factions, but it is an equivocal, hard to interpret text. It stresses the need not to neglect "the giving of causes" for observed experimental effects, because experiments in themselves are "but a dry entertainment without the indagation of causes." It therefore suggests setting up a special committee "to bring in a possible cause or causes" of the experiments that were performed before the whole Society.[47] Again, one can easily imagine Boyle agreeing with Neile that reason must be coupled with the senses "in taking notice of matters of fact," but being reluctant to spend too much time in discussing possible causes—particularly

[45] M. Hunter and P. B. Wood presented the conflicts as between the physical sciences and the life sciences; M. Feingold took the lead from Hunter but portrayed the conflicts as being between mathematicians against naturalists and experimentalists. See Hunter and Wood, "Towards Solomon's House"; Feingold, "Mathematicians and Naturalists."

[46] Molyneux to Halley, April 1686, in Birch, *History of the Royal Society*, vol. IV, p. 476 (quoted in Hunter, *Establishing the New Science*, p. 207). On Hooke, see Henry, "Robert Hooke, the Incongruous Mechanist." Henry Power's *Experimental philosophy* (London, 1664) contains elaborate hypotheses to explain the nature of the Torricellian vacuum (pp. 94–108, he denies it to be so) and magnetic experiments (pp. 153–61); on Power, see Webster, "Henry Power's Experimental Philosophy," and Johns, "Henry Power," in *Oxford Dictionary of National Biography*. A one-time secretary to Hobbes, Petty is mostly remembered for his contributions to economic thought and pioneering work on "political arithmetic." His *Discourse ... Concerning the Use of Duplicate Proportion* plainly shows Hobbes' influence; on Petty's natural philosophy, see Sharp, *Sir William Petty and some Aspects of Seventeenth-Century Natural Philosophy*.

[47] Neil's memorandum is fully reproduced in Hunter, *Establishing the New Science,* pp. 223–225.

hidden mechanical causes. The mathematician Wallis would surely disagree with the mathematician Neile's proposals. Prompted by Oldenburg, Wallis vindicated his own mathematical approach to the laws of motion—one in which he shunned the consideration of causes.[48] His *Discourse concerning Gravity and Gravitation* dismisses any discussion of the possible causes or "nature" of gravity, motion, spring of the air, etc.

> I will not dispute the Nature of Gravity or Gravitation, what or whence it is ...: But shall take for granted (what everydays Experience testifies) that there is ... such a thing as Gravity and Gravitation.[49]

He discusses whether the "positive" or actual thing is "gravity" or "levity," that is, a tendency for the bodies to move downwards or upwards (in this case, some of them would descend because impelled by lighter elements), but in no case does Wallis venture to explore the nature or origin of such tendencies. Wallis' views here are very close to Boyle's as concerns the "nature" of the spring of the air.[50]

As Michael Hunter and Paul Wood put it, in the early decades of the Royal Society there was "a broad spectrum of methodological opinion within the Society."[51] We may add that it was not only broad, but also multifaceted, so that the use of labels such as "mathematician" and "experimentalist" are too general to help us in introducing a measure of order in it. Despite Boyle's views on mathematics, he shared with the mathematician Isaac Barrow a voluntaristic theology on which both grounded a disregard for mechanical hypothesis.[52] As is well known, in the 1660s Boyle's main allies in attempting to debunk Hobbes were Wallis and the astronomer Seth Ward. In 1673 and 1674 Sir Matthew Hale (1609–1676), the Lord Chief Justice of England and Wales, anonymously published two books criticizing the explanation of the barometric experiments by the weight and spring of the air.[53] For some reason, Boyle was not willing to answer and John Wallis was commissioned to do so. Wallis dealt with the weight of fluids in the already mentioned "Discourse concerning Gravity and Gravitation, grounded on Experimental Observations," presented to the Royal Society in November 1674. The Council ordered it to be printed at its expense in January 1675. We cannot deal here with Wallis' tract except to point out that although Wallis avoids the mixed mathematics format, he provides a detailed theoretical critique of Hale's explanations of the

[48] Wallis to Oldenburg, 5 December 1668, in Oldenburg, *Correspondence*, vol. 5, p. 221.

[49] Wallis, *A Discourse of Gravity and Gravitation*, p. 2.

[50] Ibid., pp. 27–29. On the quantitative way in which he deals with the spring of the air avoiding speculation about its nature, see pp. 26f.; see Scott, *The Mathematical Work of John Wallis*, pp. 102–105. On Boyle's views on the "spring," see Clericuzio, "The Mechanical Philosophy and the Spring of Air."

[51] Hunter and Wood, "Towards Solomon's House," p. 209.

[52] Malet, "Isaac Barrow on the Mathematization of Nature."

[53] *Essay touching the Gravitation and Non-Gravitation of Fluid Bodies, and Difficiles Nuggae, or, Observations touching the Torricellian Experiment.*

barometric experiments and of new experiments adduced by Hale.[54] What seems highly relevant here is that in a matter in which Boyle was the highest authority, the Society's answer to Hale's challenge was entrusted to Wallis. This suggests that the experimentalist's and the mathematician's approach to hydrostatics and pneumatics were regarded as conflicting in no fundamental way. On the contrary it suggests that it was thought useful to combine them for answering critics. Boyle and Wallis did agree in rejecting to pay serious consideration to the investigation of the "nature" of key physical notions such as gravity or spring. Judging from the occasions in which they cooperated in crucial matters, this agreement seems to have powerfully brought them together.

7.5 Hydrostatics in the Early Scottish Enlightenment

Shortly after Wallis' *Mechanica* was published in 1671, James Gregorie (1638–1675) had a copy sent from London to St Andrews where he was the Regius professor of mathematics. He and one of the "regents" (generalist lecturers) of the University of St Andrews, William Sanders, were preparing an answer to G. Sinclair's two books on hydrostatics.[55] A somewhat contradictory figure, George Sinclair (c.1630–1696) graduated from Edinburgh University in 1649 and was appointed regent at the University of Glasgow in 1654, a position he was to resign in 1666 because of his Presbyterianism.[56] From then up to 1689 he made a living as a civil engineer and mathematical practitioner. At different times he was variously supported by the Edinburgh Town Council, which employed him to improve the town's water supply and granted him a pension in 1683. He seems to have been instrumental in making barometers widely known in Scotland, in pioneering their use as a weather instrument there, and in promoting their construction and trade. He was interested in mining and in the design of diving engines for recovering valuables from shipwrecks. After the 1689 revolution he regained his post at the University of Glasgow to become its Professor of Mathematics in 1691. Besides introductory books in mathematics, astronomy, and natural philosophy, he published *Satan's invisible world discovered* (1685), a book meant to prove the existence of devils, spirits, and

[54] Wallis, *A Discourse of Gravity and Gravitation*, pp. 10–23.

[55] Sinclair, *Ars nova et magna gravitatis et levitatis*, which deals with different mechanical topics, including hydrostatics, and *Hydrostaticks*, which also includes a long appendix on mining, "A History of Coal."

[56] On Sinclair, see Wood, "George Sinclair (d. 1696)," in Pyle, *The Dictionary of Seventeenth-Century British Philosophers*, vol. II, pp. 750f.; Emerson and Wood, "Science and Enlightenment in Glasgow, 1690–1802"; Morrison-Low, "'Feasting my Eyes With the View Of Fine Instruments'," pp. 22–24 (on Sinclair's role in making the barometer known in Scotland). Information on his birth year, education and early activities can be found in Laing, "Notice of a Scheme … for Weighing up and Recovering Ships," pp. 429–432.

witches, and indirectly to refute materialist philosophies (such as Hobbes'); it was reprinted well into the eighteenth century.[57]

Notwithstanding his publications, academic appointments, and local high reputation, Sinclair's hydrostatics was downright wrong in the eyes of both the Royal Society and the mathematically knowledgeable regents in St Andrews to whom Sinclair, seeking their endorsement, sent samples of his work in 1670 and 1671. A nasty quarrel followed in which James Gregorie helped William Sanders to write an extended critique of Sinclair's mechanics in general as presented in his *Ars nova et magna gravitatis et levitatis*, and of hydrostatics and pneumatics in particular as presented in his *Hydrostaticks*.[58] Sanders' and Gregorie's rebuke appeared in 1672 titled *The Great and New Art of Weighing Vanity* signed by "Patrick Mathers, Arch-Bedal to the University of S. Andrews."[59] The attribution of authorship to a minor university officer was meant as derogatory of Sinclair's academic and intellectual status, and was taken as such. The book contained a short appendix with Gregorie's deep, highly original solutions to pendular and projectile motion. Everything suggests that Gregorie also wrote for this occasion his own presentation of hydrostatics and pneumatics, but did so by following rather closely Wallis' "De hydrostaticis." We know that when he died in 1675 in Edinburgh (he had taken the mathematical chair there in 1674), he had finished a hydrostatical manuscript. His nephew David Gregorie occupied the mathematical chair in Edinburgh between 1683 and 1691, followed by David's brother, also called James, from 1692 to 1642. They are probably the ones who gave to James Gregorie's *primus* hydrostatics the form it has in the 1740 copy available to us (see Appendix).

We cannot discuss Sinclair's hydrostatics here, but a few particulars need to be mentioned. In his quarrel with Sanders and Gregorie, Sinclair chose to present the criticism received as a conflict between speculative mathematics and useful experimentation. His critics claimed that Sinclair's results were already in Archimedes and Stevin or could easily be deduced from them. In answer, Sinclair described his *Hydrostaticks* as "The Weight, Force, and Pressure of Fluid Bodies, Made evident by Physical, and Sensible Experiments." Sinclair's answer appeals to Boyle's

[57] The self-explanatory full title is *Satan's invisible world discovered, or, A choice collection of modern relations: proving evidently against the saducees and atheists of this present age, that there are devils, spirits, witches, and apparitions, from authentick records, attestations of famous witnesses and undoubted verity*. Besides the two books on mechanics and hydrostatics discussed below Sinclair published *Tyrocinia mathematica, Natural philosophy improven by new experiments*, and *The principles of astronomy and navigation*.

[58] For details, see the many letters crossed between Sinclair and St Andrews reprinted by the authors in "To the Reader" and "Postscript" to Sinclair's *Hydrostaticks*, and in the introduction to Mathers [pseud.], *Great and new art*.

[59] The full title is: *The Great and New Art of Weighing Vanity: or A Discovery of the Ignorance and Arrogance of the Great and New Artist*, in his *Pseudo-Philosophical Writings*. For the authorship of this book see Halkett and Laing, *A Dictionary of the Anonymous and Pseudonymous Literature of Great Britain*, vol. II, p. 1044. See also Turnbull, *James Gregory Tercentenary Memorial Volume*, pp. 238–239; Stewart, *The Academic Gregories*, pp. 36–44.

authority both to point to recent novelties in pneumatics, and more importantly, to oppose his (Sinclair's, and by implication Boyle's) "practical" approach to hydrostatics to the mathematician's "speculative" one. Sinclair opposed his "physical" demonstrations to the "geometrical" ones; the very large audience that may use his propositions, to the few number who may use the mathematician's; and the illustrations and experiments that confirm his propositions to the lack of them in the mathematical approach.[60]

In the mid 1660s Sinclair visited London. Received by Boyle, they talked about Boyle's experiments with the barometer and the air-pump. Sinclair also left a manuscript (with experiments and material that were to appear in 1669 in his *Ars nova*) with Sir Robert Moray, a Scott and prominent fellow of the Royal Society, to have it submitted to the Society if Moray thought it worthy enough—which he did not. Sinclair's 1669 Preface to the *Ars nova* claimed that that manuscript contained many results and experiments that were then new, that it was left with the "Philosophical College," that he had never heard from the Society, and suggested that he had afterwards seen in several English books "many things taken out of his Manuscript." This prompted a formal answer from the Society in the *Philosophical Transactions*. The answer made a point of not discussing the content of the *Ars nova*, but was punctilious in establishing Boyle's priority first by highlighting that "in his [Sinclair's] visits of Mr. Boyle … by his own acknowledgement, then made to that Gentleman, received much light from him," and then by stressing Boyle's publication of the *New Experiments Physico-Mechanical* in 1660.[61]

In his Prefaces and letters Sinclair always referred to Boyle and his work in a highly respectful way, although he politely but assertively disagreed with Boyle's interpretation of many experiments. In particular, Sinclair claimed (as many who opposed Torricelli's interpretation of the barometric experiment did) that "water does not weigh on water." In his *Hydrostaticks* he took issue with one of Boyle's experiments with glass bubbles from which Boyle concluded that water weighs "very near … as much in water, as the self same portion of liquor would weigh in the air." In the table of contents of *Hydrostaticks* Sinclair announced prominently his critique of Boyle ("Mr. Boyls experiment Insufficient") and devoted many pages to it and to set forth his own interpretation of the experiment, from which he concluded that "water cannot weigh in water."[62] This may have something to do with the detailed account of the doll experiment included in the 1740 "Hydrostatica" manuscript. In any case it seems relevant that the Gregories' manuscript originated as an answer to someone who was talking nonsense but disguising it as an experimentalist approach to hydrostatics.

[60] Sinclair, *Hydrostaticks*, "To the Reader," pp. 3–4.

[61] *Philosophical Transactions*, num. 50, August 16, 1669; Sinclair answered with an anonymous 8-pages leaflet, *A Vindication of the Preface of the Book Intituled… Ars Nova et Magna Gravitatis et Levitati*.

[62] Sinclair, *Hydrostaticks*, pp. 145f. (on Sinclair's claim that water does not weigh on water), 148–152 (criticism of Boyle's experiment with bubbles).

One may wonder why this manuscript was still carefully copied in 1740, when its contents were no longer new nor debatable and had turned elementary. The answer is to be found in the role hydrostatics played in the teaching of natural philosophy in early eighteenth-century Scottish higher education.[63] In 1708 Edinburgh University substantially reformed its academic organization and teaching when it abandoned the regenting system. In this system, generally adopted in Scotland in the sixteenth and seventeenth centuries, every year one of four regents, or generalist lecturers, took in charge the incoming students, and shepherded them all the way up to graduation 4 years later, each successive year teaching them a different subject—metaphysics, logic, natural philosophy, ethics.[64] In a deliberate attempt to improve the quality of the courses offered, the regents were in 1708 turned into (more or less) specialized professors of Humanities, Greek, Metaphysics (including logic), and natural philosophy (including ethics).[65] The first Edinburgh professor of natural philosophy was Robert Stewart (1675–1758, also spelled Steuart), about whom more presently.

According to student notebooks and graduation theses, since the 1690s if not before, most of the courses in natural philosophy taught in Scottish colleges were strongly influenced by Cartesianism, but they became experimental and Newtonian in the early decades of the eighteenth century.[66] Indirect evidence suggests that in Edinburgh University courses of experimental philosophy were taught at an earlier date by James Gregorie *secundus* (1666–1742), its professor of mathematics from 1692 to 1742 and likely one of the authors of the 1740 "Hydrostatica" manuscript (see Appendix).[67] In 1725, Gregorie's health problems led to the appointment of Colin Maclaurin (1698–1746) joint professor of mathematics; Gregorie did not teach after that year. One of the deepest eighteenth-century British mathematicians and natural philosophers and a dedicated teacher, Maclaurin taught three mathematics courses, elementary, intermediate and advanced, the last one including fluxional calculus and propositions from Newton's *Principia*. He also taught a separate course on experimental philosophy that included handling of experimental apparatus and telescopic observations. If Maclaurin ever followed the 1740 "Hydrostatica" manuscript in his classes, we do not know.[68] We do know, however, that the course of the

[63] Emerson, "Science and the Origins and Concerns of the Scottish Enlightenment" and "Natural Philosophy and the Problem of the Scottish Enlightenment." See also the bibliography mentioned below, n. 86.

[64] Shepherd, *Philosophy and Science in the Arts Curriculum of the Scottish Universities in the 17th Century*; "University Life in the 17th Century."

[65] Dalzel, *History of the University of Edinburgh*, vol. II, p. 306; Emerson, "Scottish Universities in the Eighteenth Century, 1690–1800"; Wood, "Science, the Universities, and the Public Sphere in Eighteenth-Century Scotland."

[66] Emerson, "Natural philosophy and the Problem of the Scottish Enlightenment"; Emerson, "Scottish Universities in the Eighteenth Century."

[67] Bower, *The History of the University of Edinburgh*, vol. II, pp. 83f.; Wood, "Science, the Universities, and the Public Sphere"; Emerson, "Scottish Universities in the Eighteenth Century."

[68] Grabiner, "Maclaurin and Newton: The Newtonian Style and the Authority of Mathematics"; Wood, "Science, the Universities, and the Public Sphere."

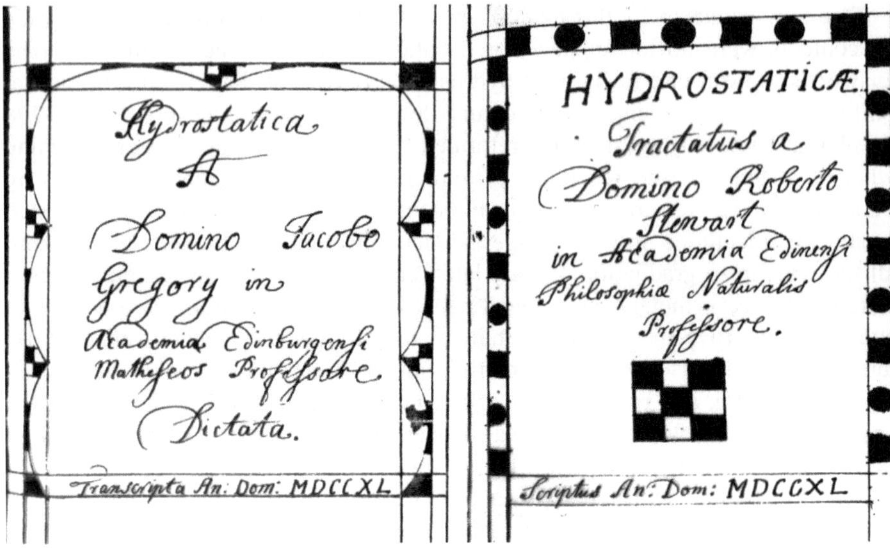

Fig. 7.6 Title pages of Gregorie's "Hydrostatica" and Stewart's "Hydrostaticae"

Edinburgh professor of natural philosophy, Robert Stewart, included demonstrations with instruments and that hydrostatics and pneumatics figured prominently in the syllabus of his courses.[69] In contents as well as in the organization of the material, the hydrostatics and pneumatics of Gregorie and Stewart show significant points in common.

The 1740 "Hydrostatica" manuscript is now bound in a manuscript volume gathering some "mixed" mathematics manuscripts, all of them copied in 1740, probably by the same amanuensis. We find in it David Gregorie's "Geometria practica" (78 pages), notes on surveying and trigonometry (18 pages), and a second mini-treatise on hydrostatics by Stewart (Fig. 7.6).[70] Sir Robert Stewart, who became regent of the University of Edinburgh in 1703 then its professor of natural philosophy from 1708 to 1742, was the son and heir of Thomas Stewart of Coltness, Baronet, and the nephew of James Steward, Lord Advocate of Scotland (1692–1709 and 1711–1713).[71] He studied at the University of Utrecht around 1700 and had the reputation of being Cartesian in his early years.[72] Apparently he was well connected with the Edinburgh Town Council, whose members were the patrons of the University.

[69] Emerson, "Natural Philosophy and the Problem of the Scottish Enlightenment," p. 256.

[70] Aberdeen University Library (AUL) Ms 2206.

[71] On Robert Stewart, see the fragmentary notices in Bower, *History*, vol. II, p. 32; Dalzel, *History*, vol. II, p. 410; Grant, *The Story of the University of Edinburgh*, vol. II, pp. 348f. See also Wood, "Science, the Universities, and the Public Sphere"; Emerson, "Natural Philosophy and the Problem of the Scottish Enlightenment."

[72] There are extant two letters from the virtuoso R. Wodrow to Stewart in Utrecht dated September 15, 1699 and 18 November 1701; see Wodrow, *Early Letters*, pp. 22, 181. On his cartesianism, Grant, *Story*, vol. II, p. 348.

Shortly after his appointment as professor of natural philosophy, in 1709 the Council granted him 50 lb sterling to buy instruments for teaching his course. In 1737 he was a founding member of the Edinburgh Philosophical Society.[73]

We know that in 1724, Stewart organized a "Physiological Library" to secure his students access to recent books in mathematics, experimental philosophy, natural history, medicine, and theology.[74] According to Michael Barfoot, one distinctive feature of Stewart's "Library" was that it accorded Robert Boyle a "transcategorical" role. By this Barfoot means not only that Boyle's works appeared cross-referenced in almost all the different categories in which the "Library" was organized, but also that Boyle himself was singled out as the foremost experimental philosopher. Moreover, his views were conflated with Newton's.[75] According to a syllabus published in 1741, Stewart's natural philosophy courses comprised chapters from Keill's *Introductio ad veram physicam*,[76] David Gregory's optics,[77] and David Gregory's "*Astronomy.*"[78] It also included "some propositions" from Newton's *Principia*, Newton's theory of colors, the study of the eye, microscopes and telescopes, and "Hydrostatics and Pneumatics from a manuscript of his [Stewart's] own writing"—this manuscript is probably the one now bound with Gregorie's hydrostatics.[79] While according to this account the course was broad enough, yet there is no evidence that the full course was ever taught. In fact, as the most recent student of Stewart's teaching found out, the only "lecture notes which have survived all refer to hydrostatics and pneumatics." This is of course consistent with the special place Boyle's understanding of hydrostatics and pneumatics occupied in Stewart's course.[80]

We have scanty and contradictory evidence about the audience and success of Stewart's course. According to the memories of Scottish church leader Alexander Carlyle, when he was a student in Edinburgh in the mid-1730s Stewart was a worn-out professor whose classes were a waste of time, although Carlyle very much enjoyed Maclaurin's three courses on mathematics and the one on experimental philosophy.[81] We do not know whether there was any competition or rivalry between

---

[73] Wood, "Science, the Universities, and the Public Sphere," p. 102; Emerson, "Natural Philosophy and the Problem of the Scottish Enlightenment," p. 256. He appears as number 14 in the list of founding members of the EPS provided in Emerson, "The Philosophical Society of Edinburgh 1737–1747," p. 190.

[74] Barfoot, "Hume and the Culture of Science in the Early Eighteenth Century," pp. 159–163.

[75] Ibid.

[76] John Keill's *Introductio ad veram physicam* was a highly popular introduction to Newtonian philosophy. On Keill's *Introductio*, see the essay by Carla Rita Palmerino in this volume (Chap. 5).

[77] Gregory, *Catoptricae et dioptricae sphaericae elementa*, translated into English as *Dr Gregory's Elements of Catoptrics and Dioptrics*. For a summary of this book, largely based on an unpublished optical manuscript by James Gregorie the Elder, see Malet, *Studies on James Gregorie*, pp. 101–115.

[78] Probably D. Gregory's *Astronomiæ, physicæ et geometriæ elementa.*

[79] The syllabus was originally published in *The Scots magazine*. It is quoted in extenso in Barfoot, "Hume and the Culture of Science in the Early Eighteenth Century," p. 152.

[80] Ibid., pp. 153, 163.

[81] Carlyle, *Anecdotes and Characters of the Times*, pp. 24–26; on Maclaurin, see pp. 16–17, 22, 26f.

Stewart and Maclaurin, although Stewart's 1741 syllabus is suspiciously similar to the content of Maclaurin's courses.[82] Be that as it may, many things suggest not only that hydrostatics and pneumatics figured prominently in the courses of natural philosophy of James Gregorie and Robert Stewart but also that they were presented with the strong interplay of mathematics and experiments to be discerned in the "Hydrostatica" manuscript.

Boyle, who in his *Hydrostatical paradoxes* presented hydrostatics as "a part of Philosophy," termed it "one of the ingeniousness Doctrines that belong to [philosophy]."[83] In Boyle's view, hydrostatics was commendable, first, for its practical dimension; then for its "delightfulness" and "subtility," as it provided "unobvious truths" that were "pure and handsome productions of reason duly exercised"; and finally, because hydrostatics provides principles to decide upon most of the controversies present "among the modern inquirers into nature."[84] There is a remarkable continuity between Boyle's views on the usefulness of hydrostatics and pneumatics and the social consideration and privileged place hydrostatics occupied two generations later in the teaching of experimental natural philosophy in Edinburgh. This place was secured by the importance of discussions about the nature of the vacuum, matter and space, the status of experimental knowledge, and the properties of fluids within contemporary general philosophical debates.[85] On the other hand, we have presented here evidence strongly suggesting that Boyle's purely experimentalist approach to hydrostatics and pneumatics was not the one that came to dominate the teaching of these subjects in Edinburgh up to the mid-eighteenth century, nor probably in the Scottish universities generally.

7.6 Concluding Remarks

Larry Stewart and others have convincingly argued the paramount importance an experimentalist approach had for British natural philosophy in the first half of the eighteenth century. We find the experimental approach to hydrostatics and pneumatics in many places, and in particular in books such as James Ferguson's *Lectures in Mechanics, Hydrostatics, Pneumatics, and Optics* (first published in 1760), which

[82] Wood, "Science, the Universities, and the Public Sphere," p. 104.

[83] Boyle, *Hydrostatical paradoxes*, in Boyle, *Works*, vol. V, pp. 189–279, at p. 193.

[84] Ibid., p. 194.

[85] For recent studies on science and moral philosophy in the Scottish Enlightenment, see Emerson, "Science and Moral Philosophy in the Scottish Enlightenment"; Wood, "Science and the Pursuit of Virtue in the Scottish Enlightenment." On the relevance of experimental philosophy for Hume's thought there is a rich bibliography, among which for our purposes see Capaldi, *David Hume, the Newtonian Philosopher*; Force, "Hume's Interest in Newton and Science"; Barfoot, "Hume and the Culture of Science in the Early Eighteenth Century"; Schabas, "David Hume on Experimental Natural Philosophy, Money, and Fluids."

we know were immensely popular and often reprinted.[86] The differences between Ferguson's hydrostatics and the manuscript here analyzed are deep and many. The former includes a mechanical description of fluids—never used for any purpose—as being made up of exceedingly small (invisible to the microscopes), exceedingly hard (incompressible), exceedingly round and smooth (so that they move most easily) particles, but lacks entirely any form of mathematical structure. The properties of fluids are demonstrated by experiments in a way that highlights the properties and the experiments to be applied to the design of springs, water-pumps and air-pumps, the "fire-engine," the determination of specific gravities, and so on. It contains many numerical tables to facilitate the application of results to practical problems.[87] Therefore, the evidence gathered here suggests that academic or learned mathematical hydrostatics and pneumatics were at variance with other versions of them that were geared to non-university audiences and emphasized its useful and experimental facets.

Far from disappearing, mathematics occupied a central role in hydrostatics and pneumatics as they were taught for decades in the University of Edinburgh. Originally the organization of hydrostatics as a mixed mathematical science was the work of mathematicians such as Wallis and James Gregorie who responded to Boyle's contributions by integrating them within a mathematical framework and idiom. As far as we know Wallis never used his *Mechanica* for teaching purposes at Oxford. It was the Gregories' reworking and enlarged paraphrase of Wallis' hydrostatics that gave it a long and new life in Edinburgh—and probably in Glasgow, Aberdeen, and St Andrews as well.[88] The teaching of hydrostatics and pneumatics in Edinburgh here reviewed put mathematics and experiments in a relation of complementarity rather than one of opposition. In this new context, what was originally part of mechanics organized and presented mathematically came to embody experimental philosophy.

Appendix: Sources and Authorship of Aberdeen University Library (Aul) Ms 2206/7

The University of Aberdeen keeps the only copy now extant (as far as we know) of a manuscript on hydrostatics and pneumatics titled "Hydrostatica" and which title page attributes authorship to "James Gregory, professor of mathematics in Edinburgh University." It is contained in one fair, 41-page long copy dated 1740 and neatly

[86] Ferguson, *Lectures on Select Subjects in Mechanics, Hydrostatics, Pneumatics, and Optics*; it was many times reprinted, the last time apparently in 1843. On Ferguson, see Millburn, *Wheelwright of the Heavens: The Life and Work of James Ferguson, FRS.*

[87] See "Lecture V. Of Hydrostatics, and Hydraulic Machines, in General," pp. 99–165, and "Lecture VI. Of Pneumatics," pp. 166–198, in Ferguson, *Lectures* (references to the 1770 edition). For the "mechanical" description of the particles of fluids, see p. 99.

[88] Wood, "Science, the Universities, and the Public Sphere," p. 112; Emerson and Wood, "Science and Enlightenment in Glasgow"; Wood, "Science and the Aberdeen Enlightenment."

written by an unknown amanuensis hand.[89] The manuscript catalog of Aberdeen University Library further identifies the author by calling him "Professor of Mathematics at St. Andrews." Although two James Gregories (or Gregorys, in what was then a characteristic English spelling) occupied the Mathematics professorship in Edinburgh, only the senior James (1638–1675) had also been professor of mathematics at St Andrews. As we shall see, there are good reasons to assume the two James were involved in the authorship of this manuscript, although the major role must have corresponded to the senior James. Upon James Gregorie's death in 1675, his friend and colleague William Sanders wrote down a list of the mathematical papers Gregorie left in a finished form. According to Sanders, Gregorie left among other things "The Theory of the whole Hydrostaticks comprehended in a few definitions and five or six Theorems."[90] The present manuscript fairly agrees with the foregoing description. It exactly contains four definitions and eleven propositions, the first six of which concern hydrostatics proper, while the last five propositions apply the former ones to explain "Torricelli's" experiment and pneumatics.

Internal references make the contents of the "Hydrostatica" manuscript mostly consistent with a date of composition around 1670 or shortly thereafter, the years in which James Gregorie the elder must have written his hydrostatical mini-treatise. As set forth above, "Hydrostatica" is a free paraphrase of Wallis' "De hydrostaticis." It simplifies and improves its original, but it is a paraphrase nonetheless. Mersenne and Boyle are more than once quoted, in particular with reference to the quantitative limits these authors provided for the atmospheric air's condensation and rarefaction. The references can be traced back to Mersenne's "Hydraulica [et] Pneumatica" (published within his *Cogitata physico mathematica* of 1644) and to Boyle's *Hydrostatical paradoxes* of 1666. As explained above, in 1671 Gregorie wrote to John Collins asking for a copy of Wallis' *Mechanica*.[91] There is therefore both internal and external evidence not to be easily dismissed that makes the senior James Gregorie author of this manuscript.

On the other hand, we know that the first two propositions of the manuscript along with their long lists of corollaries and the definition of fluid (on the whole some eight pages of text out of the 39 written pages of the manuscript) come almost verbatim from Newton's *Principia*, Book II, Section 5, Propositions 19 and 20 and their corollaries. The borrowing is explicitly acknowledged on page 10 of the manuscript: "we bring over [these two propositions] from the principles of the most illustrious Newton." The grafting of Newtonian hydrostatics into Gregorie's manuscript suggests that someone was still using it after 1687. The candidates are many, but among the most likely ones two nephews of the senior James Gregorie stand out, David Gregorie (1661–1708) and the junior James Gregorie (1666–1742), both of

[89] Aberdeen University Library (AUL) Ms 2206/7. The manuscript contains just a few amanuensis errors.

[90] University of Edinburgh Library, Dc. 1.4^{1}.129.

[91] Turnbull, *James Gregory Tercentenary Memorial Volume*, p. 325.

whom were professors of mathematics at the University of Edinburgh. We do know that David appropriated his uncle James' manuscripts without properly acknowledging his sources. If he had used his uncle's hydrostatics, it is likely that the manuscript would have reached us under David's name. Furthermore, he left Edinburgh for Oxford in 1691, before he had time to fully acquaint himself with the *Principia*. It is likely, therefore, that it was the younger James who amended and perhaps expanded the hydrostatical manuscript.[92] The younger James, of whom no written production is known, was professor of philosophy (not of mathematics) in St Andrews from 1686 to 1691 and then professor of mathematics in Edinburgh from 1692 to 1742.[93] The name in the title page of "Hydrostatica" might have been meant for him—in this case the library catalog does not identify him properly.

[92] On David's career and production, see Eagles, *The Mathematical Work of David Gregory*. On his use of his uncle James' manuscripts, see Malet, *Studies on James Gregorie*, pp. 101–115.

[93] Stewart, *The Academic Gregories*, pp. 84–87.

Part III
Mechanical Philosophy Applied

Chapter 8
From a Metaphysical to a Scientific Object: Mechanizing Light in Galilean Science

Susana Gómez

Ancient philosophers generally offered three lines of thought for conceiving of light. Classical atomism took light to be a material emission of small and swift particles; Plato spoke about light as a phenomenon charged with metaphysical tones; and Aristotle, who surprisingly only spoke of light in two brief passages in his entire corpus, asserted that light was not a substance but a quality of the medium. During the Middle Ages phenomena related to light were studied mathematically by geometrical optics, though it is well known that the mathematical approach was not, strictly speaking, a part of natural philosophy. The nature of light in that period was only studied—with few exceptions—by authors who expressed conceptions of nature that were closer to Neoplatonic cosmologies. In most cases, the study of light was nearer to metaphysics than to natural philosophy.[1]

The Neoplatonic revival of the fifteenth and sixteenth centuries was, undoubtedly, a great impulse in the direction of modern science. However, the new generation of seventeenth-century natural philosophers was not at all satisfied with the style of inquiry that came along with Neoplatonism. These natural philosophers strongly criticized the Neoplatonic understanding of occult qualities, action at a distance, sympathies, signatures, etc., and made a great effort to reinterpret those categories. As an alternative to the Neoplatonic natural philosophies of the Renaissance they proposed a conception of nature as a great machine. This mechanized conception of nature could only conceivably have arisen together with certain epistemological changes, among them the rise of experimentation and mathematization. For many seventeenth-century scientists, understanding natural phenomena in this way was essential to the process of bringing them into the domain of the genuinely scientific.

[1] About the relation of metaphysical and physical arguments in medieval and renaissance philosophies of light, see Lindberg, "The Genesis of Kepler's Theory of Light."

S. Gómez (✉)
Faculdad de Filosofia, Universidad Complutense de Madrid, Madrid, Spain
e-mail: susanagl@wanadoo.es

D. Garber and S. Roux (eds.), *The Mechanization of Natural Philosophy*, Boston Studies in the Philosophy of Science 300, DOI 10.1007/978-94-007-4345-8_8,

The development of Galileo's conception of light as it changes in the different phases of his work nicely exemplifies this effort to make a natural phenomenon an object worthy of science. In his first works Galileo tried to consolidate physical and mathematical approaches to nature. The earlier tradition of geometrical optics was not satisfactory for a natural philosopher and mathematician, who was trying to understand the nature of light. It is understandable, then, that initially Galileo was sympathetic toward those Neoplatonic thinkers who had made an effort to explain exactly what light is. Nevertheless, understood as the Neo-Platonists did, light was, fundamentally, a phenomenon charged with profound metaphysical overtones hardly reconcilable with Galileo's *sensate esperienze* and *certe dimostrazioni*. Even if at a certain moment in his scientific thought, Galileo was inclined to defend certain features of Neoplatonic conceptions of light, it later became necessary for him to articulate those views in a way he considered appropriate to science. It was, in other words, necessary for him to convert the metaphysical conception of light in Neoplatonic philosophers into a "scientific object." Atomism and a mechanical conception of nature were decisive for that transformation.

In Galileo's science, as for most seventeenth-century scientists, the development of atomism was closely connected with the project of the mechanization of nature. However, within this larger project, there were many different atomisms and mechanical philosophies.[2] These differences arose for very different reasons, some epistemological, some metaphysical, and some even theological. Among the reasons for this diversity, one of the most important was the simply difficulty of explaining certain complex natural phenomena. Light was such a phenomenon that demanded to be integrated into this new framework. As Westfall wrote:

> As its fundamental proposition, the mechanical philosophy asserted that all the phenomena of nature are produced by inert matter in motion. What about light? No philosophy of nature that ignores light can pretend to be complete, and light appears to be the least obviously mechanical of all phenomena.[3]

It seems to me that for Galileo, the explanation of what light is constituted a fundamental problem that determined the path along which his particular brand of

---

[2] Properly speaking, few seventeenth-century natural philosophers were strict mechanists, but I think we have to distinguish a proper "mechanical philosophy" in the sense Boyle attributed to it from a set of epistemological values which from the beginnings of the seventeenth century were shaping the mechanical image of the natural world. As Daniel Garber says in his article in this volume (Chap. 1), Galileo was not a strict mechanist, but I believe he really conceived the mechanical explanations as a regulatory principle of his scientific work. In fact, the development of his interpretations of the nature of light shows both his failure to achieve an absolute mechanical explanation and his efforts to achieve it. Besides the aforementioned article of Daniel Garber, which reviews the topic of the mechanical philosophy in the history of science, see Baldini, Zanier et al., *Ricerche sull'atomismo del Seicento*; Boas Hall, "The Establishment of the Mechanical Philosophy"; Clericuzio, *Elements, Principles and Corpuscles*; Festa and Gatto (eds.), *Atomismo e continuo nel XVII secolo*; Festa, Jullien and Torrini (eds.), *Géométrie, atomisme et vide*; Lüthy, Murdoch and Newman (eds.), *Late Medieval and Early Modern Corpuscular Matter Theories*; Meinel, "Early 17th-Century Atomism"; Newman, *Atoms and Alchemy*.

[3] See Westfall, *The Construction of Modern Science*, p. 36.

atomism developed. To study the nature of light required making light a scientific object in accordance with his conception of the scientific. And to accomplish this, it was necessary for him to mechanize his conception of light.

8.1 Two Galilean Atomisms: Physical and Mathematical

The distinction drawn between two kinds of atomism, physical and mathematical, in discussing the evolution of Galileo's thought is not simply a matter of our later interpretation. Some contemporaries and followers of Galileo had already spoken of the distinction. After all, the traditional division between physics and mathematics was still very strong at the time he was working. Moreover, in a sense, many historians directly inherited this division. Following this line of thought, it might seem as if Galileo's views evolved from a physical conception of atomism, one whose first signs appeared in the *Discorso intorno alle cose che stanno in su l'acqua* (1612), to a mathematical atomism carefully set out in the First Day of the *Discorsi* (1638). His physical atomism seemed tied to the classical tradition of atomism represented by the philosophies of Democritus, Epicurus, and Lucretius, each of whom explained all substances as compounds of minimal and indivisible particles characterized by their figure, size and motion. Theirs was an effort to find the ultimate causes of natural phenomena in a way alternative to the Aristotelian doctrine of forms and qualities. Such a program involves speculating, in an analogical way, about the true and real structure of bodies and about the invisible causes of observed phenomena. Galileo himself appealed to an atomic hypothesis in order to explain the results of the experiments about floating bodies in the *Discorso* and spoke of "minimal particles" in order to treat the problem of heat in *Il Saggiatore.*[4] It is interesting to note that this kind of atomism contains nothing of the mathematical. By contrast, the *Discorsi* displayed a mathematical treatment that involves an appeal to indivisibles, a form of reasoning with a long philosophical tradition beginning with the Aristotle's criticism of the concept of extended indivisibles. Opposing the idea of the actual divisibility of the continuum into an infinite number of indivisibles, Aristotle accepted the possibility of a potentially infinite divisibility. In the thirteenth and fourteenth centuries, Oxford philosophers again took up this idea of developing a mathematical atomism in which the continuum was compounded of indivisibles, though indivisibles conceived of as points without magnitude. In following

[4] About the different connotations of the concept of atom along the different phases of Galilean work, see Galluzzi, *Tra atomi e indivisibili.* I wrote this article before Galluzzi published his work, which explains why I have not taken into account some very interesting remarks and interpretations he offers, with which I mostly agree. See also Gómez, "The Mechanization of Light in Galilean Science"; Nonnoi, "Galileo Galilei: quale atomismo?"

years, this sort of atomism can be found in disputes within the domains of logic, theology, and geometry, but this mathematical atomism made no claim to offer an explanation of phenomena in material things or to replace the Aristotelian doctrine of forms and qualities.[5]

It may look as if Galileo passed from favoring the physical form of atomism to favoring mathematical atomism. But we should not underestimate the great distance separating the First Day of the *Discorsi* from medieval arguments, or even from exclusively geometrical treatments of indivisibles, as found, for example, in Galileo's contemporary Bonaventura Cavalieri. Firstly, because Galileo had in mind the project of a mathematical physics for the study of motion, his arguments about the theory of matter were not purely mathematical or logical studies, but concerned the very tools by which the real structure of nature could be deciphered. This is evidenced by the fact that Galileo was firmly convinced of the mathematical structure of nature itself. Secondly, because the First Day of the *Discorsi* was full of experiments, whether merely theoretical or carried out in practice, they were not independent of the mathematical apparatus he was developing there. An excellent example of this is the Galilean interpretation of the pseudo-Aristotelian paradox of the wheel, the so-called "Aristotle's Wheel," and its relation to the question of condensation and rarefaction of bodies.[6]

Was there really a radical break between one Galileo, the advocate of a physical atomism, and another Galileo, the defender of a pure mathematical atomism? Is it still possible to believe in a Galileo who stays strictly within the boundaries of the mathematical treatment of indivisibles, in view of the torrent of philosophical, epistemological, academic and theological problems with which physical atomism was faced?[7] My aim here is to propose that, in fact, there was no rupture, no rejection of physical atomism to opt, instead, for a purely mathematical atomism. In other words, when we speak about mathematical atomism in connection with Galileo, we have to be very careful to avoid the error of attributing to it the meaning mathematical atomism traditionally had, that is, a logical and geometrical speculation independent of the real physical explanation of natural phenomena. We must bear in mind here two very important distinctions. The first concerns the difference between physical and mathematical atomism, insofar as the former conceives of atoms as material, corporeal and spatial particles, whereas the latter defends a concept of indivisibles without magnitude, without extension, points in a geometrical sense. In this sense it can be said that Galileo began to favor mathematical atomism in *Il Saggiatore* when he asserted that the atoms of light do not have extension, and as well later in his detailed defense of the position in the *Discorsi*. On the other hand, the second distinction I have in mind involves the distance between the development of an atomism restricted to logic and mathematics—such as we find in the medieval philosophers or in

---

[5] Festa, "Repères pour une analyse historique des notions de continuum et d'atomisme"; Pyle, *Atomism and Its Critics*.

[6] For an analysis of the relation between Galileo's theories of matter and motion, see Palmerino, "Una nuova scienza della materia per la scienza nova del moto"; Palmerino, "The Isomorphism of Space, Time and Matter in Seventeenth-Century Natural Philosophy".

[7] Redondi, *Galileo eretico*.

Cavalieri—and a rather different way of using mathematical atomism in physics, as was defended by Galileo in his *Discorsi*. It is just in this sense that it would be an error to conceive of the Galilean theory of the continuum and indivisibles as a mathematical atomism understood as a renunciation of the project of explaining the real structure of matter in order to remain in the field of purely geometrical speculation.

The specific case of the nature of light provides a good example of the thread of continuity (but not identity) that connects the different phases of Galilean atomism: from the assertions regarding the sunspots to the pages of the *Discorsi* concerning the velocity of light, we find neither the renunciation of fundamental ideas or concepts, nor a replacement of the goal of understanding the nature of light through an experimental and quantitative study of its velocity. We have, instead, the development of some ideas that were, so to speak, in an embryonic state in the initial stages of Galilean science.

8.2 Neoplatonic Echoes and Materialistic Ambitions

The famous Copernican letter Galileo wrote to Monsignor Piero Dini in 1615 is one of the most controversial fragments in Galileo's thought. The main question taken up in this text is the existence in nature of a "very spirituous, tenuous and rapid substance, which in spreading out through the universe, penetrates everything without meeting resistance, warms, vivifies and renders all living creatures fertile."[8] If with these words Galileo was making reference to light—as has been claimed on many occasions—then it seems his assertion was certainly not compatible with the Aristotelian conception of light, because in this text light is not conceived of as an accident or quality of the transparent medium, as it is for Aristotle, but rather is taken to be like a *substanza spiritosissima*. It would be better to interpret those words as being representative of Galileo's affinity for the Neoplatonic view which considered light, though a substance, an immaterial one; in other words, a substantial form. Light is in this way conceived as the spirit of the world—a spiritual substance—a *vis motrix*, an animating force of all natural things, preceding all of them, as was asserted in *Genesis* to account for the creation of the world.[9]

[8] Galileo, *Opere*, vol. V, p. 301: "substanza spiritosissima, tenuissima e velocissima, la quale, diffondendosi per l'universo, penetra per tutto senza contrasto, riscalda, vivifica e rende feconde tutte le viventi creature."

[9] Light was the original substance, created on the first day and only 4 days later put into that great warehouse the Sun, as Galileo himself wrote: "Di questa luce primogenita e non molto splendida avanti la sua unione e concorso nel corpo solare, ne aviamo attestazione dal Profeta nel Salmo 73, v. 16 *Tuus est dies et tua est nox: Tu fabricatus es auroram et Solem*; il qual luogo vien interpretato, Iddio aver fatto avanti al sole una luce simile a quella dell'aurora: di più, nel testo ebreo in luogo d' *aurora* si legge *lume*, per insinuarci quella luce che fu creata molto avanti il Sole, assai più debile della medesima ricevuta, fortificata e di nuovo diffusa da esso corpo solare" (Galileo, *Opere*, vol. V, p. 302). On this, see Lindberg, "The Genesis of Kepler's Theory of Light"; Hamilton, *Three worlds of Light*; Maecling, *The Doctrine of Light*; Puliafito, "Per uno studio della *Nova de universis philosophia*."

There are in this letter, however, certain elements that lead us to be suspicious that Galileo may not completely have supported the Neoplatonic metaphysic of light. In the first place, we have to note that it is not absolutely clear in the text that this *substanza spiritosissima* can be completely identified with light, contrary to what has often been claimed. The "spiritual substance," he adds, designates not simply light but the "calorific spirit" that accompanies the light of the sun, and thanks to which light permeates and gives life to natural beings. However, if the sun's light spreads over everything with this substance, filling the universal immense spaces, we have to wonder whether what is at issue is one of the immaterial substantial forms of which Neoplatonic philosophers spoke, or if, instead, he was referring to a corporeal (and material) substance.

Traditionally, studies on the nature of light in seventeenth-century natural philosophy approach the question in terms of a struggle between only two concepts: an Aristotelian qualitative view and the substantialist conception of light. Such a dichotomy overlooks, for example, the Neoplatonic thesis about the substantiality of light, some Jesuit defenses of its substantiality, or alchemical theses about its semi-substantialist nature. The source of this error is usually the identification of "substantial" with "material," which overlooks the significant difference between the spiritual substance defended by some Renaissance philosophers and the material substance argued for by atomists.[10] In the Aristotelian tradition, corporeity is associated in some ways with the categories of form, substance and matter, but it is not exactly identical to any of them. To be corporeal means to occupy space; and therefore matter, stripped of its properties or qualities (especially those associated with magnitude and extension) is not corporeal.[11] But to be corporeal is not equivalent to being material. Moreover, while form can give matter magnitude, form is not dimensional in and of itself, and therefore corporeality cannot be identified with form either. The corporeal is the three-dimensional substance, the matter joined with the appropriate form. Finally, the substance is a being, which is *per se*, exists *per se*, and is autonomous, not dependent on anything. In nature there is no matter without form, nor are there forms that are not attached to a material substratum. The so-called "substantial forms" are such because they are that by virtue of which a thing is, in fact, that thing; they are theoretically autonomous, but in nature they are always attached to a material substratum. In nature, only the prime mover (identified with God in the scholastic version of Aristotelianism and considered outside nature) and the movers of celestial bodies are pure substantial forms, insofar as they are not material and not corporeal, and thus without magnitude. These considerations applied to the specific case of light help us to understand why in the context of the Aristotelian natural philosophy light cannot be a substance: namely, because it lacks autonomous existence, it requires a medium. Regarding its corporeity, we should say that strictly speaking

[10] On the concept of *spiritus* in Renaissance natural philosophy, see Garin, "Il termine 'spiritus' in alcune discussion fra Quattrocento e Cinquecento."

[11] Aristotle, *Metaphysics*, VII 3, 1029a20–30.

it is incorporeal because it does not occupy space in itself. However, alternatively, it could be said to be corporeal insofar as being in act it needs a material and corporeal medium. It should be noted, though, that light is conceived as the actualization of a quality in a medium, and is not considered as something with local motion; this allows Aristotle and Aristotelians to assert that the speed of light is instantaneous.[12]

The cosmological, metaphysical and theological values that Neoplatonic philosophers attributed to light were completely alien to the Aristotelian natural philosophy. Neoplatonic metaphysics of light advanced an interpretation of its nature very different from the Aristotelian one, but nonetheless conserved the Aristotelian categories of matter, form, substance and body.[13] For one thing, light—as all links along the scale of being—was autonomous in its unity, and therefore it did not need a medium; it deserved to be considered "substantial." Light existed by itself, as spirit did. It was not a quality, but a principle of activity. Secondly, because light was considered corporeal and incorporeal at the same time, we find many problems with the interpretation of sixteenth- and seventeenth-century authors who refer to Neoplatonic theses, especially those among the Jesuits. On the one hand, light was considered corporeal insofar as it was deemed to be lodged in a luminous body. Although light may seem to have existed as a form without dimensions, its dimensions could nonetheless be asserted on the basis that the body in which it is lodged has them: in this way it has a secondary corporeity, by participation in the body. On the other hand, light is instead considered to be incorporeal when it refers particularly to radiating light, that is, the light emanated spherically from luminous bodies, for this radiating light does not need any medium to pass from one place to another. It was not, in any case, material. Light was, therefore, a pure incorporeal "substantial form." Aristotelians had saved this category for the Prime Mover, and Scholastics for God. When Neo-Platonists put light in this category, it was possible for them to identify light with God.

The ideas and tone expressed by Galileo in his letter to Dini remind us of Marsilio Ficino, who in his *De lumine* had identified God with invisible and infinite light, from which finite and visible light are derived. Light was, for Ficino, as in Galileo's letter, the active power of the universe and the manifestation of the internal fecundity of all beings. However, Ficino also said that it expanded instantaneously through the universe, giving the form of magnitude to matter without form. This explains why Ficino, as like Grosseteste, Saint Buenaventura and Roger Bacon, considered light as a "corporeal form" in the sense that light gives extension or spatiality to matter, though not in the sense that light might be corporeal or spatial by itself. Two examples of this incorporeity were its penetrability and its instantaneous diffusion. Light was, therefore, a spiritual "corporeal form." Furthermore, it was possible to

---

[12] Aristotle, *Physics*, VI 3, 234a, and VII 6, 337a. See Lindberg, "The Genesis of Kepler's Theory of Light."

[13] See Lindberg, "The Genesis of Kepler's Theory of Light"; Thonnard, "La notion de lumière en philosophie augustinienne."

say that light is a substance insofar as the medium is superfluous for its diffusion.[14] Note that in this context, to say that something is a substantial and corporeal form does not mean that it is material.[15]

In spite of these fundamental differences concerning the nature of light, both Aristotelians and supporters of Neo-Platonism defended its immateriality and its instantaneous diffusion. Nevertheless, in the letter to Dini, Galileo spoke of a "very spirituous, tenuous and swift substance (*substanza spiritosissima, tenuissima e velocissima*)."[16] The problem arises precisely when Galileo attributes to this substance that expands from the Sun through the entire universe the qualification "very swift," that is, non-instantaneous spreading, and therefore having local motion. The scholastic tradition asserted that instantaneity belonged only to incorporeal beings. Some Neoplatonic philosophers attributed instantaneity to certain corporeal forms that gave dimensionality to matter (light for Grosseteste, Saint Buenaventura or Ficino). Finally, other philosophers, such as Patrizi, thought that instantaneity pertained to non-material, but substantial and three-dimensional incorporeal bodies, as for example angels or light. Therefore, the "spiritual substance" of which Galileo spoke, and to which he attributed local motion and called "very swift," coincided neither with the Aristotelian definition of light, because it was substantial, nor with Neoplatonic "corporeal forms," because it was not instantaneous. The only option was, then, to consider it as a very subtle material substance.

---

[14] On the metaphysics of light in Ficino and his conception of the Sun, see Rabassini, "La concezione del Sole secondo Marsilio Ficino"; id., "'Amicus lucis'. Considerazioni sul tema della luce in Marsilio Ficino"; id., Il vincolo dell'universo; Deitz, "Space, Light, and Soul in Francesco Patrizi's Nova de universis philosophia (1591)." Note the differences with respect to the interpretations of some eminent Neoplatonic philosophers, such as Al-Kindi or Avicebron, who thought that a medium was necessary for the diffusion of light, which was an impression on the medium, and therefore thought that radiating light was corporeal. In his synthesis of Aristotelianism and Neoplatonism, Grosseteste defended the instantaneity of light, considering it as the prime "corporeal form" because it gives extension to matter. Saint Buenaventura defended very similar ideas. The interpretations of these two later authors were picked up and developed by Roger Bacon, who continued to defend the definition of light as "corporeal form" and asserted its instantaneity, see Lindberg, *The Genesis of Kepler's Theory of Light*; id., *Theories of Vision from al-Kindi to Kepler*; id., Roger Bacon's Philosophy of Nature; McEvoy, "The Metaphysics of Light in the Middle Ages."

[15] Neoplatonic philosophy admits the existence of spatial forms separated from matter. The best examples of such non-material spatial forms are the geometrical and real entities of the Platonic World of Ideas. The way the Neoplatonic philosophers argued separately about matter and form—in addition to the idea of unity of all universal things—led to a proliferation of "corporeal forms" (corporeal insofar as they were spatial) that penetrate everywhere throughout the universe without being hindered by the resistance caused by the impenetrability of material bodies. Angels were the main protagonists of this proliferation of forms. Furthermore, the independence of spatial forms contributed to the development of a concept of space independent of the bodies that it contains, over and against the Aristotelian identification of space and place, therefore also contributing to the defense of void spaces. These elements were among the principal causes of the intersection between atomism and Neo-Platonism, as, for example, is evident in Patrizi's philosophy. On this issue, see Henry, "Void Space, Mathematical Realism and Francesco Patrizi da Cherso's Use of Atomic Arguments."

[16] "la somma velocità del muoversi per immensi spazii, essendo l'emanazione della luce come instantanea," Galileo, *Opere*, vol. V, p. 302.

However, when Galileo spoke of the diffusion of that substance, he stressed that it was not exactly light. I believe that when making this assertion, he was really trying to reconcile the animating and penetrating power of light, as found in Neoplatonic philosophies, with a materialistic explanation. Nevertheless, defending both the materiality and corporeity of this light and its capacity to extend through the universe animating all beings posed two important problems: that of the penetrability of bodies and that of opacity. Neo-Platonists avoided the problem of penetrability through their acceptance of the possible existence of spatial substantial or corporeal forms separated from matter, such as angels and light. They treated opacity in metaphysical terms, often resorting to the dichotomy between a dark matter unable to be animated by light, and another translucent matter, permeated by light. Galileo did not overlook the problem, but he approached it rather as a physicist than a metaphysician. In his letter to Dini, he made reference to the difference between the penetrability of light and that of heat, which led him to make the distinction between light and the "calorific spirit." He asserted that the animating substance, stored in the Sun, from which it expanded through the universe giving life to all natural beings, "is something more than light, in that it penetrates and spreads out through all bodily substances, even the densest ones, through which light is not able to penetrate in similar manner."[17] That "something more" was meant to indicate the calorific spirit, the true cause of life and activity of natural beings. In this letter, he did not give a satisfactory solution to this question and he only stressed the capacity of that substance to penetrate into all bodies on account of its "subtlety." It seems to me that this distinction was entirely *ad hoc*, advanced to save the problem of opacity. In fact, we will see below how years later, in the context of a correspondence with Castelli, Galileo will describe light in very similar terms to those he used to speak of the calorific spirit.[18]

Galileo had good reasons to maintain the metaphysical value of light in a letter where he was trying to reconcile the centrality of the Sun with Holy Scriptures, presenting an ode to the Sun in the purest Neoplatonic way. Nevertheless in the letter to Dini, something important changed, because for Galileo that "emanation from the Sun" was both corporeal and substantial, and the fact that it had velocity indicated that it was not one of the Neoplatonic "corporeal forms." It seems, then, that the problem is to explain how and why this significant transformation happened. Two important events connected with Galileo and the Accademia dei Lincei in the years immediately before the letter to Dini offer some insights.

In the first place, we should remember that the idea of a "calorific spirit" penetrating and animating natural beings was one of the most important theses of Bernardino Telesio. Telesio had been included in the *Index* in 1596 precisely on account of suspicions of

---

[17] Ibid., vol. V, pp. 301f.: "essere qualche cosa di più del lume, poi che ei penetra e si diffonde per tutte le sustanze corporee, ben che densissime, per molte delle quali non così penetra essa luce."

[18] Many years later, in 1637, the questions about the difference between light and heat and opacity will reappear in the letters that Castelli sent to Galileo, known as "La Mattonata" (ibid., vol. XVII, pp. 121–123, 150–155 and 156–169). These letters were published in Castelli, *Alcuni opuscoli filosofici di Benedetto Castelli*, pp. 57–79. See Campogalliani, "Mattoni al sole"; Gómez López, "Galileo y la naturaleza de la luz."

materialism in his doctrine of the *anima mundi*, according to which there was a luminous sky full of a calorific principle.[19] In 1612 Antonio Persio, a follower of Telesio's ideas, was elected—the day after his death—as a member of the Accademia dei Lincei. From that moment, the Accademia decided to edit Persio's large manuscript *De natura ignis*, where he maintained that heat was "an extremely rarefied substance or body" and included, at the end, a chapter about light.[20] Galileo, through Prince Cesi, was acquainted with Persio's work and the Accademia's editorial project. It is possible that those ideas in that period strongly influenced Galileo's ideas on light and heat.[21]

The second element relevant to the important transformation in question here is Galileo's involvement in two important controversies in the same year as Persio's election by the Accademia dei Lincei. One of them, closely connected to the intellectual life of the Accademia dei Lincei, was the dispute about sunspots. In the *Istoria e dimostrazioni intorno alle macchie solari* (also published by the Accademia dei Lincei in 1613), elements beyond the echoes of Telesio relating to the diffusion of solar heat played an important role, which resulted in a corpuscularian and materialistic conception of solar light. The other controversy concerned the problem of floating bodies. Galileo showed his sympathies for atomism on that occasion, especially in his writing about the atoms of fire (*ignicoli*) in his *Discorso intorno alle cose che stanno in su l'acqua o che in quella si muovono.*[22]

The materialistic explanation of solar light and its emission offered by Galileo in his correspondence with Castelli about sunspots sounds very different from certain ideas he expressed in the letter to Dini, as for example the one about the primitive nature of light. In the letters on sunspots, nothing is said about the distinction

---

[19] Telesio, *De rerum natura iuxta propria principia*, IV, 9–17; see Gómez, "Telesio y el debate sobre la naturaleza de la luz en el Renacimiento italiano". There are important differences with respect to content and approach between Telesio's speculation on light and the views of Ficino and Patrizi. While these later authors pointed especially to the metaphysics of light, Telesio instead conceived of his conception of light as part of physics. All of them thought that light and heat are strongly connected, but for Ficino light is the origin of all beings and heat derives from it—that is to say, heat is the result of light's activity. Telesio, instead, thought that light derives from heat, and that this explains why the sky, the part of the universe where heat dwells, would be the world of light. See also De Franco, "La teoria della luce di Bernardino Telesio."

[20] Concerning this Lincean editorial project, see Redondi, "Fede Lincea e teologia tridentina." In this essay, Redondi emphasizes the importance of an earlier work by Persio, his *Trattato dell'ingegno dell'huomo* (1576), where, though not a Copernican, Persio made an ode to the Sun, describing it as the governor of the universe, whose instrument was the light continuously spreading from it. According to Redondi, ideas like this fitted perfectly in the Lincean aim to reconcile the new science with the Bible.

[21] See letters from Bartolini to Galileo, 24 February 1612 (Galileo, *Opere*, vol. XI, p. 278); from Cesi to Galileo, 2 March 1612, where the Prince asked Galileo his opinion about the election of Persio and the publication of his manuscripts (ibid., p. 285). On 19 May 1612, Cesi sent to him "some of Persio's published things" and some days later Galileo answered that he was attending with great curiosity to the manuscripts in question, ibid., pp. 298 and 301. On 28 December 1612, Cesi sent him "le materie del Persio," ibid., p. 451.

[22] The book, published in 1612, is in vol. IV of Galileo, *Opere*, with texts connected with the debates around it in the period. On this polemic see De Ceglia, *De natantibus*.

between light and the calorific spirit. The decisive impulse toward Galileo's defining solar light as a very swift and continuous emission of material corpuscles came in a letter from Benedetto Castelli dated May 8, 1612. Castelli's argument was based on something we can observe when burning a piece of paper, for example: at first it goes black, then it changes color, and finally it catches fire and emits light.[23] Galileo received Castelli's letter some days after he had written his first letter on sunspots. Galileo's three letters on sunspots were intended to demonstrate that those observed spots were really on the surface of the sun, that they belonged to the sun. Nevertheless, it is very difficult to determine their composition and Galileo displayed great caution. However, the influence of Castelli's letter on the nature of sunspots was felt throughout the process of Galileo's responses to Scheiner. Although in his first letter, dated May 4, 1612, Galileo openly expressed the impossibility of determining the "essence" and "substance" of those spots,[24] in the last one, dated December 1, 1612, he used Castelli's argument to assert that sunspots were a kind of "bituminous substance very difficult to burn," necessary for the "restoration of the immense light continuously disseminated from so great a lamp all over the world." The problem of the continued persistence of the sun's light was raised as an apparently unwelcome consequence of the assumption that it is material: if the sun were continually emitting its matter, why does it not come to an end? In the end Galileo asserted, as suggested by Castelli, that sunspots are that "food (*pabulo*)" or "nutrition (*nutrimento*)" that serve to regenerate the solar body. In this way, a strictly material combustion process explains the origin of the sun's light. And furthermore, the nature of this process is taken to be identical to that of terrestrial fires.[25]

---

[23] Castelli to Galileo, in Galileo, *Opere*, vol. XI, pp. 294f. On the polemic about sunspots, see the Introduction by Reeves and van Helden in Galileo, *On Sunspots*.

[24] Ibid., vol. V, pp. 105f.: "Circa a cotal determinazione, io confesso a V.S non aver sin ora tanto di resoluto appresso di me, ch'io m'assicuri di stabilire ed affermare conclusione alcuna come certa; essendo molto ben sicuro, la sustanza delle macchie poter essere mille cose incognite ed inopinabili a noi, e gli accidenti che in esse scorgiamo, cioè la figura l'opacità ed il movimento, per esser comunissimi, o niuna o poca e molto general cognizione ci possono somministrare."

[25] Ibid., vol. V, pp. 230f. In a note in his second letter to Welser, Galileo had already made reference to the problem of the regeneration of the sun, see ibid., p. 140. But he was very prudent in expressing his suggestions, as shown in the sentence closing the passage just cited: "Io peró non intendo di asserire alcuna di queste cose per certa, nè di obbligarmi a sostenerla, non mi piacendo di mescolar le cose dubbie tra le cose certe e risolute," ibid., p. 231. Some pages before he had insisted on this attitude when writing: "esser ottima resoluzione il posporre ogni atto specolativo a tutte le altre nostre occupazioni. Perchè, o noi vogliamo specolando tentar di penetrar l'essenza vera ed intrinseca delle sustanze naturali; o noi vogliamo contentarci di venir in notizia d'alcune loro affezzioni. Il tentar l'essenza, l'ho per impresa non meno impossibile e per fatica non men vana nelle prossime sustanze elementari che nelle remotissime e celesti," ibid., p. 187. The problem of solar regeneration is also present in the letter to Dini of 1615, where Galileo will give an explanation very close to some of the ideas expressed by Ficino in his *De vita*. There Ficino related the luminous and spiritual radiation from stars to a kind of circular breathing of the universal spirit, a *leitmotif* of the Hermetic tradition. On the metaphysics of light in Ficino and his conception of the Sun, see Rabassini, "La concezione del Sole secondo Marsilio Ficino"; Vasoli, "Su alcuni temi della filosofia della luce nel Rinascimento."

Does the materialistic conception of light asserted by Galileo in 1612 raise suspicions about the authenticity and sincerity of his sympathies regarding the Neoplatonic conception of light expressed in the letter to Dini in 1615? Or should we infer that in this letter of 1615 he is retracting ideas he defended 3 years before? While we shouldn't underestimate the obvious rhetorical attractiveness in that context of an ode to the central role of the sun in the universe, I think we should not completely mistrust Galileo's sympathies for a Neoplatonic conception of light.[26] We historians of science tend too frequently to suppress contradictions and to look for the hidden key that, supposedly, explains the perfect coherence of the problem under study. But perhaps in this case it would be more appropriate to consider this period of Galileo's investigations into the nature of light more as a creative moment that comes with its dose of confusion, the fluctuation between different alternatives, a moment when Galileo was still trying to reconcile explanations containing elements in principle mutually incompatible. Paraphrasing Copernicus, Galileo himself recognized in the letter to Dini that his speculation was "just a minor work, in need of being put into better form, washed off and cleaned up with affection and patience, since it is only sketched out and suitable for being filled out with limbs appropriate to its shape, but for the moment it is unruly and rough. If I have the chance, I will put it into better shape."[27]

Almost all studies of the structure of matter in Galileo's science agree that by 1612, Galileo had already expressed his preference for atomism and had rejected an Aristotelian natural philosophy founded on forms and qualities. Never, as far as I know, has Dini's letter been cited as a chapter in Galileo's atomism. The Neo-Platonism of this Copernican letter is only interpreted as related to cosmological or scriptural questions. But is it really possible that Galileo had set aside his atomistic and mechanistic tendencies, expressed both in his writings about floating bodies and in his texts on sunspots in 1612, to endorse a Neoplatonic speculation full of immaterial and spiritual entities? Or, instead, should we think that he was trying to reconcile Neoplatonic metaphysics of light with an atomistic conception? Shouldn't we, perhaps, speak of an attempt to mechanize light, which until then had been treated only metaphysically by Neoplatonic philosophers or mathematically by geometrical optics? Everything seems to point to a desire on Galileo's part to keep the cosmological advantages of conceiving light as the world spirit, as in the Neoplatonic philosophies, without, at the same time, giving up on his aim of offering a mechanical explanation of natural phenomena. The goal seems to be to explain light in terms of a material and mechanical natural phenomenon, that is, to convert the luminous spirit into atoms. Certain pages in Galileo's *Saggiatore* are the key to understanding this transformation in his thought.

---

[26] *Le Soleil à la renaissance.*

[27] Galileo, *Opere*, vol. V, p. 30: "un piccol parto, bisognoso d'esser ridotto a miglior forma, lambendolo e ripulendolo con affezione e pazienza, essendo solamente abbozzato e di membra capaci sì di figura assai proporzionata, ma per ora incomposte e rozze: se averò possibilità, l'anderò riducendo a miglior simmetria."

8.3 From a Spiritual to a Material Light: *Il Saggiatore*

The traditional account of Galileo's theory of matter points to *Il Saggiatore* as the main starting point of his atomism. In particular, it highlights the passage where Galileo made a reduction of sensible or secondary qualities to the interaction of small particles, defined and differentiated entirely in terms of size and motion, with the sense organs of animals. The traditional account recognizes that Galileo had already spoken of very small corpuscles, *minimi* or *ignicoli* in the *Discorso intorno alle cose che stanno in su l'acqua*. Nevertheless, it is not clear that in this work Galileo is referring to authentic indivisibles; it would be more accurate to hold that he had had material particles in mind that in fact could not be divided further.[28] This traditional reconstruction of Galileo's theory of matter, though, seems scarcely coherent with the explicit use of the word "atom," accompanied with the term "democriteo," both in the *Discorso* and in the defense that Castelli wrote on that occasion. I believe, instead, that we should rather hold that even though Galileo had already expressed his philosophical sympathies for atomism in these earlier writings, he had still not worked out the concept of the indivisible that would allow him to go beyond the pure explication of some natural phenomena in terms of material and mechanical causes. Such an analysis came to light for the first time in *Il Saggiatore*, a work where we may observe a problematic but crucial step in the development of Galilean atomism that years later will lead toward the mathematical formulation of the theory of indivisibles set out in the First Day of the *Discorsi*.

In *Il Saggiatore*, Galileo speaks of the minimal extended particles which make up all substances, while the specific concept of "atom" as the "ultimate resolution of matter," *indivisible*, is saved for the particles of light, thus stressing the difference between the particles of light and the extended particles, such as for example those that produce the sensation of heat. Between both kinds of particles there is a difference "in the same proportion of excellence that is found between the finite and the infinite, the temporal and the instantaneous, the extended (*quanto*) and the indivisible, light and darkness."[29] Atoms of light are indivisible, that is, without extension.

---

[28] Nonnoi, "Galileo Galilei: quale atomismo?" finds an example of Galileo's attitude in this sentence of Castelli in his defence of Galileo (in the *Errori di Giorgio Coresio*): "gli atomi sono così detti non perchè siano *non quanti*, ma perchè, sendo i minimi corpuscoli, non se ne danno altri minori da i quali possano essere divisi," Galileo, *Opere*, vol. IV, p. 281. See also Redondi, "Atomi, indivisibili e dogma."

[29] Galileo, *Opere*, vol. VI, pp. 351f.: "con quella proporzione d'eccellenza qual è tra il finito e l'infinito, tra 'l temporaneo e l'istantaneo, tra 'l quanto e l'indivisibile, tra la luce e le tenebre." I think that it is erroneous to attribute to Galileo an atomistic conception of the secondary qualities. When he speaks about tactile sensations, odors or flavors, it should be noted that he does not use the word "atom," but expressions like *minimi sottillissimi*, *minimi ignicoli*, *minimi del fuoco*, all of which refer to extended particles with specific figures and motions. Nothing is said about their indivisible nature. The extension of these corpuscles implies a defense of a classical atomistic theory of qualities and sense perception only if Galileo had conceived them (as he did in the *Discorso* when speaking about atoms of fire) as the smallest division of matter. But Galileo did not do so in *Il Saggiatore*. Therefore, it would be better to speak of a corpuscularian rather than atomistic explication of sense perception and secondary qualities.

This seems to be perfectly coherent with the assertion according to which the motion of fire particles is "temporally (temporaneo)" finite, while the diffusion of light is instantaneous:

> And perhaps when such attrition stops at or is confined to the minimal extended [particles] [*minimi quanti*], their motion is temporal and their action calorific only; but when their ultimate and highest resolution into truly indivisible atoms is arrived at, light is created. This may have an instantaneous motion, or rather an instantaneous expansion and diffusion, rendering it capable of occupying immense spaces by its—I know not whether to say its subtlety, its rarity, its immateriality, or some other property which differs from all these and is nameless.[30]

Certainly, this is one of the most mysterious passages in Galileo's writings, and it raises many questions. First of all: is it possible that matter, when it reaches its "highest resolution," stops being material by turning itself into a non-material being on account of the fact that it is without extension? Or was Galileo skeptical of the scholastic maxim that naturally ties the materiality of nature with extension? And secondly, in addition, how was the instantaneity of atoms of light to be interpreted in comparison with the "very swift" spreading of corpuscles of light in Galileo's correspondence with Castelli in 1612, or with the "very swift" substance of the letter to Dini in 1615?

Galileo had described light as an emission of material corpuscles both in the passages discussed earlier on sunspots and in the correspondence with Castelli. In the letter to Dini he had flirted with the definition of light as a spiritual and very swift substance, from which it was deduced that it was also extended and material. Nevertheless, when he had to deal with the problem raised by the non-penetrability of light in opaque bodies, while at the same time wishing to attribute to this luminous substance an animating force, he had to appeal to a distinction between light and the calorific spirit. However, to state that the calorific spirit is a spiritual, very subtle substance, does not get to the heart of the problem. It seems necessary to keep both obstacles and to explain the meaning of that subtlety, that spiritual nature. It seems to me that Galileo had certain sympathy towards some Neoplatonic cosmological ideas, but epistemologically, he professed to be convinced about the need for scientifically certain demonstrations. And these kinds of demonstration must be free of supposedly occult properties whose powers Renaissance Neoplatonic philosophers had made the central *explanans* of their cosmologies. A good solution that would allow saving the animating and vital force of light, without falling back on the obscure Neoplatonic definition of spirit as a substantial form, involved the mechanization of spirit. In addition, this way of proceeding might reach reconciliation with atomism, as long as its classical versions were suitably revised.

In *Il Saggiatore* there was no longer any allusion to the distinction between light and the calorific spirit. In this text, Galileo opts for a light that penetrates all bodies, overlooking the problems of opacity and transparency, to which he will return in the *Discorsi*. The absolute penetrability of light, which puts it on the level of spirit, is

---

[30] Ibid., vol. VI, p. 352, trans. in *Discoveries and Opinions of Galileo*, p. 278: "E forse mentre l'assottigliamento e attrizione resta e si contiene dentro a i minimi quanti, il moto loro è temporaneo, e la lor operazione calorifica solamente; che poi arrivando al'ultima ed altissima risoluzione in atomi realmente indivisibili, si crea la luce, di moto o vogliamo dire espansione o diffusione instantanea, o potente per la sua, non so s'io debba dire sottilità, rarità, immaterialità, o pure altra condizion diversa da tutte queste ed innominata, potente, dico, ad ingombrare spazii immensi."

explained here by its composition out of indivisibles *non quanti*, without extension. The definition of the indivisibles of light as non-extended entities permits Galileo to do without the distinction between light and a "calorific spirit." The spiritual substance of the letter to Dini is now light itself, and it is translated into the language of indivisibles. Nevertheless, this description of the nature of light has a consequence that conflicts with certain ideas contained both in the letter to Dini and in the passages on sunspots, in particular, what Galileo says there about the velocity of light. If the indivisibles that make up the luminous substance do not have extension, then they are not corporeal, and therefore they could not have velocity. In fact, Galileo says in *Il Saggiatore* that light is "an instantaneous motion, or rather an instantaneous expansion and diffusion (*espansione e diffusione instantanea*)." Note here the precision of Galileo's language when he suggests that the word "motion (*moto*)" is not adequate to make reference to a non-extended indivisible. Although Galileo recognized explicitly that this description of the nature of light could lead him to assert its immateriality ("*non so s'io debba dire sottilità, rarità, immaterialità*"), he did not seem willing to renounce its material nature, especially after the great effort he had made to offer an explanation of spiritual substance in terms of the highest resolution of material bodies. Overcoming this problem required developing the atomistic theory, which he had not yet developed; he would reveal it in the First Day of the *Discorsi*, which led him to conceive spatial and material magnitudes compounded of an infinity of an non-extended indivisibles (*infiniti atomi non quanti*). Nevertheless, in the period when he was writing *Il Saggiatore*, Galileo could only imagine the later consequences of his own evolution. If years earlier, in the period of the letter to Dini and the polemic on sunspots, he had to admit that if something is material its velocity must be finite, now, in *Il Saggiatore*, saying that light was the ultimate resolution of matter and that its atoms were not extended led him to conclude that its diffusion was instantaneous. What remained was explaining how corporeal magnitudes could be composed of infinite atoms *non quanti*, and how temporal magnitudes could be composed of "infinite instants." It would be then when he would return to the problem of the velocity of light.[31]

The connection and continuity between the atomic composition of light outlined in *Il Saggiatore* and the arguments of the *Discorsi* on the same issue becomes clear thanks to Galileo's rhetorical strategy, about which very little comment has been made. In *Il Saggiatore*, immediately after the passage about indivisibles of light and their instantaneous diffusion, the passage on light closes with the following words:

> I do not wish, Your Excellency, to engulf myself inadvertently in a boundless sea from which I might never get back to port, nor in trying to solve one difficulty do I wish to give rise to a hundred more, as I fear may have already happened in sailing but this little way from shore. Therefore I shall desist until some more opportune occasion.[32]

[31] Ibid., vol. VIII, p. 201.

[32] Ibid., vol. VI, p. 352, trans in *Discoveries and Opinions of Galileo*, p. 279, italics are mine: "Io non vorrei, Illustrissimo Signore, *inavvertentemente ingolfarmi in un océano infinito*, onde io non potessi poi ridurmi in porto; nè vorrei, mentre procuro di rimuovere una dubitazione, dar causa al nascerne cento, sì come temo che anco in parte possa essere occorso per questo poco che mi sono scostato da riva: però voglio riserbarmi ad altra ocasion più oportuna."

The *Discorsi*'s dialogue about the velocity of light ends with these lines:

> But in what seas are we inadvertently engulfing ourselves, bit by bit? Among voids, infinites, indivisibles, and instantaneous movements, shall we ever be able to reach harbor even after a thousand discussions?[33]

The occasion Galileo for which was waiting in 1623 seems to have arrived on the First Day of the *Discorsi*.

8.4 The Geometrical Matter of the Spiritual Light

At the beginning of 1638 the *Discorsi* were ready for print. Fifteen years had passed between the two passages discussed above during which many things had happened, both in Galileo's life and in the larger world.

In 1637 Descartes published the *Discourse on Method* and his three famous works on optics, meteorology, and geometry. The treatise on optics was the result of the researches he had carried out starting in 1625 and which, in a way, reflected the polemics and scientific exchanges that had been and were to be produced in this field among such principal scientists of the period as Constantijn Huygens (Christiaan's father), Isaac Beeckman, Pierre Gassendi, Ismael Boulliaud, Marin Mersenne and Thomas Hobbes. The mathematical approach to the study of the phenomena of light that caused them so many problems did not dispel speculations concerning the physical nature of light. The tendency to conceive light as a mechanical phenomenon, produced by the exchange and transmission of motion among material particles, became clearer and clearer.

Meanwhile the critics of the Aristotelian qualitative view were progressively put in a corner. By the 1640s, the main problem was to elucidate whether light was a fluid, that is, an emission of particles moving in the void or any other very subtle fluid, or an effect produced by the transmission of motion in a continuous medium. Both the conversion of Hobbes from his theory of light as the emission of particles to a theory of the continuous medium and the Cartesian conception of light as an instantaneous transmission of motion in a continuous medium perfectly reflected the new state of research in optics.[34] These two authors laid the foundations of modern research on light. Yet it would be misleading to think that there was only one dominant view in the period. Besides Hobbes or Descartes, we find defenders of very different views, like for example van Helmont, determined to support the semi-substantial nature of light[35]; the biblical alchemical animism of Robert Fludd, ever entangled with the mystical terminology of light and darkness; a generation of new chemists who began to flirt with corpuscularian doctrines and thought that light was a material, active and vital

[33] Ibid., vol. VIII, p. 89, trans. in *Two New Sciences*, p. 51: "Ma in quai pelaghi ci andiamo noi inavvertentemente pian piano ingolfando? tra i vacui, tra gl'infiniti, tra gli indivisibili, tra i movimenti instantanei, per non poter mai, dopo mille discorsi, giugnere a riva?"

[34] Giudice, *Luce e visione*.

[35] See the letter from van Helmont to Mersenne, 15 January 1631, in *Correspondance*, vol. III, pp 34f. About the semi-corporeal nature of light defended by some Paracelsians and Helmontians, see Clericuzio, *Elements, Principles and Corpuscles*, pp. 61f., 90–92, 180f.

principle of extreme subtlety, a thought which surfaced many times in esoteric debates about its semi-corporeal nature[36]; semi-materialistic interpretations developed by authors well integrated in the scientific circles of the period, such as Ismael Boulliaud, who in his *De natura lucis*, wrote "light is the middle proportion between substance and accident;"[37] calculations and speculations about the velocity of light compared with that of angels, as that described to Mersenne by his correspondent Gabriel Thibaut.[38] Even Mersenne, ever well informed about scientific innovation, in some of his works found it difficult to accept completely the mechanistic view of light.[39]

Talking about the complex map of the conceptions of light, we shouldn't forget the Jesuits. Although the great majority of their works on optics took place from the 1650s, presumably in reaction to the confrontation with atomism and mechanistic views, they had already expressed their insistence on denying the materiality of light throughout the first half of the century.[40] Gassendi deserves a special mention,

---

[36] It suffices here to recall Theodore Deschamps, William Davidson, and some years after, in the 50s, Helmontians as Noah Biggs. See Clericuzio, *Elements, Principles and Corpuscles*, pp. 53, 61, 91.

[37] The book had already been finished in 1635 (then under the title *De lumine et coloribus*). In November of that year, Boulliaud sent the manuscript to Peiresc, who gave it to Gassendi. It was published in Paris in 1637, and on 30 October of that year, Boulliaud sent it to Galileo. It may be inferred from his answer (dated 1 January 1638) that he did not like it very much, Galileo, *Opere*, vol. XVII, pp. 207f., 245f. On this issue he agreed with Descartes; see Descartes' letter to Mersenne, 11 October 1638, in Mersenne, *Correspondance*, vol. VII, p. 11.

[38] Letters from 25 January 1647 and 24 May 1648, in Mersenne, *Correspondance*, resp. vol. XV, pp. 163–164 and vol. XVI, pp. 329f.

[39] While Mersenne's ideas on light changed over the years, his point of view was never completely clear. At the beginning of his scientific career, in his *Quaestiones in Genesim*, he defended the scholastic view defining light as a visible quality. He reintroduced the question in the *Harmonie Universelle*, I, prop. IX, p. 17, and he included the *Tractatus opticus* of Hobbes in his *Universae geometriae mixtaeque mathematicae synopsis*. In his *Cogitata*, he asserted that light was a motion of the ether or subtle matter. See Mersenne, *Correspondance*, vol. XIV, p. 681, note 2.

[40] The main source for the Jesuit view of light is found in the *Commentarii Collegii Conimbricensis In quatuor libros de coelo*, chap. VII. This text discussed the different conceptions of light and their criticisms, and concluded with a clear reluctance to accept the materiality of light, with the defense of its instantaneity, and with the distinction between a divine light (with a clear symbolic value) that pertains to "glorious bodies" and natural light. It is clear that there are cosmological reasons for maintaining these ideas connected with the defense of the Aristotelian distinction between celestial and terrestrial worlds, stressing the incorruptibility of the heavens. Just one of the central points arguing against the materialistic conception of light is the fact that celestial bodies would become depleted and would be extinguished if their light were a material emission, ibid., p. 381. Among the Jesuits who studied the nature of light from the late 1640s on, we find Zucchi, *Optica philosophica experimentis et ratione constituta*; Maignan, *Cursus philosophicus*; id., *Perspectiva horaria, sive de horographia, gnomonica libri*; Kircher, *Magnes, sive de arte magnetica*, id., *Ars magna lucis et umbrae*; Schott, *Magiae universalis naturae, & artis. Partem priman opticam, sive thaumaturgum opticum*; id., *Physicam curiosam, sive mirabilia naturae*; [Eschinardi], *Microcosmus physico-mathematicus*; id., *Dialogus opticus*; id., *Centuria problematum opticorum, seu Dialogi optici*; Fabri, *Dialogi de lumine;* id., *Physica, id est scientia corporearum*, id., *Synopsis optica*; Traber, *Nervus opticus sive tractatus theoricus, in tres libros opticam, catoptricam, dioptricam distributus*. Exceptions to the general trend of Jesuits on this problem are Étienne Noël, a French Jesuit who in his *Physica vetus et nova* (1648) defended a materialistic theory of light, and Grimaldi, *Physico-mathesis de lumine* (1665), who though he did not actually assert the material composition of light, developed his arguments as if it were so, using an analogy with fluids.

because although the publication of the first version of his atomism took place after Galileo's death (the *Animadversiones* were published in 1649), it is he to whom the atomists among Galileo's followers will largely refer in highly favorable terms.[41] In Galileo's correspondence, there is no trace of any exchange with Gassendi about the structure of matter and the nature of light, but we have to remember that Gassendi (1592–1655) was much younger than Galileo. Only in 1640, when it was perhaps too late for the Pisan scientist, did Gassendi take part in the polemic that had broken out between Fortunio Liceti and Galileo as a result of the publication of the *Litheosphorus sive de Lapide Bononiensi*, a debate which directly touched the problem of the nature of light.[42] Gassendi was one of the main defenders of the corporeal and material nature of light; he conceived it as a very rarefied substance, a kind of very tenuous fire composed of particles that spread in the air or even in the void. Furthermore, if light was an emission of corpuscles or atoms, and not the transmission of a motion, then it was necessary to conclude that its velocity is not instantaneous.[43] No less important was the fact that in Gassendi's particular version of atomism, atoms are conceived of as material particles with an internal active principle, opposing strict mechanical philosophy and helping to reconcile the modern materialistic interpretations of nature with Renaissance naturalistic philosophies. Although, as already noted, the publication of Gassendi's works occurred after the death of his admired Galileo, a letter he wrote in 1635 shows that both men shared a problem about the nature of light: the difficulty reconciling the mathematical and physical approaches on the assumption of the atomic nature of light. In that letter, Gassendi was reportedly very concerned about the apparent contradiction between a geometrical optics that spoke of non-extended points and a physical optics that conceived light as a compound of spatial atoms.[44]

While this rich variety of ideas, debates, philosophical and epistemological approaches were in the course of developing, Galileo was attempting to resolve the conflicts that decisively conditioned both his work and the larger Italian scientific world. The publication of *Il Saggiatore*, under the patronage of the Accademia dei Lincei, came at the same time as the ascent to the papal throne of Urban VIII, a papal lover of sciences and arts who seemed ready to open the doors to the new scientific and philosophical ideas of the period. It was a moment of optimism for those intellectuals who, after the anti-Copernican decree of 1616, had been constrained to stay within the limits of extreme prudence. In *Il Saggiatore*, Galileo himself did not speak explicitly about the centrality of the Sun and the movement of the Earth, although it is clear that both his interpretation of the nature of comets and his defense of their rectilinear path were only compatible

---

[41] The bibliography on Gassendi is immense and is updated in Murr (ed.), *Gassendi et l'Europe*. See also Bloch, *La philosophie de Gassendi*. On Gassendi's influence and the differences between his matter theory and strictly mechanistic philosophies, see Clericuzio, *Elements, principles and Corpuscles*, pp. 63 and ff.

[42] The correspondence of Gassendi on this issue is in Galileo, *Opere*, vol. XVIII.

[43] Gassendi, *Opera Omnia*, lib. VI, chap. 11, vol. I, pp. 422f.

[44] Letter to Mersenne, 13 December 1635, in Mersenne, *Correspondance*, vol. V, pp. 532–537.

with a moving Earth.[45] However, he was very confrontational with respect to the Jesuits, arguing astutely and with a characteristically ironic and at times arrogant tone. The book immediately kindled the wrath of the Jesuits. Galileo underestimated their irate reaction and their inevitable revenge as he made ready to continue the great cosmological reform he had started with the discovery of the Medicean stars and stopped—at least publicly—with the decree of 1616. In 1624 he decided to use the already projected "Dialogo sopra il flusso e riflusso" again to reply to Ingoli. Both Ingoli's works, *De motu et quiete Terrae disputatio* (1616) and the *Replicationes ad J. Kepleri impugnationes* (1618) had been written to be standard bearers for the anti-Copernican crusade. Galileo answered them with a "Letter to Ingoli" (1624), where he not only defended the motion of the Earth, but also made many comments about the infinite, the center (or centers) of the world and gravity.[46] The "Letter" was never printed because when, in April of 1625, as soon as Galileo was ready to give it to the printer, a letter from Mario Guiducci informed him that an official of the Holy Office Congregation intended to prohibit or correct *Il Saggiatore*, accusing it of defending the motion of the Earth.[47] This letter from Guiducci gave rise to the thesis of Redondi, according to which the true danger that the Holy Office and the Jesuits felt in Galileo's ideas, that which triggered off many of the vicissitudes which led to the condemnation of 1633, was his support of an atomistic theory of matter openly in conflict with the dogma of the Eucharist, or more exactly, in conflict with the Aristotelian-scholastic grounding of the sacrament, as mandated by the Council of Trent.[48] The Aristotelian distinction between substance and accident made it possible to hold that the accidents of the sacred host (color, flavor and other appearances that pertain to bread) remain after consecration, while at the same moment, transubstantiation had taken place, and bread is no longer bread but Christ's body. Atomism undermines this explanation because it identifies accidents with substance; that is to say, it asserts that accidents (the so-called secondary qualities) do not exist *per se*, but that they are only sensible qualities that result from the shape, size and motion of the material particles. Therefore, if accidents remained, it would be impossible for any change of substance to have occurred. According to Redondi, Guiducci's letter really referred to this problem as deriving from the passages Galileo wrote in *Il Saggiatore* about the corpuscularian explanation of sensations, as we discussed earlier. This seems to be confirmed by the discovery in the *Archivio delle Sacra Congregazione per la dottrina della Fede* of an anonymous document asking for an ecclesiastical report concerning the compatibility or incompatibility between Galileo's view on secondary qualities and the Eucharist. Much had been written on Redondi's thesis, and although it is true that most historians agree in denying that the principal problem that Galileo had with the Inquisition was atomism and not his defense of the motion

[45] Besomi and Camerota, *Galileo e il Parnaso Tychonico*, especially pp. 1–31; Solís, "Los cometas."

[46] Bucciantini, *Contro Galileo*.

[47] Guiducci to Galileo, 18 April 1625, in Galileo, *Opere*, vol. XIII, p. 265.

[48] Redondi, *Galileo eretico*.

of the Earth, nearly all of them, and I include myself among these historians, recognize the value of Redondi's thesis for having raised the theological issues surrounding the rehabilitation of atomism in the seventeenth century.[49]

The period of renewed optimism for the defenders of the new science opened with the papacy of Urban VIII, and the publication of *Il Saggiatore* in 1623 concluded in 1632 with one of the most important achievements of the baroque science: the publication of the *Dialogo sopra i due massimi sistemi*. But it also ended with its bitter aftertaste: Galileo's complex trial and the condemnation of 1633. Throughout those years it does not seem that the theory of matter occupied a privileged place in Galilean thinking, very busy defending the motion of the Earth and replying to the objections to his cosmological view put by Aristotelian cosmology and physics. The ban on speaking "*amplius quovis modo de mobilitate terrae nec de stabilitate solis*" and the house arrest in Arcetri were well used by Galileo for returning to and developing many physical questions that had occupied him in his youth. The result was the *Discorsi intorno a due nuove scienze* in 1638, the First Day of which became one of the key elements of the debates on matter theory that signaled the origins of modern science.

Despite suggestions concerning the philosophical and theological danger of defending atomistic theories, there is no evidence that Galileo withdrew from atomism, nor is there any reason to suppose, in any case, that such was his desire. We only know that after 1623, he opted for public silence and that when he returned to the subject (firstly in the *Postille alle Esercitazioni filosofiche* of Rocco in 1634 and years later in the First Day of the *Discorsi*), his matter theory no longer resembled the atomism of Democritus or Epicurus to which he had been so close in earlier years, especially in the period of the *Discorso* on floating bodies. In the *Discorsi*, Galileo defended a particular version of atomism based on the concept of extended magnitudes composed of infinite non-extended indivisibles, a kind of atomism that seems closer to certain geometrical atomisms of the medieval tradition than to the pagan Democritean doctrines. In *Il Saggiatore*, geometry was made conspicuous by its absence while physical atoms were defended. In the *Discorsi*, geometrical arguments were the main protagonists of a theory of matter difficult to apply to the explanation of real physical phenomena. One might think, then, that this change was due to the suspicion of heterodoxy that fell on atomism, especially if we remember that for the Aristotelian point of view speaking in mathematical terms did not necessarily imply speaking about the physical realm.[50] If the Copernican doctrine

---

[49] This is not the place to give a complete account of the different views on Redondi's thesis. I refer to two examples: Bucciantini in his *Contro Galileo* and some passages of Camerota, *Galileo*, pp. 389 and ff. The discovery of a new document in 1999 in the Archives of the Holy Office has reopened the polemic on this question. See Martínez, "Il manoscritto ACDF"; Artigas, "Un nuovo documento sul caso Galilei"; Mateo-Seco, "Galileo e l'Eucaristia"; Shea, "Galileo e l'atomismo."

[50] The Society of Jesus prohibited the teaching of atomism in their schools in 1632. It is significant that only a year later, the Aristotelian Antonio Rocco in the *Esercitazioni filosofiche* advanced a criticism of atomism, which Galileo answered in manuscript notes, where he had already defended his concept of bodies composed of infinite non-extensive indivisibles. See Galileo, *Opere*, vol. VII, pp. 682f., 745f., and Redondi, "Atomi, indivisibili e dogma"; Nonnoi, "Galileo Galilei: quale atomismo?"

could be maintained on the condition that it was used exclusively as a mathematical hypothesis, as Galileo has been asked to do, why not do the same with atomism, putting it on the plane of the geometrical speculation without implicating the problems of a physics of Democritean origins? I do not exclude the possibility that theological and ecclesiastical considerations had a certain weight in the transition from the atomism of *Il Saggiatore* to the theory of matter developed in the *Discorsi*, but I do not think, in any case, that this change should be understood as an abandonment of physical atomism (insofar as it referred to the physical realm of nature) in favor of a purely mathematical atomism. Nor can it be seen as the result of a renunciation of some ideas he had defended earlier, motivated by dogma. Regardless of the evident impact of anti-atomistic currents that derive from theological considerations, Galileo had to face up to some purely philosophical and scientific problems, problems that required a further development of his original atomistic and Democritean sympathies. The main question arose when Galileo characterized the particles of light as indivisibles in the passage of *Il Saggiatore* discussed above; even more, when he had suggested that he conceived them as non-extended entities, from which it follows that their propagation has to be instantaneous. Certainly, it was a conception of indivisibles incompatible with the Democritean idea of solid and spatial atoms, a conception scarcely useful to elaborate a mechanical interpretation of natural phenomena. Therefore, it was just necessary to re-elaborate a new version of atomism, a new physical-mathematical atomism.

In contrast with the view according to which Galileo rejected the physical version of atomism he advocated in *Il Saggiatore* and defended instead a non-physical mathematical theory of indivisibles in the *Discorsi*, I believe that one can establish some continuity between those two seemingly different works with regard to this issue. Specifically, I claim that there was not a radical rupture in Galileo between a physical and a purely geometrical atomism, and that the indivisibles of the *Discorsi* were not essentially different from the atoms of *Il Saggiatore*. What Galileo developed in his last work was a necessary revision of a kind of atomism with which he wanted to explain the problems of rarefaction and condensation, while eluding at the same time the obstacles represented by the penetrability of bodies and the infinity of the world. The fact that in 1623, 15 years before the publication of the *Discorsi*, Galileo limited the idea of indivisibility to light particles, thus permitting light to spread instantaneously throughout "immense spaces," makes credible Galileo's own declaration in a letter to Fulgenzio Micanzio in 1634 that in 1616 he had already begun to think about a new physic of matter.[51] In that letter, Galileo made reference to a note

[51] "La nota del nostro q. comun padre e maestro poteva esser circa la condensazione e rarefazione, come punti da me più tosto stimati difficilissimi che resoluti, non vi havendo in quei tempi altro che difficoltà; ma ben poi circa 18 anni sono, ritrovandomi alla villa con il Salviati del Dialogo, mi cadde nella mente una mattina, mentre eramo a messa, un pensiero, nel quale poi più profondamente internandomi, mi vi son venuto confermando, et a me è parso poi sempre ammirando come per modo stupendo di operar della natura, secondo il qual modo (e credo in nessun altro) si possa distrarre e rarefare una sustanza in immenso senza ammettere in essa veruno spazio vacuo, et all'incontro in immenso condensarla senza alcuna penetrazione di corpi: pensiero, credami, assai peregrino, il quale insieme con moltissime altre novità spero che ella vedrà sparse nelle opere

by Paolo Sarpi, famous and persecuted defender of atomism with whom he had shared ideas and discussions from the time they first met in Padua in 1600. The "very difficult" problems to which he alluded were condensation and rarefaction, two crucial phenomena in the debate between Aristotelians and atomistic natural philosophers. According to the atomists, rarefaction is explained by way of the spatial voids existing among atoms—spatial voids that the Aristotelians did not accept in their plenist physics. Furthermore, to assert that condensation is explained in terms of an increase of matter with respect to volume, but without accepting inter-atomic voids, implied the acceptance of the penetrability of bodies, which Aristotelians also denied.[52] Starting from this problem, Galileo sets out to elaborate a version of atomism that overcomes Aristotelian objections, that is to say, an atomism that could maintain the definition of rare and dense in terms of matter and volume but without accepting either the existence of extensive void spaces or the penetrability of substances. The concept of atoms of light spreading instantaneously through the universe already pointed in this direction, but Galileo still needed a theory of the continuum and indivisibles that gave a satisfactory answer and that could be generalized to all material substances. As he said in a letter to Micanzio in 1634, the idea he had while at mass in 1616 had to be developed and improved throughout the years before finalizing the matter theory we read in the *Discorsi*. And along the way, and due to that fact, some details changed, among them the conception of the motion of light as being either instantaneous or temporal.

Therefore, it is no great surprise that in 1638 Galileo ended the passage of the *Discorsi* about light by repeating exactly the same words with which he had closed his speculation on light 15 years earlier: "But in what seas are we inadvertently engulfing ourselves, bit by bit? Among voids, infinites, indivisibles, and instantaneous movements, shall we ever be able to reach harbor even after a thousand discussions?"[53] These words closed the dialogue about light in the First Day of the *Discorsi*. Those pages led some of Galileo's followers to think that he had given up, and not extended his atomism to the case of light but limited himself only to speculate about its velocity. In fact, the conversation between the three protagonists of the dialogue started with an assertion by Salviati about the light's power

che mi restano da mandar fuora, le quali penso di ridurre al netto in questa vernata per mandarle poi alla P. V., acciò ne faccia il suo volere," Galileo to Fulgenzio Micanzio, 19 November 1634, in Galileo, *Opere*, vol. XVI, pp. 162f. In the winter following that letter, Galileo sent Micanzio the draft of the First Day, as can be inferred from a letter of Micanzio. In a missive of 3 February 1635 he wrote: "Ho ricevuti altri tre fogli del Dialogo, in tutto 6; li ho anco letti con l'avidità che non posso esplicare: ho necessità di meditarli a verso per verso. La novità delle cose, le ragioni e demostrationi di problemi non più sentiti, mi mettono in un nuovo mondo. L'intento mio mi portava tutto al punto della rarefattione e condensatione, ma m'accorgo che non ci si può ben arrivare che per li passi precedenti." And on 10 February 1635, he made clear the relation of that draft to his *Postille* to the work of A. Rocco, see ibid., vol. XVI, p. 203 and pp. 208f.

[52] Aristotle, *Physics*, IV 9, 216b25–217b30.

[53] Ibid., vol. VIII, p. 89, trans. in *Two New Sciences*, p. 51: "Ma in quai pelaghi ci andiamo noi inavvertentemente pian piano ingolfando? Tra i vacui, tra gl'infiniti, tra gli indivisibili, tra i movimenti instantanei, per non poter mai, dopo mille discorsi, giugnere a riva?"

to liquefy some substances: "I cannot believe that the action of light, however pure, can be without motion, and indeed the swiftest." In response to this the Aristotelian Simplicio answered: "Daily experience shows the expansion of light to be instantaneous."[54] Following on this was Galileo's famous experiment for calculating the velocity of light. It is obvious that the rudimentary character of his proposed experiment could not succeed in establishing a definite value, but Galileo certainly did not doubt that the motion of light was *in tempore*. When Galileo had said in *Il Saggiatore* that the smallest parts of light are indivisible, he differentiated them from spatial *minima* with finite velocity. Atoms of light were non-spatial and therefore, so he then thought, their velocity had to be instantaneous. In the *Discorsi* he continued to defend the idea that all motion requires space, but with the fundamental difference that in the latter work he had managed to demonstrate how all spatial and material magnitudes are composed of non-spatial, but material, indivisible parts. And light was nothing but a spatial fluid composed of infinite non-extended atoms. To this we must add that Galileo applied to temporal magnitudes the same arguments he employed to assert that spatial bodies are composed of *infiniti atomi non quanti*. In any temporal interval, there are infinite instants, and the velocity of any body is the sum of its instantaneous velocities. It was therefore possible to attribute velocity to the indivisibles composing the material fluid of light.[55] This helps us to understand Galileo's change of opinion about the velocity of light between *Il Saggiatore* and the *Discorsi*. It also shows continuity between those two apparently different works: it proves that the indivisibles of the *Discorsi* were not essentially different from the atoms of *Il Saggiatore*. In other words, it demonstrates that there was no rupture between a physical and a purely geometrical atomism. Galileo had thus revised the conception of atomism needed to explain the problems of rarefaction and condensation, while at the same time eluding the obstacles raised by the penetrability of bodies and the infinity of the world.[56] Therefore, it is not surprising that in the 1638 book, Galileo ended that passage repeating exactly the same words with which he had closed his speculation on light 15 years earlier.

---

[54] Galileo, *Opere*, vol. VIII, p. 87, trans. in *Two New Sciences*, p. 49: "Io non saprei intendere che l'azzione della luce, benché purissima, potesse esser senza moto, ed anco velocissimo.... mostra la esperienza quotidiana, l'espansione del lume esser instantanea."

[55] Galileo followed almost literally the criticism Aristotle made to Zeno's paradoxes in his *Physics*, VI, 232b and ff., intending to deny the possibility of motion in an infinitely divisible continuum space. But in this case, he wanted to demonstrate the possibility of motion in a space composed of infinite indivisibles. Aristotle, assuming that "time and space are divided according to the same rules and divisions," refuted Zeno's paradoxes against motion, asserting that both time and space are infinitely divisible continuous magnitudes. If Galileo wanted to assert the possibility of motion in a continuous medium composed of *infiniti atomi non quanti*, he had to do the same with time. He did just that in the Third Day of the *Discorsi*.

[56] This is how Galileo expounded it in the previously mentioned letter to Micanzio in 1634, and repeated it using almost the same words in the *Discorsi*, just after the passage on light: "il comprender come stia il negozio della rarefazzione e della condensazione, senza incorrer per causa di quella nell'inconveniente di dovere ammettere spazii vacui, e per questa la penetrazione de i corpi: inconvenienti, che amendue mi pare ch'assai destramente vengano schivati con l'ammetter detta composizione d'indivisibili," Galileo, *Opere*, vol. VIII, p. 93.

These considerations also prove that for Galileo to speak about the velocity of light did not mean that he overlooked the problem of its atomic composition, because both questions were interrelated. This becomes clearer if we look at the kind of problems that the three protagonists of the *Discorsi* were discussing before and after those brief pages on light: namely, the composition of fluids, condensation and rarefaction. In fact, after describing the diffusion of light as a phenomenon of expansion, Galileo went on in the First Day treating the questions of condensation and rarefaction.[57] His aim there was—as it always had been—to explain materialistically the rarefaction of substances. Classical atomism was his best option, but it raised a serious problem because rarefaction took place through the proliferation of void spaces among spatial atoms. Moreover, one of the principal criticisms of atomism was precisely that the existence of such extended voids could lead to an acceptance of an infinite universe. Galileo's reticence regarding this possibility (or his fear of being accused of defending an infinite universe) was clear in some passages of the First Day, where he stressed that both atoms and voids had to be necessarily *non quanti*.[58] However, I think that Galileo, subtly and rhetorically, had already expressed this fear in the passage of *Il Saggiatore* discussed earlier, where, after speaking of a luminous substance that spreads through immense spaces in reaching its highest degree of rarefaction (*rarità*), he said: "I do not wish, Your Excellency, to engulf myself inadvertently in a boundless sea." Was this not precisely the infinite ocean, the universe of infinite atoms described by Lucretius in his *De rerum natura*? The new theory of indivisibles expounded in the *Discorsi* overcame that problem.

[57] "Veramente non l'ho sperimentata, salvo che in lontananza piccola, cioè manco d'un miglio, dal che non ho potuto assicurarmi se veramente la comparsa del lume opposto sia instantanea; ma ben, se non instantanea, velocissima, e direi momentanea, è ella, e per ora l'assimiglierei a quel moto che veggiamo farsi dallo splendore del baleno veduto tra le nugole lontane otto o dieci miglia; del qual lume distinguiamo il principio, e dirò il capo e fonte, in un luogo particolare tra esse nugole, ma bene immediatamente segue la sua espansione amplissima per le altre circostanti; che mi pare argomento, quella farsi con qualche poco di tempo; perché quando l'illuminazione fusse fatta tutta insieme, e non per parti, non par che si potesse distinguer la sua origine, e dirò il suo centro, dalle sue falde e dilatazioni *estreme*," ibid., vol. VIII, pp. 89 and 104f.

[58] The effort to resolve the problem of condensation and rarefaction without admitting either infinite void spaces or the penetrability of bodies reflected Galileo's effort to overcome two of the principal criticisms that Aristotelians launched against atomism, because as he himself had said in the *Discorsi*, he really did not have any problem accepting void spaces and penetrability, ibid., vol. VIII, p. 105. He expressed the same in his correspondence with Baliani, who after reading the *Discorsi*, wrote to Galileo about his suspicions in view of the subtle strategies Galileo used to elude void spaces and penetrability. Baliani, who was a convinced atomist and did not have any problem accepting the existence of void spaces, and did not see any obstacle in also accepting their penetrability. Galileo replied by saying that he, too, had no problem with them, but he was trying to please the Aristotelians: "Quanto a quello che ella mi dice della opinione sua circa alla condensazione e rarefazione, cioè che ammette la penetrazione dei corpi l'uno con l'altro, già ho io scritto (come ella può vedere) che chiunque tale operazione volesse ammettere, io gli concedo quanto li piace, non havendo io hauto intenzione di scrivere quanto in tal proposito ho scritto se non in grazia di quelli che negano la penetrazione e gli spazii vacui potersi dare in natura," 1 September 1639, ibid., vol. XVIII, pp. 93–95. See also the letters from Galileo to Baliani of 1 August and from Baliani to Galileo of 19 August 1639, in ibid., vol. XVIII, resp. pp. 75–79 and 86–88.

Only then, with this new sort of atomism, could it be explained that matter, in reaching its highest degree of rarefaction and being converted into indivisibles, became a luminous substance able to "occupy immense spaces (*imgombrare spazii immensi*)." But it did so without having to support an infinite universe. Furthermore, it was a way to deal with the doubts one might have had 15 years earlier about Galileo with respect to the materiality of light insofar as it is composed of non-extended atoms, a reasonable doubt if one believes that in nature there is no matter without extension, however conceptually independent they may be.

As regards penetrability, we have seen that the question was already present in the letter to Dini in 1615, where Galileo had shown his ambiguity in making a vague mention of the subtlety of the spiritual substance. With the non-extended atoms of *Il Saggiatore* the question was partially resolved, but at the risk of denying the materiality of light. With the new atomism of the *Discorsi* Galileo seems to have found a solution. If light is composed of a perfect fluid of non-spatial atoms, it could penetrate into all dense bodies and dissolve them without falling into the problems raised by the penetrability. In other words, light could penetrate into all bodies as spirit does; however, now this spirit is really corporeal and material, neither a "corporeal form" nor a "substantial form" as angels and light were in Neoplatonic philosophies and in some Jesuit works.

Chapter 9
Causation in Descartes' *Les Météores* and Late Renaissance Aristotelian Meteorology

Craig Martin

9.1 Introduction

Over 70 years ago, Étienne Gilson showed the parallels between Descartes' *Les météores* and the Coimbrans' textbook that was based on Aristotle's *Meteorology*. The topics treated in Descartes' work follow those found in the frequently-taught Jesuit textbook. They both discussed the formation of clouds, rain, rainbows and other lights in the sky, minerals and salts, and the cause of winds and earthquakes.[1] The similarities do not end at the structure and topics treated that Gilson pointed out but extend to large portions of the treatises' content. To be sure, differences appear, but many Aristotelian meteorological concepts are found throughout Descartes' treatise without being changed at all or only in a minor way. Descartes' *Les météores* was neither revolutionary, nor was it intended to be revolutionary.

Les météores was first published in 1637 together with the *Discours de la méthode, La géométrie,* and *La dioptrique.* Some recent studies on Descartes' physics and the *Les météores* have emphasized the eighth discourse where Descartes explained the rainbow through a geometrical analysis of refraction.[2] This emphasis has given the appearance that his study on meteorology was part of Descartes' larger goal of applying mathematics to natural philosophy as he did in the accompanying *La dioptrique*. Other studies, however, have emphasized the physical aspects of his account on the rainbow and how the deductive method Descartes used relies on observation and experience.[3] Moreover, while Descartes appeared to

[1]Gilson, "Météores cartésiens et météores scolastiques," pp. 102–137.

[2]For example, Gaukroger, *Descartes' System of Natural Philosophy*, pp. 25–28.

[3]Garber, "Descartes and Experiment," pp. 94–104.

C. Martin (✉)
History Department, Oakland University, Rochester, MI, USA
e-mail: martin@oakland.edu

D. Garber and S. Roux (eds.), *The Mechanization of Natural Philosophy*, Boston Studies in the Philosophy of Science 300, DOI 10.1007/978-94-007-4345-8_9,

be rightfully proud of his treatment of the rainbow, it should be kept in mind that this discourse is meant to be part of natural philosophy, not mixed mathematics, and is much different from the previous seven discourses, which rely on descriptive accounts of the movements of corpuscles.[4] Concentrating on the problem of the rainbow distorts the meaning of the entire treatise because most of *Les météores* is not an attempt to ground meteorology on mathematics but rather is a discussion of how a wide range of sublunary phenomena might be explained using only matter and local motion.

In the first discourse of *Les météores*, Descartes announced his method for this subject. He would explain meteorological phenomena without recourse to substantial forms or real qualities, not because he denied their existence *tout court,* but because they were superfluous. He wrote:

> Then, know also that in order to keep my peace with the philosophers, I have no desire to deny that which they imagine to be in bodies in addition to what I have given, such as their *substantial forms,* their *real qualities* and the like; but it seems to me that my explanations ought to be approved all the more because I shall make them depend on fewer things.[5]

Here Descartes' unwillingness to reject outright the existence of substantial forms and real qualities was likely a matter of delicacy. In a 1642 letter to Regius, Descartes suggested that this tactic was meant to illustrate that these concepts were of no use without incurring the anger of Regius's colleagues by directly arguing against their existence.[6] Thus *Les météores* played a role in his attempt to eliminate substantial forms from physics.

This elimination is widely regarded as central to Descartes' critique of scholastic natural philosophy; for example, in *Le monde* and in the *Principia* he similarly suggested that matter and motion sufficiently account for the natural world.[7] Moreover, the reliance on matter and motion is a hallmark not only of Descartes' physics but an identifying feature of the new natural philosophies of the seventeenth century. What is not recognized, however, is that the removal of substantial forms and final causes was already common in some Aristotelian treatises on meteorology. Granted, in *Le monde* Descartes attempted to root out teleology and formal causation in all of physics, while Aristotelians were more likely to remove them just from the field of meteorology, leaving these kinds of explanation intact for psychology, biology, and other fields. Aristotelian commentaries were diverse; and it cannot be assumed that they slavishly followed Aristotle. For example, one contemporary commentator on Aristotle, the Jesuit Niccolò Cabeo, used the field of meteorology as paradigmatic for all of natural philosophy and thereby eliminated substantial forms from his explanations of all natural phenomena except for

[4] See Descartes, *Œuvres*, vol. I, p. 370.

[5] Descartes, *Œuvres*, vol. VI, p. 239. Translation in Descartes, *Discourse on Method, Optics, Geometry, and Meteorology*, p. 268.

[6] Descartes, *Œuvres*, vol. III, pp. 491f.

[7] *Principia* IV 187, in Descartes, *Œuvres*, vol. IX, p. 309. On this point and on the possibility that portions of *Les météores* came from early drafts of *Le monde*, see Gaukroger, *Descartes: An Intellectual Biography*, pp. 226f.

the human intellect. Understood in this context, *Les météores* appears less a herald for the new sciences than a treatise that was participating in contemporary debates on explanation in meteorology.

Because of his tendency to deny the existence of influences, assessing Descartes' knowledge of his contemporaries' work is tricky. Nonetheless, we know that he learned some scholastic philosophy at the Jesuit college at La Flèche.[8] What impact his studies had on him is unclear. In a letter to Mersenne in 1640, 3 years after the publication of *Les météores*, Descartes requested the names of authors of Jesuit textbooks in philosophy because he could only remember those of the Coimbrans, Francisco Toletus, and Antonio Rubio.[9] Of those three, only the Coimbrans wrote on meteorology. In the same year he praised Eustachius a Sancto Paulo's *Summa philosophica*, a textbook on the entire range of philosophy including meteorology, which he said he had just recently purchased. Descartes' description of his limited memory of earlier readings may very well be true, but nevertheless should not be taken as proof of his total ignorance of contemporary Aristotelians in the 1630s. The correspondence between *Les météores* and other Aristotelian meteorological works is evidence of at least a minimal amount of familiarity with one or more of these books.

Furthermore, there is additional evidence that Descartes was familiar with the content of other treatises on meteorology, in particular ones that did not rely on formal causation. Soon after the publication of the *Discours*, Libert Froidmont (1587–1653), a professor of theology and philosophy at Louvain and author of the well circulated and frequently reprinted *Libri sex meteorologicorum* (1627), criticized Descartes on a number of grounds: his philosophy was too close to atomism, had unacceptable implications about the human soul, and did not utilize teleology.[10] In a lengthy letter to Plempius, Descartes responded to a number of Froidmont's points. With regard to *Les météores*, Froidmont had written that Descartes' description of the composition of bodies by their parts and shapes was "too gross and mechanical"[11] and he complained that Descartes "hopes he will explain too many things by position and local motion, which cannot be understood without some real qualities."[12] In sum, Descartes' meteorology suffered by its use of only matter and motion, without recourse to formal causation. Descartes defended himself not by arguing that "real qualities" were unnecessary or superfluous but by

[8]On Descartes and Jesuit instruction at La Flèche, see Rodis-Lewis, "Un élève du collège jésuite de La Flèche: René Descartes," pp. 25–36; Giard, "Sur la compagnie de Jésus et ses collèges vers 1600," pp. 199–225.

[9]Descartes, *Œuvres*, vol. III, p. 185. For a discussion of this request see: Ariew, *Descartes and the Last Scholastics*, p. 26.

[10]Armogathe argues that Descartes was familiar with observations found in Froidmont's *Meteorologica*. See Armogathe, "The Rainbow: A Privileged Epistemological Model," p. 252.

[11]Descartes, *Œuvres*, vol. I, p. 406.

[12]Ibid., p. 408.

contending that his work treated similar problems as other meteorological tracts had done. He wrote:

> But if one should wish to list the problems which I explained only in the treatise *De meteoris,* and compare them with what has been done up until now by others on the same subject, in which he [Froidmont] is very versed, I am confident that he would not find such a great occasion for condemning my somewhat bloated and mechanical philosophy.[13]

Perhaps he was bluffing about his knowledge of contemporary works on meteorology. Nevertheless, in his defense of himself, Descartes maintained that his meteorology addressed the questions typical of the state of the field, thereby suggesting he had some idea what the state of the field was and that he was aware that others were not using formal and final causes as explanations, that they too were "gross and mechanical."

What Descartes and Froidmont meant by the word "mechanical" is unclear and it seems likely they did not share a similar definition for the word.[14] Nevertheless, a comparison of *Les météores* with the Aristotelian meteorological tradition shows that Descartes was to a certain degree correct about a number of his Aristotelian contemporaries if we accept that he meant "mechanical" to mean a reliance on material causation and a limited application of formal and final causation. An analysis of their work suggests that a debate over whether final and formal causes should be part of meteorological explanations had already begun before Descartes and continued into the middle of the seventeenth century. Aristotelian commentators on the meteorology cannot be thought of as a homogenous group of authors; rather, their views varied widely. The supposed novelty of eliminating substantial forms from meteorology, however, was in fact no novelty at all. Descartes' meteorological theories should not be understood as revolutionary but rather as a continuation of earlier debates. To understand why the field of meteorology differed from other parts of natural philosophy it is necessary to return to both Aristotle and his commentators.

9.2 Aristotle's Meteorology

The seventeenth-century creators of new philosophies that competed against Aristotelian models prided themselves on their dependence on fewer causes. Even though few historians now trust the accuracy of claims of the complete independence of the promoters of such novel natural philosophies—they contain caricatures rather than portraits of scholastic thought—it is still generally accepted that Aristotelian natural philosophy privileged final and formal causation over material and efficient causation. This privileging is stated explicitly in the *Physics*, among other places,

[13] Ibid., p. 430. On this letter and the meaning of the word "mechanical" in Descartes and Froidmont, see Gabbey, "What was 'Mechanical' about 'The Mechanical Philosophy' Chap. XX" p. 18. The last two lines of the above translation are taken from Gabbey.

[14] See Gideon Manning's article (Chap. 10) in this volume for a discussion of the exchange between Froidmont and Descartes. See Daniel Garber's contribution for a general discussion of the term "mechanical philosophy" in the seventeenth century.

and is apparent in numerous treatises, such as the biological and psychological works.[15] For example, sensation and intellection are understood in terms of form and as actualities of potentialities; and the parts of living beings are considered with respect to their being "for the sake of something." Nevertheless, despite the overall emphasis on forms and ends, Aristotelian works also discussed material and efficient causation; and according to Aristotle it is possible to give explanations, perhaps not always complete explanations, of large portions of the natural world using just matter and motion. In fact Descartes was aware of the significant roles played by material and efficient causation in Peripatetic philosophy, as he contended that his use of shape, motion, and size in physics corresponded to some of the principles that Aristotle employed.[16]

For Aristotle there were limits to teleology just as there were limits to material explanations of nature. A detailed explanation of the varying virtues of the causes is found in *Meteorology* IV 12.[17] Contemporary scholars, as well as numerous medieval and early modern commentators, have considered this chapter as well as the entirety of *Meteorology* IV to be an introduction to biology, a bridge between discussions of the elements, qualities, and the formation of homeomerous substances to discussions of the functions that these substances have in animate beings.[18] Mary Louise Gill understands *Meteorology* IV 12 as delineating to what extent unqualified (*haplos*) necessity can explain the natural world. Gill equates Aristotle's "unqualified necessity" to a "material necessity [that] is grounded in the natures of materials and in general laws of material causation."[19] In *Meteorology* IV 12, Aristotle contended that there exists a hierarchy of substances, starting from the elements at the bottom, going to the homeomerous substances such as flesh and bone, to the anhomeomerous substances such as organs, and finally to entire organisms at the top. The level of a substance within this hierarchy corresponds to the kind of causation that should be used to explain it. Thus, organs are known with respect to final causality, with respect to the "for the sake of something" *tou heneka,* that is, their function within an organism, even though they are simultaneously composed of an underlying matter and are the matter, which composes the entire organism. The characteristics of the homeomerous bodily parts—the flesh, blood, and bone, that is, the matter of these organs—however, can be known through what Gill calls material necessity because, "these things come to be by heat and cold and their combined motions."[20]

---

[15] *Physics* II 9, 200a32–b3.

[16] *Principia* IV 200, in Descartes, *Œuvres*, vol. VIII–1, p. 323.

[17] The authorship of *Meteorology* IV has been and perhaps still is questioned. I treat the book as authentic. There were extremely few doubts expressed on its authenticity before 1915. For a summary and bibliography of most germane scholarship on this question see Baffioni, *Il IV libro dei "Meteorologica" di Aristotele*, pp. 34–44; 386–392.

[18] Furley, "The Mechanics of *Meteorologica* IV: A Prolegomenon to Biology"; Gill, "Material Necessity and Meteorology IV 12."

[19] Gill, "Material Necessity and Meteorology IV 12," pp. 146–147.

[20] 390b2–14.

The preceding 11 chapters to *Meteorology* IV discuss this material necessity, the motions caused by the hot and the cold. Similarly to Gill, David Furley stresses this book's reliance on matter and motion,[21] on "unqualified necessity," identical to Gill's "material necessity," as the primary explanation.[22]

Even though *Meteorology* IV 1–11 explores material properties as the result of matter and motion, the creation of primarily but not exclusively animate homeomerous substances is seen through the prism of teleology. The actions of the hot and cold participate in the process of concoction, whereby an unformed substance attains its perfected form, or, in the terms of medieval and Renaissance scholars, becomes a perfect mixture. Concoction is a type of *teleiōsis*, and although this book explains how the hot and the cold cause physical transmutation, these transmutations are often seen with regard to specific ends. The fact that even the "material necessity" of *Meteorology* IV is subordinate, at least partially, to final causes does not mean that there are no limits to Aristotelian teleology. Although *Meteorology* IV explains why and where there are limits to teleology, the three preceding books of the *Meteorology*, the three books that actually treat meteorological themes, are a better place to examine Aristotelian natural philosophy that has little recourse to final and formal causes.

The first three books of the *Meteorology* discuss changes in the sublunary region, a region that, according to Aristotle, is filled with irregular and episodic changes. Sublunary change results from the eternal motions of the celestial bodies that drive the transformation and cyclical motions of the four elements. Aristotle described the proximate cause of meteorological phenomena as being two exhalations that move in continual cycles between the surface of the earth and the uppermost limit of the terrestrial region. These two exhalations are a vaporous exhalation, which is wet and cold, and a smoky exhalation, characterized by dryness and heat. The movements of the dual exhalations provide a unity of explanation for Aristotle, as they give an account for a wide variety of phenomena, including many phenomena that are now considered to be beyond the scope of the atmospheric sciences, such as the apparently fiery paths of comets and the flickering light of the Milky Way. Additionally, according to Aristotle, an analogous pair of exhalations circulates beneath the surface of the earth and explains geological and hydrological phenomena such as earthquakes, hot springs, and the features of the sea and rivers.

---

[21] Furley, "The Mechanics of *Meteorologica* IV." Gill and Furley are by no means the first to emphasize matter and motion in *Mete.* IV. For example, Federico Pendasio (ca. 1600), a professor of philosophy at Padua and Mantua, claimed that the opinion that this book treated primarily matter and motion was widespread. See his *Lectiones in quartum librum meteorologicorum*, f. 1r: "Principium autem hoc statuo quod apud omnes est compertissimum, librum hunc, partem esse naturalis philosophiae, tractat enim quae concernunt materiam et motum."

[22] Although Furley refers to *Meteorology* IV's "mechanics," I refrain from this label because Aristotle does not use machines or actual mechanisms as models or analogies. Calling this "mechanical" is anachronistic. For this precise definition of "mechanical philosophy" in antiquity, see Berryman, "Galen and the Mechanical Philosophy," pp. 235–253.

As for most of Aristotle's natural philosophy, the purpose of meteorology is to provide causes. Aristotle did not rely on the supernatural to explain extreme examples of weather or other catastrophes and his naturalism was not subordinate to the ethical goal of removing fear of the Gods, as it was for Epicureans.[23] Rather the *Meteorology* is dedicated to explanations *via* material and efficient causation. The material cause of these atmospheric and subterranean changes are the elements and the two exhalations composed of them; the efficient cause is the motions of the celestial bodies, in this case the sun and the moon.[24] Final and formal causes are not part of his explanations for these subjects, because the matter of meteorological phenomena is perpetually imperfect, being partial transformations of the elements, as numerous medieval and Renaissance commentators noted. Moreover because these partially transformed elements are inanimate, as Olympiodorus argued, they do not participate in the formation of organs and organism, which have clear purposes and ends.[25] Thus it is perhaps not surprising that in Theophrastus's *Metaphysics*, where he attacks those who proclaim that "all things are for the sake of an end and nothing is in vain," his first counter example is meteorological, namely "the incursions and refluxes of the sea, or droughts and humidities, and in general, changes, now in this direction and now in that, and ceasings-to-be and comings-to-be."[26]

Granted, some meteorological phenomena are endowed with purpose. Seasonal rains ensure the availability of crops; even though climatic and seasonal weather patterns exist,[27] specific rains, snows, earthquakes, and floods are without order or clear purpose. Moreover, Aristotle's meteorology, as nearly all ancient meteorology, emphasized rare and irregular phenomena, such as meteors, comets, various fires in the sky, cyclones, and so forth. As a result, proper knowledge of these topics is difficult, he wrote: "Of these things some puzzle us, while others admit of explanation in some degree."[28] The arguments for intractability for the field of meteorology

[23] See Book six of Lucretius's *De rerum natura* and Epicurus's *Letter to Pythocles.*

[24] *Meteorologica*, I 1, 339a20–33, trans. by E. W. Webster, in *The Complete Works of Aristotle*, vol. 1, p. 555.

[25] On *Meteorology* I–III as about inanimate homeomerous substances see Olympiodorus, *In Aristotelis meteora commentaria*, p. 273, 20f. For other discussions of the limits of teleology in Aristotle, see *De generatione animalium*, V 1, 778a29–778b7; *De partibus animalium*, I 1, 642a2–3.

[26] Theophrastus, *Metaphysics*, IX 28–29. On the fact that meteorology was a prime example of dysteleology, see Vallance, "Theophrastus and the Study of the Intractable: Scientific Method in *De lapidibus* and *De igne*," pp. 28f. For the view that Theophrastus' position on the limits of teleology was common to Aristotle, see Recipi, "Limits of Teleology in Theophrastus?" pp. 182–213; but, for the view that Theophrastus was attacking Aristotle, see Lennox, "Theophrastus on the Limits of Teleology," pp. 143–151.

[27] See *Physics* II 8, 198b16–21. Whether Aristotle actually endorsed a teleological position in this case has been a matter for debate. For a discussion of this issue see Furley, "The Rainfall Example in *Physics* II.8," pp. 115–120.

[28] *Meteorologica*, I 1, 339a2–3.

are even more pronounced in Theophrastus' *Meteorology*, where he provided not one cause but a multitude of possible causes.[29] In Aristotle's eyes and those of other ancient Peripatetics, because meteorology was a field dedicated to the part of the natural world that lacks clear order, being composed of the elements that have been partially but not completely transformed, it was best understood by material and efficient causes rather than by formal and final ones, which in turn give us only probable or hypothetical knowledge when they provide any explanation at all. Because much of meteorology is distant and thus difficult to observe our knowledge of it is provisional, or as Aristotle wrote in *Meteorology* I 7, "we consider a satisfactory explanation of phenomena inaccessible to observation to have been given when our account of them is free from impossibilities."[30]

9.3 Medieval and Renaissance Aristotelian Meteorology Commentaries

For much of the Middle Ages and Early Modern period, the most common method to discuss meteorology was to write a commentary or textbook based on Aristotle's writings.[31] Because over 200 commentaries on meteorology were written during this period, I will limit my considerations to those of some of the most famed Renaissance Peripatetic authors, such as Agostino Nifo (1469–1538) and Pietro Pomponazzi (1462–1525); authors of text-books that might have been available to a young Descartes, such as the Coimbrans', Eustachius a Sancto Paulo's, John Poinsot's, and Daniel Sennert's; and the more significant works written in the years surrounding Descartes' composition of *Les météores*, including those of Libert Froidmont, Francesco Resta, and Niccolò Cabeo.

By the late Middle Ages the field of meteorology in these commentaries and textbooks became defined by its dysteleology. Averroes (1126–1198) retained the exhalations in his description of the scope of this work; whereby, in his view, the

---

[29] Daiber, "The *Meteorology* of Theophrastus in Syriac and Arabic Translation," pp. 166–293, with an English trans. of treatise, pp. 261–271.

[30] *Meteorologica*, I 7, 344a5–7, trans. by E. W. Webster, in *The Complete Works of Aristotle,* vol. 1, p. 562. Cynthia A. Freeland contends that *Meteorology* I–III relies on abduction rather than dialectics or syllogisms. See Freeland, "Scientific Explanation and Empirical Data in Aristotle's *Meteorology*," pp. 67–102. For a discussion of the lack of teleology in the *Mete*. See: Liba Taub, *Ancient Meteorology*, pp. 80–84. This lack is not always recognized, see Meinel, "Les *Météores* de Froidmont et les *Météores* de Descartes," p. 107.

[31] By meteorology, I limit myself to the field that considered the causes of atmospheric and subterranean events and do not consider the prognostication of weather via signs. By limiting myself to this field, I am following Aristotle's definition of meteorology, which was understood as such by the large part of practitioners of natural philosophy in the Aristotelian tradition. For Aristotle's definition see *Meteorologica*, I 1, 338a19–339a5.

first three books treat the accidents of the dual exhalations and the final book homeomers in general.[32] Thus the four books, following the general schema of Aristotelian intellection, start with accidental particulars and end with universal statements. Albertus Magnus (1193–1280) believed that the first three books of the *Meteorology* treated substances that were in state of becoming simple mixtures and the final book discussed simple mixtures.[33] Thus for him, the *Meteorology* followed the priority implied in the act of becoming and the completion of this act. Later medieval scholars, such as Jean Buridan (1300–1358) and Blasius of Parma (ca. 1400) contended that the differing scopes of the first three books and the fourth one match the change from imperfect mixtures to perfect mixtures.[34] It was this view that was to dominate throughout the Renaissance and well into the seventeenth century.

During the Renaissance, the intractability and imperfection of meteorological phenomena was a basis for emphasizing the conjectural nature of natural science (*scientia*). Two of the most famed philosophers of the early sixteenth century, Nifo and Pomponazzi, put forth this position. Nifo, who was a professor of philosophy in several Italian universities, used meteorology and Aristotle's confession of the inability to understand all causes in order to distinguish the natural sciences from the mathematical. He wrote: "It must be said that natural science is not a science *simpliciter*, such as the mathematical sciences are, but is a science that explains the why (*propter quid*). It is the science of finding the causes which can be held through a conjectural syllogism that gives the *propter quid* of the effect." This account of the effect however is not definitive. He supported this position by his use of Aristotle's meteorology, and argued that, "Aristotle in the book of the *Meteorology* concedes that he does not provide the true causes of natural effects, but that which is possible through conjecture."[35] For Nifo, knowledge of meteorology is uncertain.

Nifo's contemporary and rival Pietro Pomponazzi's take on the intractability of meteorological phenomena led to an even more skeptical view of the nature of natural

[32] Averroes, *In quartum librum meteorologicorum,* in *Opera,* vol. IV, f. 460r.

[33] Albertus Magnus, *Liber quartus meteororum,* in *Opera omnia* vol. IV, p. 705.

[34] Buridan, *Expositio libri meteororum*, f. 103r; Blasius of Parma, *Expositio in libros meteorologicorum*, ff.1r; 49r. This view was followed in the sixteenth century by, among others, Agostino Nifo, Konrad Gesner, Francesco de Vieri, Agostino Pallavicini, Joannes Hawenreuter, Jacques Charpentier.

[35] "Dicendum, scientiam de natura non esse scientiam simpliciter, qualis est scientia mathematica, est tamen scientia propter quid: quia inventio causae, quae habetur per syllogismum coniecturalem, est propter quid effectus. per haec delentur obiectiones, quae contra haec fieri solent: Prima quidem delentur ex eo, quia non est circulus in demonstratione, cum primus processus sit tantum syllogismus, secundus vero demonstratio propter quid. deletur etiam Secunda obiectio, quia effectus semper est notior ipsa causa in genere notitiae quia est. nunquam enim causa potest esse ita certa quia est, sicut effectus, cuius esse est ad sensum notum. Ipsum vero quia est causae, est coniecturale, utrum tale esse coniecturale est notius ipso effectu, in genere notitiae propter quid. nam posita inventione causae semper scitur propter quid effectus. unde & Aristo., in libro Meteororum concedit se non tradidisse veras causas effectuum naturalium, sed quo erat sibi possibile coniecturabiliter" (Nifo, *Expositio super octo Aristotelis Stagiritae libros de physico auditu*, f. 6v).

philosophy. According to his view, the idea of a complete science is held only by fools; meteorology is proof that we will never be able to have an accurate account of the entire natural world. Pomponazzi maligned both religious thinkers who argued that meteorological events, disasters in particular, were the result of the will of God, and those "stupid philosophers" and "Peripatetics," the latter being a category in which probably most of his contemporaries included him, who want to know everything, and proclaim that all events can be traced back to "movement of the heaven."[36] Moreover he contended that unlike Seneca, Aristotle did not believe that earthquakes and winds have final causes.[37] For Pomponazzi, meteorology is evidence for the absence of determinism and purpose in the universe and for the existence of limits for human knowledge.

9.4 Descartes and Seventeenth-Century Aristotelian Meteorology

While a number of medieval and Renaissance scholars tried to follow Aristotle's intention of leaving out final causation from meteorology, not all Early Modern Aristotelians followed this position. Throughout the sixteenth century and the first half of the seventeenth century, commentators on the *Meteorology* and authors of textbooks in natural philosophy and of other meteorological tracts became divided over the question if the field of meteorology relied only on material and efficient causes and could be explained through the motion of corpuscles and the processes of rarefaction and condensation. On the one hand, some commentators continued in the tradition of Nifo and Pomponazzi and explicitly denied that this subject could be explained by final and formal causes; and, on the other hand, several scholars introduced this field by giving its final and formal causes.

During the first decades of the seventeenth century, the role of formal and final causes distinguished meteorological treatises in much the same way that disputations about the location of comets and the division of the cosmos into distinct sublunary and supralunary regions did. There is even a partial correspondence between those

[36] "Peripatetici autem et alii stulti Philosophi qui volunt omnia scire, dicunt ex necessitate motus Coeli haec evenire" (Pomponazzi, *In libros meteororum*, f. 167r).

[37] "Quoniam Aristoteles non posuit causam finalem terrae motus, Seneca autem in suis quaestionibus ponit finem, quia fiant terrae motus, et ego quia promisivo his in hoc libro dicturum de causa finali omnium effectuum, qui in his quatuor libris determinantur, ita etiam observabo. loquamur ergo de fine extrinseco, utrum terraemotus habeat utilitatem aliquam pro fine in universo propriam, et sic etiam de ventis; nulla enim res abstracta est in mundo quae non conveniat naturae ad aliquid, et propriam habeat utilitatem in universo, et in suo genere sit maxima bona: Deus enim secundum Philosophos est auctor optimus et sapientissimus, cum autem universum sit opus Dei, oportet ergo quod perfectissime hoc fecerit, ut Plato posuit in Thimeo [sic]" (Pomponazzi, *In libros meteororum*, f. 190v).

scholars who did not use formal and final causes in meteorology and those who accepted the existence of the transmutation of the heavens, even though the two issues are not connected in terms of argument (see chart on p. XX). While the debates about causation in the meteorological world were neither as heated nor as potentially dangerous as ones regarding the possibility of change in the heavens, nevertheless they were divisive and included what might be characterized as extreme reformulations of Aristotelian physics, especially that of Niccolò Cabeo (1585–1650), which eliminated any role for metaphysical entities in the realm of physics.

At the start of *Les météores* Descartes followed the traditional distinction made in commentaries on Aristotle's *Meteorology* that substances can be classified as perfect or imperfect mixtures. Whereas in Aristotle's work, matter theory is not treated in depth until the final book, Descartes began his treatise with an exposition on the subject, making it the conceptual foundation for his exposition. He described the traditional elements as being composed of small irregular particles that join together, although never perfectly. Smaller particles that move more quickly than larger ones fill up any spaces between the pores of bodies made up of these larger particles, which move more slowly but have more impetus and thus can agitate other particles easily. The motions, combinations, shapes, and positions of these particles give rise to the various types of substances as well as their transformations. While rejecting traditional explanations of the elements, he retained the more typically Aristotelian terminology of vapors and exhalations. For Descartes, vapors are those bodies composed of fine material that are present within the pores of terrestrial bodies; exhalations are closely related to these vapors but are more regular in their shapes, being composed of particles with a shape similar to those which constitute water but are only finer. He likens exhalations to "spirits or brandies." The sun agitates the vapors and exhalations causing their irregular but cyclical motion throughout the atmosphere.[38] These vapors and exhalations are a constant resource in the *Les météores* and are the composing matter of winds, clouds, and lightning, among other things.

Nearly all, if not all, early seventeenth-century meteorological treatises employed the terms vapor and exhalation to distinguish Aristotle's two exhalations. Typically *vapor* was the wet and hot exhalation, while *exhalatio* was hot and dry, as well as being frequently characterized as smoky. A number of authors of meteorological treatises, such as the Coimbrans, Eustachius a Sancto Paulo, Francesco Resta, John Poinsot, and Daniel Sennert divided the material cause of meteorology into proximate and remote causes: the former being the *vapor* and *exhalatio*, the latter the elements, or in the case of the Coimbrans just two elements, earth and water (see chart on p. XX). While Descartes did not use the term material cause and *a fortiori* did not distinguish between proximate and remote material causes, nonetheless there are parallels between his description of matter and those of the Aristotelians. In effect the irregular and fine particles correspond to the remote material cause while his vapor and exhalation are the proximate ones.

[38] Descartes, *Œuvres*, vol. VI, pp. 239–241.

Moreover, Descartes' appeal to the sun as the cause of the motion of the vapor and exhalations mirrors Aristotelian positions. Granted, Descartes hedged his identification of the sun as the mover, by adding the phrase "or some other cause," and by making an analogy to light, which he argued was the result of the motion of fine particles as propelled by luminous bodies.[39] The Aristotelian textbooks for the most part state that the force and motion of the celestial bodies, particularly the sun, are the efficient causes of atmospheric change. Eustachius was an exception, as he believed that "Deus Optimus Maximus" is the efficient cause.[40] Alternatively, the Coimbrans distinguished the instrumental cause of the force and motion of the celestial bodies from their heat.[41] While Descartes did not employ a "quality" in his explanation, the concept of the heat of the sun as efficient cause does appear in Jean-Baptiste Duhamel's *De meteoris* (1660), a work that melded Aristotelian and Cartesian ideas with other novel philosophies.[42]

While the elaborate taxonomy of causes and explanations that is found in some of these texts books (proximate, remote, instrumental, *per se*, and *per accidens*) probably aggravated a number of thinkers who wished to leave behind Aristotelian philosophy, the work of John Poinsot (also known as John of St. Thomas, 1589–1644) might have given them cause to rethink their antagonism. His "Tractatus de meteoris" in his 1634 *Cursus philosophicus thomisticus* begins with the proclamation that he will not apply final and formal causes, only material and efficient ones.[43] The efficient cause is divided into two: *per se,* which is the power (*virtus*) of the sun, stars, and celestial bodies that comes in the form of heat, and *per accidens,* which is antiperistasis. Poinsot utilized corpuscular motifs to explain how heat acts as an efficient cause. According to Poinsot, *vapor* is composed of subtle aqueous parts. Heat causes evaporation by lifting these subtle parts to the higher regions, where they in turn fall, causing precipitation. The dry smoking *exhalationes*, however, are affected by the heat and *virtus* of the sun and stars, which cause them to rise and eventually flame up causing winds, thunder, lightning, and comets. The power of heat acts by thinning out (*subtilizando*) both of the exhalations and by separating (*segregando*) the more subtle parts from the thick ones. After these subtle parts have reached higher levels above the surface of the earth they either burst into flames or precipitate depending on whether they are smokey or watery.[44] Thus meteorological phenomena are caused by the separation and motion of small particles without recourse to final or formal causation.

Against Poinsot, a number of authors of meteorological treatises argued that meteorological phenomena could be understood by formal or final causes. The

[39] Ibid., p. 240.

[40] Eustachius, *Summa philosophiae quadripartita,* p. 155.

[41] Collegium Conimbricense, *In libros meteorologicos,* pp. 4f.

[42] Duhamel, *De meteoris et fossilibus*, p. 11.

[43] Poinsot, *Cursus philosophicus thomisticus: Tomus tertius philosophia naturalis*, p. 129.

[44] Ibid., pp. 129f.

Coimbrans, in a lengthy discussion, concluded that the double exhalations do not have new substantial forms but retain the forms of the element. Eustachius contended that the formal cause of atmospheric change is located in the forms of the elements.[45] Libert Froidmont identified the substantial form of fire as the formal cause of meteors and maintained that the formal cause of wind was the form of the exhalations.[46] Francesco Resta divided formal causes into proximate and remote in his 1644 *Meteorologia*. For him, the remote formal cause of rain was the substantial form of water and the proximate the form of the drops.[47] In most of these treatises, once the formal cause was established there are few applications of it in further explanations, although it could be used to deduce the secondary properties of the exhalations; *vapor* has the properties of water, *exhalatio* has those of fire. Nevertheless, the paucity of applications of formal causes is justified by a strict reading of Aristotle. Thus, both Poinsot and Sennert, who denied that meteorological substances, being imperfect, have their own *forma misti*, seem more than reasonable in their interpretations. In this context, Descartes' contention that he did not utilize formal causes does not seem particularly bold.

If Aristotelian meteorology without formal causes is easy to imagine, the same can be true for final ones. Neither the Coimbrans mentioned them, nor did Sennert. Eustachius, however, wrote that the final cause of meteorology was the moderation of the weather, the perfection of the universe, and the manifestation of divine power and wisdom.[48] Froidmont argued that winds were useful to mankind by stopping putrefaction and making the world temperate.[49] Similarly, Resta contended that winds thin out the air and make the weather more temperate.[50] Given that much of the subject of meteorology was disaster-provoking weather and geological events, others followed Pomponazzi who lambasted those who thought everything had a purpose. This line of thought did not have to wait until Voltaire and the Lisbon earthquake of 1755. Furthermore, the conviction of those who held that there were entelechies for the weather does not appear to be strong. Froidmont, who wrote scholia on Seneca's *Naturales quaestiones*, a work that attempted to demonstrate the connectedness of the world to divinity through a number of meteorological examples, did not press particularly hard on this issue in his attack on *Les météores*. He took issue with the lack of teleology in Descartes' discussions of organisms but not in his polemics against his meteorology.[51]

[45] Eustachius, *Summa philosophiae quadripartita*, p. 154.

[46] Froidmont, *Libri sex meteorologicorum*, p. 41.

[47] Resta, *Meteorologia de igneis aereis aqueisque corporibus*, p. 795.

[48] Eustachius, *Summa philosophiae quadripartita*, pp. 154f.

[49] Froidmont, *Libri sex meteorologicorum*, pp. 196–198.

[50] Resta, *Meteorologia de igneis aereis aqueisque corporibus*, p. 363.

[51] Descartes, *Œuvres*, vol. I, pp. 402–409.

9.5 Cabeo's Meteorology

Although Descartes contested some aspects of the Aristotelian meteorological tradition and Froidmont was unhappy with his work, Gilson and Gaukroger reasonably point to the conservative nature of *Les météores* despite its novel treatment of the rainbow and the enumeration of an explicitly corpuscularian position.[52] The relative conservativeness of this work can be better appreciated by a close examination of Cabeo's commentary on the *Meteorology.* Cabeo, a Jesuit who lived in Northern Italy, is best known for his *Philosophia magnetica*, a work that attacked Gilbert and posited that corpuscular effluvia cause magnetic attraction. His interest in corpuscular philosophy continued in his *Commentaria in libros meteorologicorum* (1646), a four-volume tome to which Cabeo devoted much of the 1630s. Although Cabeo was familiar with the work of many proponents of novel natural philosophies, such as Galileo, Tycho Brahe, Paracelsus, and Kepler, he did not refer to the works of Descartes and does not appear to have been influenced by his writings.

The *Commentaria* is broad in scope and, although it includes a literal exegesis of Aristotle's text and maintained that Aristotle's writings are the starting place for the study of natural philosophy, the treatise went well beyond Aristotle's words and even the field of meteorology. It is in effect a comprehensive study of natural philosophy that delves into cosmology and chemistry as well as meteorology. For Cabeo, Aristotle's *Meteorology* was the ideal vehicle for investigating the natural world because it avoided the metaphysical speculation that dominated the rest of Aristotle's work. Most importantly, he thought that *physica* should avoid substantial forms as an explanation and rely only on truly physical causes and should be understood through the detailed observation of their physical effects, as Aristotle had done in the *Meteorology*.[53]

Cabeo separated *physica,* as he called it, from metaphysics and mathematics, both of which he considered speculative.[54] For him *physica* is concerned with the sensible, that is, "all of the effects of those things that can be perceived and are actually perceived," and "the sensible causes of all effects, which can be perceived by external sensation," while "those [causes] that cannot be perceived do not pertain to *physica*."[55]

---

[52] See supra, p. XX [1].

[53] Cabeo shared his concern over the non-physical nature of substantial forms with other contemporary Jesuits, namely Honoré Fabri. See Roux, "La philosophie naturelle d'Honoré Fabri (1607–1688)."

[54] "Supponendum igitur est tres iam communiter ab omnibus distingui scientias totales speculativas Methaphysicam, Physicam, & Mathematicam, quae dicuntur scientiae totales," Cabeo, *Commentaria in libros meteorologicorum*, vol. I, p. 6.

[55] "Omnes ergo illi effectus, qui sensu percipi possunt, & de facto sensu percipiantur horum omnium effectuum cognoscendi ratio spectabit ad Physicam, & ex complexione cognitionum harum proprietatum, & effectuum integrabitur Physica, quae tota versatur in hoc ut ostendat causas sensibiles omnium effectuum, qui sensu externo percipi possunt, & quae sic percipi non possunt non spectabunt ad Physicam," ibid., vol. I, p. 9.

Because Aristotle was too occupied with metaphysics, dependence on his writings caused other Peripatetics to ignore sensible objects or to analyze nature using metaphysical concepts. Thus many of Aristotle's views were not authoritative because he was "more accustomed to metaphysical speculation, than physical observation."[56] Within the category of metaphysical speculation, he included abstractions and indiscriminate applications of logic that reduced things (*res*) into universal categories, differences, and divisions, all three of which he believed to have no physical reality. In his view, metaphysical entities are chimerical because they are not material, sensible, or physical.[57] Thus he intended to correct the "many Peripatetics occupied in these metaphysical subtleties [who] do not read these books [i.e, the *Meteorology*]."[58] Cabeo commented on the *Meteorology* because it allowed him to create a natural philosophy based on physical bodies and not metaphysical concepts, while still maintaining allegiance to Aristotle.

Substantial forms were the main target of Cabeo's attack on metaphysics. The common conception of form as essence was mistaken, according to Cabeo; rather, forms are real, physical, material entities, namely spirits and vapors that have powers and virtues. He realized that this was not the accepted interpretation of Aristotle, as he wrote: "And thus perhaps the substantial form is a metaphysical essence and formula according to Aristotle; it is not a physical entity."[59] Nevertheless, because of substantial forms' non-sensible nature Cabeo believed they should have no place in natural philosophy. Moreover, he went on to reject the twin concepts of form and privation, a foundation of Aristotelian physics, because "one of which is nothing, the other which is metaphysical."[60] Instead, Cabeo defined form as active matter. He wrote:

> This is a form truly physical, this is a vapid and subtle spirit; for it is that, which gives determined being to each thing. For a thing is such because it is animated by this kind of spirit. From this [spirit], there is an active force, so great and of such kind; and just as the

[56]"Sed etiam hic videtur Aristoteles magis metaphysicis speculationibus assuetus, quam physicis observationibus," ibid., vol. IV, p. 418; *Commentaria* IV, 79–80: "unde cum Aristoteles physicum agit, omnino antiquos sequitur, sed quia iste Philosophus maxime pollebat ingenio metaphysico, & apprime arridebat philosophari per metaphysicas abstractiones, reducendo semper res ad universalissimas, & metaphysicas rationes, ut constat in tota eius physica; imo & in tota morali, & poetica, & rhetorica ipsa; semper enim res deducit ad differentias, divisiones, & metaphysicas abstractiones" (ibid., vol. IV, pp. 79–80); "omnino Aristotel. ingenium erat ad subtilitates metaphysicas, & abstractiones: non concrescebat illa subtilitas ingenii, ut concrescunt physica," ibid., vol. IV, p. 351.

[57]"sed videant ne physicam reliquant philosophiam, ut chimaeras sectentur metaphysicas," ibid., vol. I, p. 114; "illud est materia, non chimaerica, sed physica," ibid., vol. III, p. 406.

[58]"Sed istos libros non legunt multi peripatetici occupati in illis subtilitatibus metaphysicis," ibid., vol. IV, p. 352.

[59]"& sic fortasse forma substantialis, est essentia & ratio metaphysica apud Arist. Non entitas physica," ibid., vol. IV, p. 80.

[60]"non forma, & privatio, quorum alterum nihil est, alterum quid metaphysicum," ibid., vol. I, p. 406.

> diversity of the sublunary objects comes from these diverse spirits, which are implanted in them, the diversity of faculties, properties and virtues comes from these. This spirit is true act, it is true form, not a metaphysical formula conceived in the mind, but a physical principle of a faculty.[61]

What Aristotle called form, and what some considered metaphysical, is in fact a specific type of body that unifies a substance. It is a spirit, a vapor that consists of small particles of matter and contains active forces that order the world.[62]

Cabeo's commitment to using only what he thought were physical entities greatly altered the traditional way of understanding causes not just for meteorology but for the entire natural world; in his account of substantial change, generation and corruption, and the nature of mixture both final and formal causes are conspicuously absent. For Aristotle, homeomerous substances were mixtures of the four traditional elements. When fully mixed, a new form either supervened upon or replaced the forms of the elements. Medieval and Renaissance scholastics thought mixtures were perfect, when the form was fully actualized and imperfect when only partially.[63] Cabeo, however, discussed generation, corruption, and the perfection of mixture in terms of the position and motion of three kinds of matter. Generation results when particles of spirit become fixed in a medium; destruction occurs when these particles disperse or fly away from the grosser particles, causing their binds to dissolve. "This," wrote Cabeo, "is a true physical mixture, and it becomes perfect, when there is a concoction of its wetness, by which the spirituous parts are joined with the fixed."[64] Perfection is not the realization of their form or end but rather the result of the power of the binding: "All of the perfect compositions [*perfecta compositio*] of sublunary substances are such because their parts are joined with a perfect link." The stronger the link, the more perfect the mixture.[65]

---

[61] "forma vero physica est ille, spiritus vapidus, & subtilis, ille enim est, qui dat rei unicuique determinatum esse. Ideo enim res est talis, quia tali spiritu animatur. Ab isto est vis activa, tanta, & talis; & sicuti diversitas harum rerum sublunarium provenit a diversis istis spiritus, qui rebus inditi sunt; ita diversitas facultatum, proprietatum, operationum, virtutum, ab iisdem prodit. Hic vero verus actus, haec vera forma, non metaphysica, mente concepta ratio, sed physicum principium facultatum," ibid., vol. III, p. 4.

[62] Arnim, *Stoicorum veterum fragmenta*, § 439–462.

[63] For Aristotle's theory of mixture and combination, see: Joachim, "Aristotle's Conception of Chemical Combination," pp. 72–86. For early modern debates over the distinction of these terms see Lüthy, "An Aristotelian Watchdog as Avant-Garde Physicist: Julius Caesar Scaliger."

[64] "Haec est vera generatio physica, de quae hic Philosophi, quod nimirum partibus fixis; iterum volatiles aliae separatae adiungantur, & convenienti humore adglutinentur, & haec vera physica mixtio, & perficitur, ut constabit ex infra dicendis, concoctione illius humidi, quo partes spiritosae, cum fixis coniunguntur, & tota perfectio," Cabeo, *Commentaria in libros meteorologicorum*, vol. IV, p. 84.

[65] "Tota rerum sublunarium perfecta compositio in eo consistit, ut partes sint perfecto vinculo copulatae, & quo magis coniunctae fuerint, & minus separabiles, etiam ab efficaciori agente, diceretur certe res magis perfecta, in ratione unius, & compositi; istam autem partium compositionem, seu colligationem, dixi iam saepe fieri in humido," ibid., p. 98.

Thus, the teleological principle becomes equivalent to a physical state. On the other hand, the corruption of substances comes from the weakening and dissolution of the links, which causes the "spirits and subtler parts to separate from the corporeal and the thick and still ones to fall, while the subtle ones go up into the air."[66] Neither generation nor corruption truly creates or destroys anything; generation and corruption are merely the division and union of parts.[67]

In his efforts to remedy the metaphysical tendencies of his Aristotelian predecessors and peers, Cabeo argued that *physica* should be based only on physical principles. These principles largely came from the *Meteorology*. The spirits and vapors that traditionally explained aerial and subterranean change become the model for the entire natural world up to but not including the human soul.[68] Cabeo made the *Meteorology* a starting point in order to undermine metaphysical accounts of the natural world based on formal and final causes. These compositions, the linking, and the dissolution of the bindings of the elements explain the diversity of substances. Even though his philosophy used active principles, the place and position of elemental bodies were seen as crucial to creating new substances. His adoption of physical principles taken from his reading of the *Meteorology* ensured that substantial forms would have no explanatory role in his understandings of physical change, generation, and corruption.

9.6 Conclusion

While a number of followers of the Aristotelian tradition from antiquity until Cabeo's time understood meteorology as a field that relied on material and efficient causation, Cabeo went further and used Aristotle's *Meteorology* as a model for all of natural philosophy. Cabeo's interpretation is no more faithful to Aristotle than the Coimbrans' or Eustachius's that admits some role for substantial forms and final causes in meteorology. Nevertheless, his commentary shows that different directions could be pursued in interpreting Aristotle's primary work that does not use final and formal causation. Eustachius inserted God as the efficient cause and contended that the weather is the manifestation of divine wisdom. To the contrary, Poinsot explicitly denied entry to formal and final causes in his meteorological discussions, and Cabeo seized upon the fact that substantial forms are not involved in the *Meteorology* and

[66] "Dico ergo, ut saepe indicatum est, & non semel etiam fusius explicatum, rem aliquam corrumpi, nihil aliud esse, quam ex attentuatione humidi, quasi ex dissolutione vinculi separari spiritus, & partes subtiliores, a corporalibus: & crassas, & consistentes concidere, subtiles in auras abire," ibid., vol. IV, p. 80.

[67] "ut in corruptione nihil deperditur, sed quae erant unita dividuntur; in generatione nihil producitur, sed quae erant divisa uniuntur," ibid.

[68] For his view on the soul, see ibid., vol. IV, p. 82.

thus made it the basis for a *physica* that did not rely on metaphysics. Aristotle and many of his followers recognized that formal and final causes were not needed for most of meteorology, but both Cabeo and Descartes did not limit themselves to this field. Rather, they thought they could use the material and physical principles of meteorology as a foundation for explanations of natural phenomena in general. The kinds of explanation previously held by meteorology became the model for all of physics.

Although it seems unlikely that Descartes and Cabeo influenced each other, their approaches demonstrate that in portions of physics, such as meteorology, both Aristotelians and their critics were capable of coming to similar conclusions regarding the insignificance of substantial forms in natural philosophy. For Cabeo, Aristotle's *Meteorology* with its emphasis on material and efficient causation became the basis of all of physics. Descartes' *Les météores* relied on similar tactics whereby the failure to use formal causation would not provoke anger among schoolmen, because the field traditionally utilized this kind of explanation only to a limited degree. Therefore meteorology was an ideal topic for Descartes to unfold his physics in a non-controversial manner. The innovations of *Les météores* are not to be found in Descartes' removal of substantial forms and real qualities, but in other areas, such as the use of images as a means of visual persuasion.[69]

Daniel Garber has argued that while the *Discours* was attacked, it was not perceived as revolutionary by conservative Aristotelians, such as Froidmont.[70] It does not appear that Descartes perceived it as revolutionary as well; he did not appear to expect anything but widespread acceptance of the contents of *Les météores* among those teaching in Jesuit colleges.[71] While Descartes suppressed *Le monde, Les météores*, focused on the inanimate terrestrial world, was a less controversial vehicle to present his larger goal of a physics that had no recourse to final and formal causes. Furthermore, because the subject matter of meteorology did not demand discourses on cosmology, which had proved dangerous to Galileo among others, and because a certain camp of Aristotelian natural philosophers—not just Cabeo, but Poinsot and Sennert as well—was already in agreement with his exclusive reliance on efficient and material causation, *Les météores* was a less dangerous vehicle than *Le monde* to provide a new model for natural philosophy.

---

[69] Christoph Lüthy, "Where Logical Necessity Becomes Visual Persuasion: Descartes's Clear and Distinct Illustrations," pp. 101–103; Claus Zittel, *Theatrum philosophicum: Descartes und die Rolle ästhetischer Formen in der Wissenschaft*, pp. 187–230.

[70] Garber, "Descartes, the Aristotelians, and the Revolution that did not Happen in 1637," pp. 471–486.

[71] Descartes, *Œuvres*, vol. I, p. 455.

Name and date	Material	Efficient	Formal	Final	Place of comet
The Coimbrans, 1608 [1592]	Proximate: vapor & exhalatio Remote: Earth & water (p. 1)	Force and motion of Celestial bodies Instrumental: Heat (pp. 4–5)	Does not have new substantial forms but retain forms of elements (p. 5)	No mention	Highest region of the air (p. 28)
Eustachius a Sancto Paulo, O. Cist., 1609	Vapor & exhalatio (p. 154)	Deus Opt. Max. (p. 155)	Forms of the elements (p. 154)	Moderation of the weather; perfection of universe; manifestation of divine power and wisdom (p. 154–155)	Sublunary
Libert Froidmont, 1639 [1627]	Vapor & exhalatio (p. 22)	The Sun (p. 30)	Formal cause of Meteors: Substantial form of fire (p. 41); Formal Cause of wind: substantial form of vapor and exalatio	Final cause of wind: utility; fights disease by stopping putrefaction; makes world temperate (pp. 196–198)	Some are sublunary (pp. 116–117)
Francesco Resta, O. M., 1644	Proximate: Halitus (p. 1)	Sun, Celestial Bodies	Remote formal cause of rain: substantial form of water Proximate: division of water into drops; form of drops. (p. 795) Formal cause of wind: motion. (p. 360)	Final cause of wind: thins out air, makes weather more temperate. (p. 363)	Sublunary (p. 14)

(continued)

(continued)

Name and date	Material	Efficient	Formal	Final	Place of comet
Daniel Sennert, 1632 [1618]	Proximate: Vapor & exhalatio Remote: Four elements (p. 250)	The sun, subterranean heat, winds (p. 250)	Does not have *forma misti* (p. 251)	No mention	Some comets sublunary, some supralunary (p. 256)
John Poinsot (John of St. Thomas, O. P.), 1638 [1634]	Proximate: Vapor & exhalatio (pp. 129–130)	Per se: Virtus of sun, stars, and celestial bodies. Per accidens: Heat that separates small particles from large ones. (p. 130)	Do not pertain (p. 129)	Do not pertain (p. 129)	Sublunary (p. 137)
Niccolò Cabeo, S. J., 1646	Earth, water, exhalations; mercury, sulfur, salt; wet medium, spirits, fixed parts	Power of heat, spirits	Do not pertain	Do not pertain	Supralunary
René Descartes, 1637	Vapor, exalatio	Sun	Unnecessary	Unknowable	Supralunary
Jean-Baptiste Du Hamel, 1660	Vapor, exhalatio, fumus (p. 2)	Heat of sun; subterranean heat (p. 11)	Unnecessary	No mention	Supralunary (Astronomia physica, p. 70ff.)

Chapter 10
Descartes' Healthy Machines and the Human Exception

Gideon Manning

10.1 Introduction

In responding to Libertus Fromondus' criticism of *Les météores* as "excessively gross and mechanical [*mechanica*]," Descartes made the now-famous concession that in natural philosophy he utilized "shapes, sizes and motions, as happens in mechanics [*Mechanica*]," and even went so far as to characterize his work as a "mechanical philosophy [*mechanica philosophia*]."[1] On reflection it seems likely that Fromondus and Descartes were speaking past one another in their exchange, as Alan Gabbey has persuasively argued, with each meaning something rather different by *mechanica*.[2] Still, one wonders why, notwithstanding Descartes' concession, the medical claims from Discourse Five did not prompt Fromondus' "mechanical" accusation.[3]

I would like to express my gratitude to Sophie Roux and Daniel Garber for inviting me to contribute to this volume. I also wish to thank Melissa Pastrana, Benjamin Goldberg, Justin Smith and especially Helen Hattab for their encouragement and constructive suggestions during the writing process.

[1] Descartes, *Œuvres*, vol. I, pp. 406, 420 and 430 respectively. It is interesting to note that in the first seventeenth-century French edition of Descartes' correspondence from 1659, Descartes is not quoted as referring to his "mechanical philosophy" but to "ma Philosophie, toute grossiere & Mechanique qu'elle est" (Descartes, *Lettres de M. Descartes*, p. 50).

[2] Gabbey, "What Was 'Mechanical' about 'The Mechanical Philosophy'?," pp. 11–24.

[3] Fromondus did voice concern that the operations traditionally associated with the vegetative and sensitive soul could not be explained by "so ignoble and brutish a cause" as the fire in the heart, Descartes, *Œuvres*, vol. I, p. 403. If his calling *Les météores* "mechanical" only served to emphasize that Descartes' preferred causes were "excessively gross," like those associated with a lowly workman's craft, then Fromondus implied Discourse Five was mechanical by citing Descartes' "ignoble" causes. Be this as it may, there is no explicit mention of Discourse Five being "mechanical" in Fromondus' letter.

G. Manning (✉)
Division of the Humanities and Social Sciences, California Institute of Technology,
Pasadena, CA, USA
e-mail: gmax@hss.caltech.edu

D. Garber and S. Roux (eds.), *The Mechanization of Natural Philosophy*, Boston Studies in the Philosophy of Science 300, DOI 10.1007/978-94-007-4345-8_10,

In Discourse Five we are told, for example, that (1) “laws of mechanics are … the same as those of nature,” (2) we should regard the human body—a natural object—“as a machine” and (3) the heart beat follows “just as necessarily as the motion of a clock from the force, position and shape of its counterweights and wheels.”[4] Although Discourse Five does not draw equally on these three claims, their presence surely warrants calling Discourse Five “mechanical.”[5] It may even be the most mechanical of all the parts of the *Discours* and its companion essays.

A cohort of Dutch physicians, including the Utrecht rabble rouser Henricus Regius and the Leiden-based Cornelis van Hogelande, was drawn to the mechanical philosophy after reading the *Discours* and they appear to have found (2) especially important. In 1641, Regius published a student's medical disputation that declared: “The world, therefore, is a great machine and each of its parts is a small machine.”[6] Several years later, in 1646, Van Hogelande similarly embraced (2): “we are of the opinion that all bodies, however they act, are to be viewed as machines, and their actions and effects … are to be explained only according to mechanical laws.”[7] Among Cartesian physicians there are many passages like these in which we are advised to consider natural bodies as machines.

But what is a machine for Descartes and the Cartesians?[8] Remarkably, nowhere in Descartes' medical and natural philosophical writings does he ever explicitly tell us.

---

[4] Descartes' assimilation of the laws of motion to the laws of mechanics and his identification of natural bodies with machines are frequently cited to support the conclusion that he is a mechanical philosopher. For elaboration on this point and further references, see Daniel Garber's contribution to this volume.

[5] As we will come to see, Descartes continued to endorse (2) in the *Meditationes*, *Passions* and *Description*. In the *Meditationes*, for example, the Meditator speaks of “this whole machine of limbs [*totamque hanc membrorum machinam*]” and “the machine of the human body [*machinamentum humani corporis*].” In the *Passions* he wrote about the composition of “the machine of our body [*la machine de nostre corps*]” and in the *Descriptions* he claimed to describe “the whole machine of our body [*toute la machine de nostre corps*].”

[6] Regius, *Disputatio medica prima [–tertia] De illustribus aliquot quaestionibus physiologicis*, para. ix; cited and translated in Clarke, “Henricus Regius,” *The Stanford Encyclopedia of Philosophy (Summer 2011 Edition)*, Edward N. Zalta (ed.), URL=<http://plato.stanford.edu/archives/sum2011/entries/henricus-regius/>. I have been unable to consult the original.

[7] “[O]mnia corpora quocunque modo agentia, tanquam machinas consideranda, eorundemque actiones atque effectus … non nisi … secundum leges mechanicas, explicandos … existimamus,” Cornelis Van Hoghelande, *Cogitationes* …, p. 174f.

[8] For some recent answers to this question, see Des Chene, *Spirits and Clocks* and Garber, “Descartes, Mechanics, and the Mechanical Philosophy,” pp. 185–204. Garber's work is particularly relevant with its goal of illuminating what the learned tradition of mechanics came to mean for Descartes. Although I am sympathetic with much of what Garber has to say, if I understand him correctly he endorses the view that machines are just corporeal substances. He writes, “For Descartes, I suggest, a machine has become simply a collection of parts whose states are determined by the size, shape, and motion of those parts, as well as by the collisions among them,” ibid., 198. The parts are themselves just bits of extended matter on Garber's view, as is the machine itself. My disagreement with these conclusions hinges on whether machines must have a nature over and above their nature as corporeal substances. That they require such a nature is especially clear in the medical case. The possible exception is the corporeal world as a whole (see notes 3 and 71 below). For Descartes' description of the universe as a machine see Descartes, *Œuvres*, vol. VIII-1, p. 315, transl. in *The Philosophical Writings of Descartes*, vol. II, p. 279; Descartes, *Œuvres*, vol. V, p. 546, transl. in *The Philosophical Writings of Descartes*, vol. III, p. 213, the latter of which is cited later in this paragraph.

Instead, in the *Principia* he claimed not to "recognize any difference between artifacts," such as machines, "and natural bodies except that the operations of artifacts are mostly performed by parts [*peraguntur instrumentis*] which are large enough to be easily perceivable by the senses."[9] Indeed, Descartes' substance dualism between minds and bodies would seem to require that anything lacking thought and reason must be an extended corporeal substance—if not one substance then, by exclusion, the other. As Descartes once advised a correspondent, he "preferred to write 'The Universe is composed of matter, like a machine' or 'all the causes of motion in material things are the same as in artificial machines'."[10] Here Descartes presumed that machines are identical to extended corporeal substances.[11] His contribution is to add that the universe and natural objects are extended corporeal substances too, from which it follows that natural objects are machines.[12]

When Descartes' *Traité de l'homme* was finally published in its original French in 1664, its commentator, the French physician Louis de La Forge, introduced a sophisticated definition of a machine in an effort to help convey what Descartes was attempting to do when Descartes compared natural objects to machines. La Forge explained that a machine is:

> [A] body composed of several organic parts, which being united, work together to produce several movements, of which they would not be capable if they were separated. I call organic parts all sorts of simple or complex bodies, which being united together are able to help through their structure, shape, movement, rest, and location, in the production of the motions and functions of the machine of which they are parts.[13]

[9] Descartes, *Œuvres*, vol. VIII-1, p. 326, transl. in *The Philosophical Writings of* Descartes, vol. I, p. 288. I have modified the translation.

[10] Descartes, *Œuvres*, vol. V, p. 546, transl. in *The Philosophical Writings of Descartes*, vol. III, p. 213. Whereas living things had arguably been the paradigm substances before Descartes, he treated machines as the paradigm substances in his physics. This represents a genuine shift.

[11] John Schuster and Stephen Gaukroger explain why Descartes might have been able to take this for granted: "No mechanic would appeal to teleological processes, occult virtues or immaterial causes to account for the functions of a simple mechanical device. Explanations in the mechanical arts rested on the appeal to a clear picture of the structure and interactions of the constitutive parts of the apparatus," Gaukroger and Schuster, "The Hydrostatic Paradox," p. 552. Schuster and Gaukroger maintain that these kinds of explanations were exemplars for Descartes. I would add that they likely set the minimal conditions for what a corporeal substance needed to be.

[12] If a machine is identical with a corporeal substance, and therefore wholly defined by the modes of extension, then I would argue that only the universe as a whole is a machine. I elaborate on this point in the conclusion below and in note 71.

[13] "[U]n Corps composé de plusieurs parties organiques, qui estant unies, s'accordent à produire quelques mouvemens, dont elles ne seroient pas capables, si elles estoient separées. J'appele parties organiques, toutes sortes de Corps simples ou composez, qui estant unis ensemble, peuvent aider par leur conformation, figure, mouvement, repos, et situation, à la production des mouvens, et fonctions de la Machine, dont ils sont parties," Descartes, *Traité de l'homme* p. 173. In citing "mouvens" and "fonctions" it is hard to know if La Forge is just saying the same thing: functions are just movements or effects of the interactions of a machine's parts. This would seem to be Descartes' own view when he discusses "movement" and "functions" in the opening of the *Traité de l'homme*, and it would be consistent with the common association at the time between function and actio. But there is an alternative: La Forge might be using "fonctions" as equivalent to the Latin "utilitatis" or "usus." Though this might seem an academic point, what is at stake here is the possibility of assigning not just functions but also malfunctions to machines. Do they genuinely deviate from their nature? Do machines have a nature that can be corrupted? More on this point in sections three and four below.

This is more than Descartes ever told us, and there is a great deal worth highlighting in this definition, including its now foreign sense of "organic" and its sophisticated emphasis on coordinated interaction of parts.[14] Yet, if we treat this definition as a characterization of what a machine was for Descartes and the Cartesians—the idea of which lent credibility to associating natural philosophy with mechanics and mechanical explanation—what is of greatest interest is what La Forge left out. Specifically, we can immediately see that Descartes and the Cartesians failed to take account of what a machine was for, of what purpose it served (to say nothing of its maker). In other words, they have left out a traditional way of individuating a machine in terms of its nature.

In the pages to follow I will be questioning Descartes' apparent assimilation of machines to corporeal substances. I will argue that where natural philosophy overlaps with medical theory, Descartes was committed to machines having "natures" that could not be explained entirely in terms of the vocabulary of corporeal substance.[15] Although this is at its most obvious when Descartes turns his attention to medical question, I will further argue that the same considerations that show Descartes abandoning this austere ontology of machines in medical theory generalize to all the machines in his mechanical philosophy.

I begin by showing that Descartes consistently endorsed the strategy of importing principles from natural philosophy into theoretical medicine. In keeping with this strategy, he had good reason to show the compatibility between his medicine and his mechanical philosophy and as a result medical questions are essential to his scientific and philosophical project In the next three sections I discuss Descartes' appeal to machines in his bid to create a scientific medicine. Focusing on the opening to the *Traité de l'homme* and the vindication of the teachings of nature in the *Meditationes*, I emphasize the rationale Descartes offered for characterizing the human being's body, when viewed as a machine, as healthy or diseased. The main innovation of the paper lies is showing that Descartes justifies talk of healthy and diseased states by

---

[14] The question of what makes a particular body organic in the seventeenth-century is a complicated one, as suggested by Anne Conway's reference to "a mere Organical body like a Clock," Conway, *Principles of the Most Ancient and Modern Philosophy*, p. 222. Being "organic" or having "parties organiques" refers to organization. "Organic," in this sense, is compatible with "mechanical." For a brief history of "organic," see Cheung, "From the organism of a body to the body of an organism: occurrence and meaning of the word 'organism' from the seventeenth to the nineteenth centuries," pp. 319–339.

[15] Dennis Des Chene and Lisa Shapiro also see medical questions as potential problems for Descartes and the mechanical philosophers more generally, Des Chene, *Spirits and Clocks* and "Life and Health in Cartesian Natural Philosophy," pp. 723–735; Shapiro, "The Health of the Body-Machine?," pp. 421f. At least since the 1950s French scholarship has been sensitive to this issue. In particular, see Gueroult, *Descartes' Philosophy According to the Order of Reasons* and Canguilhem, *La connaissance de la vie.* Gary Hatfield, who first discussed Descartes and medicine in Hatfield, "Descartes' Physiology and its Relation to his Psychology," pp. 335–370, has returned to the issue in Hatfield, "Animals," pp. 404–425. Hatfield's latest discussion provides an account of "natures" that distinguishes among physical kinds, although he does not develop the metaphysical underpinnings for such an account. I believe Descartes provides the necessary metaphysical underpinnings in Meditation Six as I detail in sections three and four.

appeal to the technical scholastic terminology of "extrinsic denomination." Prompted by the claims in sections two, three and four I conclude the paper by discussing the broader implications of my conclusions for the ontology of machines, the mechanical philosophy's contribution to medicine, and Descartes' position on the reality of an animal's healthy and diseased states.

10.2 Natural Philosophy and Medicine

Prior to discussing Descartes' view of health and disease, a more general historical point needs to be made. The main protagonists in medical theory during Descartes' lifetime, in spite of all that happened in the sixteenth-century and was beginning to happen in the seventeenth-century, remained Aristotle of Stagira, Galen of Pergamum and their respective followers.[16] Not only is there every indication that Descartes was familiar with their medical work, from his correspondence and personal notes we know that he was also familiar with more recent figures in medicine, such as Andreas Vesalius,[17] Jean Fernel,[18] Casper Bauhin,[19] Hieronymus Fabricius,[20] and William Harvey.[21] It has recently even been suggested that Descartes' medical interests are on display in nearly 20% of his surviving corpus, including large portions of the *Traité de l'homme*.[22] Whether this is an overstatement or not, it is clear that many of

[16] Hippocrates, Avicenna and Paracelsus need to be mentioned as part of this background, but from the standpoint of institutional learning in France, Italy and the Netherlands—the main venues through which Descartes became acquainted with learned medicine—Aristotle and Galen reigned supreme. It is little surprise then that the physician Plempius initiated his exchange with Descartes over the medical content of Discourse Five by drawing a battle line between the Aristotelians and the Galenists, a battle line which Descartes took in stride and rightly saw as a standard trope in medical disputes (Descartes, *Œuvres*, vol. I, p. 497).

[17] Descartes, *Œuvres*, vol. II, p. 522, transl. in *The Philosophical Writings of Descartes*, vol. III, p. 134.

[18] Descartes, *Œuvres*, vol. I, p. 533.

[19] Ibid., vol. II, pp. 591 and 592.

[20] Ibid., vol. IV, p. 555.

[21] It has been further argued that beside "scholastic commentaries and manuals, such as those of Riolan … [and] Cureau de la Chambre, Descartes was surely acquainted with.... Aselli, Bartholin … Van Helmont, Silvius, Fallopius, Colombus, [and] Paracelsus" (Georges-Berthier, "Le mécanisme cartésien," pp. 43f.). For more on Descartes' sources see the references cited in note 23.

[22] The 20% claim is made by Vincent Aucante in his *La philosophie médicale de Descartes*. Also worth noting is the work of Anne Bitbol-Hespériès who, in her edition of the *Traité de l'homme* and chapter two of her *Le Principle de Vie Chez Descartes*, traces the origin of many of the physiological descriptions Descartes offers. Bitbol-Hespériès even concludes that Descartes had at his disposal all the sources Harvey consulted in forming the theory of the circulation of the blood in the early decades of the seventeenth-century (Bitbol-Hespériès, "Cartesian Physiology," p. 362). Overall she reinforces the impression Thomas Hall gives, that Descartes was adapting the work of many others to fit his own theoretical interests, Hall, "Descartes' Physiological Method: Position, Principles, Examples," pp. 53–79. Also worth mentioning is Aucante's separate edition of Descartes' private notes, which provides the additional service of documenting the extensive number

Descartes' contemporaries saw him as actively engaged with medical questions. Not only did the University of Bologna attempt to hire Descartes as a professor of theoretical medicine in 1633, the subtitle to the first volume of Descartes' letters in 1657 announced that they considered "the most important questions of morality, physics, medicine and mathematics."[23] In 1645, with a retrospective air, Descartes even wrote to the Marquess of Newcastle that "the preservation of health has always been the principal end of my studies".[24]

Descartes' interest in medicine and medical questions should not surprise us. By the seventeenth-century it was a long established practice throughout Europe to first obtain a degree in the arts, which included the study of natural philosophy, before advancing to candidacy in medicine. Likewise, teaching careers still often began in logic and natural philosophy before advancing to a higher paying professorship in medicine where this faculty was present.[25] Practicing physicians, whether academically employed or not, were thus expected to be competent natural philosophers. And students drawn to natural philosophy who wanted to pursue their interests within the academy had little choice but to attend medical school, as opposed to seeking a degree in Law or Theology. As a result, natural philosophy and medicine were closely associated, both institutionally and in the lives of many students.

The example of Isaac Beeckman, who exerted a strong influence on Descartes' intellectual development, is a case in point. Beeckman had received a medical degree from the University of Caen just a few months prior to meeting Descartes in 1618. This much is well known, but less well known is the fact that corollaries Beeckman added to his medical thesis included a defense of atomism against medical polemics.[26] Beeckman's interest in atomism predated his medical studies, but his commitment to atomism took mature form only during his medical studies. And it was this atomism that Beeckman would share with Descartes in 1618. Insofar as

---

of traditional medical questions Descartes considered. Connecting Descartes' thinking with contemporaries like Kenelm Digby and Gassendi, as well as immediate predecessors such as Ambrose Paré, Jean Fernel, Jean Riolan the younger and the elder, Aucante argues convincingly that Descartes' near constant attention to medicine resulted in his medical views changing over time (this is also a theme in Aucante's *La philosophie médicale de Descartes*).

[23] Early modern philosophy's explicit and implicit connections to medicine have been slow to emerge in discussions among historians of philosophy. A prime example of the neglected connection between medicine and philosophy is Bologna's invitation to Descartes, which I detail in Manning, "Descartes and the Bologna Affair," in preparation.

[24] Descartes, *Œuvres*, vol. IV, p. 329.

[25] The literature on the history of Western universities up to the seventeenth century is vast and the details will vary by country and religious affiliation. For a general study of Italy see Grendler, *The Universities of the Italian Renaissance* and for France see Brockliss, *French Higher Education*.

[26] Beeckman, *Theses de febre tertiana intermittente*. Only the title page and corollaries to Beeckman's thesis survive, portions of which are reproduced in Beeckman, *Journal tenu par Isaac Beeckman*, vol. I, pp. 200–201. For more on Beeckman see the biographical entry in Berkel et al. *A History of Science in The Netherlands*, pp. 410–413. Beeckman's specific influence on Descartes is the subject of de Buzon's contribution to this volume (Chap. 6).

Descartes' later ambition to replace the natural philosophy of the schools owed something to Beeckman's atomism therefore, it owed something to medicine.

The rationale for this bond between the disciplines, which would have been labeled as the "subalternation" of medicine to natural philosophy, can be traced to the medical protagonists noted earlier.[27] It "behooves the natural philosopher" Aristotle wrote near the beginning of *De sensu*, "to obtain a clear view of the first principles of *health* and *disease* .... Indeed we may say of most natural philosophers .... that [they] complete their works with a discussion on medicine."[28] This very passage led Aquinas to comment that:

> It is the job of the natural philosopher to investigate the primary and universal principles that control health and illness; it is the physician's to put these principles into practice, following the idea that he is the maker of health ... [The physician] should not limit himself to making use of medicines, but should also be able to reflect on the causes [of health and illness]. To this end, the good physician begins his training [with] natural philosophy.[29]

Countless other natural philosophers and physicians active in the years between Aristotle and Aquinas and Aquinas and Descartes could be cited advocating the same disciplinary dependence, implying that the best philosopher of necessity discusses medicine and that *The best doctor is also a philosopher*, as the title of one of Galen's works proclaimed.

Descartes shared this strategy for supporting medicine. As a result, his effort to infuse mechanical thinking into natural philosophy also involved addressing fundamental questions in medicine; questions about the nature of health and disease in particular. Descartes' often cited tree metaphor from the letter preface to the *Principia* conveys precisely the relationship between medicine and natural philosophy that Aristotle envisioned:

> Thus the whole of philosophy is like a tree. The roots are metaphysics, the trunk physics, and the branches emerging from the trunk are all the other sciences, which may be reduced to three principal ones, namely medicine, mechanics and morals.[30]

---

[27] Some physicians and philosophers would reject this formal association and the idea that medicine could be a science on a par with natural philosophy. For an account of the disciplines in the medieval period see Weisheipl, "The Nature, Scope and Classification of the Sciences," pp. 461–482. For medicine and natural philosophy in particular see Ottosson, *Scholastic Medicine and Philosophy* and French, "Where the Philosopher Finishes, the Physicians Begins," pp. 75–106. For a discussion of the relationship between medicine and natural philosophy during the sixteenth-century, see Schmitt, "Aristotle among the Physicians," pp. 1–15.

[28] Aristotle, *Complete Works*, 436a 17–22; emphasis added and slightly amended. See also ibid., 480b 22–30 for an equally perspicuous statement of the same view.

[29] "ad naturalem philosophum pertinet inuenire prima et universalia principia sanitatis et infirmitatis; particularia autem principia considerare pertinet ad medicum, qui est artifex factiuus sanitatis ... [medici] non solum experimentis utentes sed causas [sanitatis et egritudinis] inquirentes ... et hec est ratio quare medici bene artem prosequentes a naturalibus incipiunt" (Aquinas, *De sensu*, 8.277–9.316).

[30] Descartes, *Œuvres complètes*, vol. IX-2, p. 14; transl. in *The Philosophical Writings of Descartes*, vol. I, p. 186.

Medicine grows out of natural philosophy or "physics." But the *Principia* are short on details about this organic relationship.[31] An earlier and considerably longer passage from the *Discours* provides greater clarity. Describing what his physics will ultimately accomplish, Descartes claimed:

> [W]e could use this knowledge—as the artisans use theirs—for all the purposes for which it is appropriate, and thus make ourselves, as it were, the lords and masters of nature. This is desirable not only for the invention of innumerable devices (*artifices*) ... but also, and most importantly, for the maintenance of health, which is undoubtedly the chief good and the foundation of all the other goods in this life .... [A]ll we know in medicine is almost nothing in comparison with what remains to be known, and ... we might free ourselves from innumerable diseases, both of the body and of the mind, and perhaps even from the infirmity of old age, if we had sufficient knowledge of their causes and of all the remedies that nature has provided.[32]

Descartes' view is roughly this: the knowledge we gain in physics is causal knowledge that can be put into practice to greatly improve our lives. It does this primarily by better equipping us to manipulate the world to produce desirable effects. Just as artisans are able to manipulate the world with their artifacts, we will know precisely what needs to be done to produce the outcomes we desire. In medicine, in particular, what we gain is knowledge of the causes of health and disease, which we can then exploit to create a more efficacious regimen and therapeutics. In short, for Descartes it is just as Aristotle and Galen and Aquinas and the Schools had indicated it should be.

10.3 The Human Being as If a Machine

Understanding now why Descartes took such a strong interest in medicine, we can return to his mechanical philosophy with greater appreciation of what it had to accomplish: it had to preserve the link between natural philosophy and theoretical medicine by, at the very least, providing insight into the nature of health and disease. It is in the *Traité de l'homme* that the operations of the human body—the physician's traditional object of study—take center stage for the first time in Descartes' writing, and indeed, the *Traité de l'homme* bears several hallmarks of a medical text, including especially its emphasis on anatomical knowledge.[33]

[31] The forth to last proposition of the *Principia* strikes a similar cord, speaking of "medicine and mechanics" and "all the other arts which can be fully developed with the help of physics," Descartes, *Œuvres*, vol. VIII–1, p. 327, transl. in *The Philosophical Writings of Descartes*, vol. I, p. 289.

[32] Descartes, *Œuvres*, vol. VI, pp. 61–62, transl. in *The Philosophical Writings of Descartes*, vol. I, pp. 142f.

[33] From a text translated into Latin in 1523, Galen explained that the physician studied "the human body," Galen, *Galen's Hygiene*, p. 5. Prior to the sixteenth-century, the most important definition of medicine and its subject matter in the Latin west came from Avicenna's *Canon*, where "medicine is the science by which we learn the various states of the human body," Avicenna, *Canon*, p 715.

In the opening to the *Traité de l'homme* Descartes introduces the supposition that bodies, and specifically the human body, should be viewed as machines whenever we try to explain their actions. This is a pivotal claim for Descartes, reiterated later by van Hogelande, explained in detail by La Forge and embraced by other supporters of the mechanical philosophy. In the remainder of this section I will concentrate on this hypothesis and what it tells us about machines and the human body. In the next section I will examine what the *Meditationes* has to tell us about health and disease.

The *Traité de l'homme* begins with Descartes' announcement that he intends to:

> [S]uppose the body to be nothing but a statue or machine [*machine*] made of earth, which God forms with the explicit intention of making it as much as possible like us. Thus God not only gives it externally the colors and shapes of all the parts of our bodies, but also places inside it all the parts required to make it walk, eat, breathe, and indeed to imitate all those of our functions which can be imagined to proceed from matter and to depend solely on the disposition of our organs.
>
> We see clocks, artificial fountains, mills, and other such machines [*machines*] which, although only man-made, have the power to move of their own accord in many different ways. But I am supposing this machine to be made by the hands of God, and so I think you may reasonably think it capable of a greater variety of movements than I could possibly imagine in it, and of exhibiting more artistry than I could possibly ascribe to it.[34]

The "supposition" or hypothesis about bodies being machines is the touchstone for everything Descartes has to say about the human body in the *Traité de l'homme*. His analogy between God's fabrication and ordinary and familiar machines also serves to clarify and make plausible this fundamental supposition. I will return to these familiar machines in a moment but notice, as Canguilhem labored to emphasize, that before placing the human machine into their company Descartes' supposition leads him to specify a material, efficient, final and then a formal cause for the

---

The *Traité de l'homme*'s affinities with medical texts is noted in Cunningham, "The Pen and the Sword," pp. 631–665. For a more positive characterization of the inroads anatomical knowledge had made in *De anima* commentaries at the beginning of the seventeenth century than the one I offer, see Edwards, "Digressing with Aristotle," pp. 127–170. See the official prohibition against teaching portions of *De anima* directly related to anatomy and medical subjects in the 1599 *Ratio studiorum*'s "Rules for the professor of philosophy."

[34] Elsewhere Descartes implies he adopted the strategy of supposition because he had yet to develop an account of embryological development that began with an initial chaos of matter in motion subject only to the laws of nature, *Œuvres*, vol. II, p. 525, transl. in *The Philosophical Writings of Descartes*, vol. II, p. 99. The prospects for such an account never looked good to Descartes' followers, who, like Malebranche, tended toward preformationism. Historians of science and philosophy such as George Canguilhem, Jacques Roger and Geneviève Rodis-Lewis have also found such an account unappealing. Roughly, they argue Descartes is conceptually confused where he insists mechanical principles can be used to explain the creation of a machine and not just its operations; mechanical principles can do the latter, but not the former, Canguilhem, *La connaissance de la vie*, pp. 115f.; Roger, *Les sciences de la vie*, pp. 151f.; Rodis-Lewis, *L'anthropologie cartésienne*, pp. 149–167. My own view is that the ontology of corporeal substance does preclude what these scholars claim cannot be done, but that once we have the right understanding of a machine's ontology the mechanical philosophy can explain the creation of a machine. Such an explanation is reconstructed on Descartes' behalf in Hatfield, "Animals," pp. 440–425.

human machine.[35] The material cause is "earth," a component of the matter theory developed in *Le monde*. The efficient cause is "God," whose intention—and here the final and formal causes are introduced—is to create a machine as much like a genuine human being's body as possible.

While La Forge seems to have taken Descartes to be minimizing the role of creators and their purposes in his appeal to a machine—they are not mentioned in La Forge's definition of a machine when commenting on this passage—both are actually present in the *Traité de l'homme*, and both do genuine work. Citing God as the machine's creator allows Descartes to bypass concerns that no artisan could make such a variously moving machine capable of imitating a human being's body.[36] Only a perfect maker would suffice. The comparison with familiar machines also calls attention to a maker's purpose. All the machines Descartes cites are literally geared to produce their effects.[37] In the case of the human machine, God's purpose is "to imitate all those of our functions which can be imagined to proceed from matter." This final cause does nothing less than initiate the other causes; it is the cause of the other causes in the *Traité de l'homme*. Thus, in supposing that the human being's body is a machine in the *Traité de l'homme*, Canguilhem is right that Descartes specified all four Aristotelian causes. This point has been obscured by many factors, however, not least La Forge's failure to include any explicit mention of final causes in his definition of a machine and Descartes' own now famous repudiation of final causes beginning in the 1630s.[38]

There is one final point about the *Traité de l'homme* that needs to be made. As Descartes details the supposition, God's skill as a craftsman has no limits aside from those imposed by the pre-existing material used in his creation.[39] Even with this limitation, however, Descartes believes the fabricated machine "capable of a greater variety of movements than I could possibly imagine" owing to God's perfection. As his readers would know from *Genesis*, our bodies come from the earth too, so an obvious question is whether there is any difference between our human bodies and these machine bodies, these human machines. In the final paragraph of the *Traité de*

[35] Canguilhem, *La connaissance de la vie*, pp. 112f.

[36] Something similar can be said about the "fable" introduced in *Le monde*.

[37] Cf. Descartes' very intriguing exchange with Pierre Gassendi where Descartes insists God should be esteemed only as an efficient cause in light of his complex and beautiful creations, *Œuvres*, vol. VII, pp. 374–375. The exchange with Gassendi highlights the asymmetry between God's machine, which is the corporeal world itself, and manmade machines, whose creators can surely be praised for their intelligence and good intentions.

[38] Descartes' public objections to the use of final causes in physics do not begin until the composition of the *Meditationes*, but they become a consistent feature of his repertoire after that. The nature of his objections varies, however, between our lacking access to God's purposes and purposes in general lacking a role in proper explanations. See Descartes, *Œuvres*, vol. IV, p. 113 and ibid., vol. V, p. 158 respectively.

[39] Of course Descartes' God makes the matter as well. Yet, once the matter has been created, the immutable character of God's activity is such that God is no less constrained by his initial act of creation than we are. To this extent it seems right to speak in a Platonic fashion about the God and the constraints under which he works in the *Traité de l'homme*.

l'homme Descartes claimed that "they imitate perfectly the movements of a real man." Of course, imitation is not identity, but the only apparent difference between human bodies and human machines has to do with the manner of their creation. Whether this difference qualifies as a significant one, such that the human machines found in the *Traité de l'homme* cannot be the bodies of human beings is not one that Descartes answered in the *Traité de l'homme*.[40] Instead, he simply proceeds as though perfect imitation implies identity. He moves from treating the human being's body *as if* it is a machine according to his initial supposition to concluding that the human being's body is a machine because of what mechanical explanations suffice to explain.[41] Having accomplished all this by explicit appeal to a maker's skill and purpose, we are left to wonder why La Forge provided such a limited characterization of a machine in his commentary to the *Traité de l'homme*. A possible answer will emerge in the next section.

10.4 Prelude to the Healthy Human Body When Viewed as a Machine

Over the course of his life Descartes studied many diseases and proposed numerous therapies to correspondents.[42] If the interview recorded by the Dutch student Frans Burman is to be trusted, Descartes even subscribed to the view that "nature herself works to effect her own recovery."[43] This remark came on the heels of Descartes having criticized physicians, who he thought would be well advised to follow nature's lead. The clear implication of a remark like this, even if it does not attribute

[40] One might think of the language test from Discourse Five in this context, but there the issue is not whether fabricated human bodies are identical to ours because their lack of identity is taken for granted. At issue instead is whether a mere body can do what a soul-infused body can do. In fact, the test is not about finding minds at all but about identifying the limits of mechanical intelligence. I argue for this claim against alternative interpretations in Manning, "Descartes, Other Minds and Impossible Human Bodies," in preparation.

[41] Descartes is often criticized for his tendency to move from a hypothesis made plausible by analogies and mechanical explanations to the conclusion that he had demonstrated his hypothesis. For some recent discussion, see Clarke, "Hypothesis," pp. 249–271; McMullin, "Hypothesis in Early Modern Science," pp. 7–37; Ariew, "The New Matter Theory and Its Epistemology," pp. 31–48. In the present case the potentially dubious move is from a hypothesis that the human body is a machine to the conclusion that the two are identical. I attempt to sort out this and related issues in Manning, "Analogy and Falsification in Descartes Physics," in preparation.

[42] For a detailed account of Descartes' therapeutics, see Aucante, *La philosophie médicale de Descartes*, pp. 375–416.

[43] Descartes, *Œuvres*, vol. V, p. 179, transl. in *The Philosophical Writings of Descartes*, vol. III, p. 354. This account of the physician's role in reestablishing health developed from the Hippocratic claim in *Epidemics* 6.5: "The body's nature is the physician in disease." For a discussion of "nature" as used by physicians, see Neuburger, *The Doctrine of the Healing Power of Nature Throughout the Course of Time*.

knowledge to nature, or our individual natures, nevertheless assigns an end to her teachings. In particular, what nature teaches us is the best way to preserve our health or reestablish it when it is lost. In other words, our nature is directed to our health or preservation and it is our role, and the role of the physician, to learn from and aid our nature in her pursuit of this end. The immediate question for Burman, in light of Descartes' published repudiation of final causes as well as his apparent identification of machines with corporeal substances, should have been whether, and in what sense, any nature can have such an end. Burman also might have inquired about the status of nature herself, which for Descartes would seem to be matter or extension. Burman might also have asked relative to what norm or standard should our health and illness be judged? Fortunately, Descartes outlines answers to these and related questions in his comments about the teachings of nature in Meditation Six. It is there that his metaphysics, natural philosophy and medicine most productively come together in an account of health and disease.[44]

In order to discuss the relevant passage from Meditation Six it is necessary to set the scene. Recall that by this point in the *Meditationes*, Descartes has rehabilitated our sensory perceptions after their fall from grace in Meditation One. Instead of serving as guides to the way the world is itself, the senses are now characterized as guides to "what is beneficial or harmful for the composite of which the mind is a part."[45] There are teleological and normative considerations manifest in this claim. Just a few lines later Descartes explicitly brought out the teleological and normative aspects of sensory experience when he noticed "it is not unusual for us to go *wrong* even in cases where nature does urge us towards something." The point is clear: our nature, given through the medium of the senses, can misinform us and thereby misguide us as to what is beneficial and harmful. The specific example Descartes goes on to cite is that of dropsy.[46] When suffering from dropsy our desire to drink leads us into error because water, which the senses inform us is beneficial and desirable, is not actually good for us. In spite of the theodicy from Meditation Four then, the errors we commit owe something to God, who endowed us with a nature that includes the sense-based desires that are in error. As Descartes puts it, a "sick man is no less one of God's creatures than a healthy one, and it seems no less a contradiction to suppose that he has received from God a nature which deceives him."[47]

To resolve this challenge to God's goodness, Descartes initially suggested we go wrong in such cases because our "nature is corrupted [*corrupta*]." This, however, will not do without significant qualification. I will say more about this in a moment, but, as

[44] The *Passions* is also a good place to find Descartes' metaphysics, natural philosophy and medicine coming together, though in the service of his moral philosophy. Whether the claims found in the *Passions* and Meditation Six are compatible is a question I do not explore.

[45] Descartes, *Œuvres*, vol. VII, p. 83, transl. in *The Philosophical Writings of Descartes*, vol. II, p. 57. For a recent discussion of the teleological implications of this claim in English, see Simmons, "Sensible Ends," pp. 49–75.

[46] Dropsy is now known as edema, a disorder in which the desire to drink, if fulfilled, can result in the exacerbation of symptoms and possibly even in death.

[47] Descartes, *Œuvres*, vol. VII, p. 84, transl. in *The Philosophical Writings of Descartes*, vol. II, p. 58.

with nearly every answer Descartes rejected in the *Meditationes*, it is worth pausing to consider whether it might have already been proposed, and by whom. In this case, as in so many others, Aristotle is one of Descartes' likely targets. In his *Metaphysics* Aristotle characterized disease as the privation or corruption of a healthy nature.[48] Beyond just Aristotle, however, the presence of a "corrupted nature" is most frequently cited in the learned medical tradition. In late antiquity Galen would insist that disease is "the opposite [of health], i.e. ... some constitution contrary to nature" and this view persisted among physicians long after Descartes.[49] The terminology of Galen's medicine was so entrenched, in fact, that most seventeenth-century physicians continued to cite contra-naturals (*res contra naturam*) in discussing health and disease. Even Cartesian physicians like Henricus Regius continued to use Galen's terminology well into the second half of the seventeenth-century.[50]

Returning to Descartes' appeal to a corrupted nature, he goes on to argue that without further clarification thinking of disease or illness as a corruption has unacceptable consequences. Specifically, it still implies that God is responsible for our errors. As he said, "this [appeal to a corrupt nature] does not remove the difficulty" of casting God as a deceiver because it is our very nature that misinforms us, when it becomes corrupted. With this claim, Descartes is hovering very close to an attack on the accepted view of learned physicians and a cornerstone of medical practice. Descartes will go on to argue, on the basis of experience, that we can reconcile our errors with God's perfect goodness.[51] In the end then, Descartes' will accept that our nature is sometimes corrupted—just as the physician's claim.

Remarkable as this eventual about face may seem in light of what Descartes had written up to this point, his agreement with the physicians is of less interest to us than the initial move he makes to distance God from the deceptions of our corrupt nature. Descartes wrote:

> Perhaps it may be said that they go wrong because their nature is corrupted [*corrupta*], but this does not remove the difficulty .... A clock constructed with wheels and weights observes all the laws of nature [*leges omnes naturae observat*] just as closely when it is badly made and tells the wrong time as when it completely fulfils the wishes of the clockmaker.

On Descartes' view, the nature of an ordinary machine, like a clock, always respects the laws of nature. Insofar as such machines frustrate our purposes it is because the laws of nature do not bend to our will. Our challenge when building

---

48 Aristotle, *The Complete Works*, 1044b 29–34.

49 Galen, *On Diseases and Symptoms*, pp. 134f. There are too many texts to cite, but the challenge would be to find alternative accounts of disease among learned physicians.

50 Laurence Brockliss and Colin Jones observe that medicine in early modern France was "until about 1690 ... in most respects fundamentally Galenic," Brockliss and Jones, *The Medical World of Early Modern France*, p. 99. For more on Descartes' and Regius' traditional medicine, see Manning, "Naturalism and Un-Naturalism Among the Cartesian Physicians." For Regius' medicine and more generally his relationship with Descartes, see also Bitbol-Hespériès, "Descartes et Regius: leur pensée medicale," pp. 47–68 and Verbeek, *Descartes Among the Dutch*, pp. 13–33.

51 Descartes, *Œuvres*, vol. VII, pp. 87–88, transl. in *The Philosophical Writings of Descartes*, vol. II, pp. 60–61.

machines, in fact, is always to realize our purposes within the constraints set by nature and her laws. It is because of this that Descartes believed a better physics would produce better machines and, eventually, a better medicine.

Yet this analysis of a clock jeopardizes the very idea that an ordinary machine can have a corrupted nature. Descartes' standard for corruption is located entirely into the mind of the watchmaker. The machine itself is never corrupted because qua machine it is always doing what the laws of nature require. Fortunately for Descartes, he goes on to say much more about the reality of a human machine's healthy and diseased states as well as its "nature". The passage needs to be quoted in full:

> In the same way, if I should consider [*ita, si considerem*] the body of a man as a kind of machine equipped with and made up of bones, nerves, muscles, veins, blood and skin in such a way that, even if there were no mind in it, it would still perform all the same movements as it now does in those cases where movement is not under the control of the will or, consequently, of the mind. I can easily see that if such a body suffers from dropsy, for example, and is affected by the dryness of the throat which normally provides in the mind the sensation of thirst, the resulting condition of the nerves and other parts will dispose the body to take a drink, with the result that the disease will be aggravated. Yet this is just as natural as the body's being stimulated by a similar dryness of the throat to take a drink when there is no such illness and the drink is beneficial. Admittedly, when I consider the purpose of the clock, I may say that it is departing from its nature when it does not tell the right time; and similarly when I consider the mechanism of the human body in relation to the movements which normally occur in it, I may think that it too is deviating from its nature if the throat is dry at a time when drinking is not beneficial to its continued existence [*conservationem*] .... As I have just used it, 'nature' is simply a denomination [*denominatio*], which depends on my thought; it is quite extrinsic [*extrinseca*] to the things to which it is applied, and depends simply on my comparison between the idea of a sick man and a badly made clock, and the idea of a healthy man and a well-made clock. But by 'nature' in the other sense I understand something which is really to be found in the things themselves; in this sense, therefore, the term contains something of the truth .... When we say, then, with respect to the body suffering from dropsy, that is has a disordered nature because it has a dry throat and yet does not need drink, the term 'nature' is here used merely as an extrinsic denomination [*denominatio extrinseca*]. However, with respect to the composite, that is, the mind united with this body, what is involved is not a mere denomination, but a true error of nature, namely that it is thirsty at a time when drink is going to cause it harm.[52]

There are obvious similarities here to the opening paragraph from the *Traité de l'homme*. Nevertheless, far more intricate claims are being made in Meditation Six than in the earlier work and we should proceed slower.

Before explaining Descartes' remarks about health and disease, it will help to begin with something recognizable: Descartes' appeal to ordinary and familiar machines. We saw that in the *Traité de l'homme* Descartes began by supposing a machine made to perfectly imitate our bodies "and all those functions which can be imagined to proceed from matter." After having specified all four Aristotelian causes in the creation of this machine—material cause: earth; efficient cause: God; formal cause: human form; final cause: imitation—Descartes proceeded to cite a number of more familiar machines to

[52] Descartes, *Œuvres*, vol. VII, pp. 84f., transl. in *The Philosophical Writings of Descartes*, vol. II, pp. 58f., slightly modified.

make his supposition plausible. He specifically emphasized what a supremely gifted craftsman could do with the right parts. In the *Meditationes* Descartes proceeds in the reverse order. The introduction of a familiar machine, this time just a clock, takes place prior to supposing the human being's body is a machine. The point of juxtaposing familiar machines with the human being's body, when viewed as a machine, is the same in both cases, however. Descartes wants us to imagine less of an ontological gap between nature and art than we may have initially been inclined to see.

A less obvious similarity between the two works is Descartes' retention of the initial *as if* strategy from the *Traité de l'homme*. Descartes uses the Latin subjunctive *considerem*, but in keeping with his earlier slide into identifying our bodies with machines in the *Traité de l'homme*, he fails to match *considerem* with another imperfect subjunctive later in the *Meditationes*. The human body might as well be a machine is Descartes' message; the two might as well be identical for natural philosophical purposes. The material cause is also much more elaborate in Meditation Six. No longer just made of "earth," the human being's body, when viewed as a machine, is reportedly composed of homoeomerous parts: "bones, nerves, muscles, veins, blood and skin." By contrast, the formal cause in Meditation Six remains exactly the same as in the *Traité de l'homme*. It is still the pre-existing human body, for the new machine is characterized as capable of performing all the "same movements" that the human being's body "now does." Moreover, the states of the machine, such as dryness in its throat, are still referred back to the human being's body and what "normally provides in the mind the sensation of thirst."

What of the other causes from the *Traité de l'homme*? For example, the ones La Forge failed to mention in his definition of a machine? Although Descartes cites a clockmaker, there is no explicit appeal to God as the efficient cause of the human machine, as there had been in the *Traité de l'homme*. Perhaps because deception is not an option after Meditation Three, God's role and the final cause of imitation shift almost entirely to our choice to "consider the body of a man as a kind of machine." In other words, what we have in the *Meditationes* is a straightforward supposition that our body is a machine without God figuring as the agent who makes the supposition plausible. In spite of this, God reenters as the guarantor that our supposition is of a genuine possibility in Meditations Three through Five because God alone can assure the veracity of our power of conceiving. Most importantly, however, and it bears repeating, because of our choice to consider the human being's body as a machine the final cause of the machine remains imitation.

10.5 The Healthy Human Body When Viewed as a Machine

Unfortunately, what we have just learned does not take us very far into the long passage from Meditation Six cited in the previous section. In fact, in the portions I have yet to discuss there are no obvious parallels to draw to the *Traité de l'homme*. Nevertheless, Descartes' most explicit discussion of the metaphysical basis for ascriptions of health and disease along with the attribution of natures to machines is found in Meditation Six. Prompted by a question about our corruptible nature, Descartes contrasts our nature with the nature of mere machines in terms of "extrinsic denominations." In the remainder of this section I will attempt to focus on these aspects of the text for, properly understood, the long passage from Meditation Six shows how Descartes used machines to connect his medicine and his mechanical philosophy. By the end of this section we will also be equipped to see how a machine differs from a mere corporeal substance, a point I will address in my concluding remarks.

Returning to the long passage from Meditation Six, recall that if we judge the healthy and diseased states of a human being' body, when viewed as a machine, by what Descartes says about the clock and its nature, then health and disease look to be entirely mind-dependent. This is a serious misinterpretation of Descartes' text, however, to say nothing of being an entirely implausible view. Still, it is an interpretation with some textual support, not the least of which is Descartes' dualism between mind and body. To bring this out more clearly, consider the case of our mental health. The mind is an immaterial substance and it is usually assumed that the mind cannot be discussed within a mechanized natural philosophy that concerns itself solely with bodies in motion.[53] In other words, the mind qua mind is the wrong kind of ontological entity for the Cartesian natural philosopher to study. However we end up describing mental health, because the mind is immaterial neither its healthy nor its unhealthy states can be adequately grasped given the resources of corporeal substance alone.

A human being's body, on the other hand, presents a different but no less significant challenge where its healthy and diseased states are concerned. In keeping with the paradigm case of the clock, the human being's body qua body is subject to no other laws than the laws of nature, i.e. of motion or mechanics. This is precisely Descartes' point in connection with the claim regarding clocks that fail to tell the right time. They "go wrong because their nature is corrupted, but this does not remove the difficulty [because] … a clock constructed with wheels and weights observes all the laws of nature just as closely when it is badly made." Assuming we would be able to identify any corruption of nature where the laws of nature are concerned, in such a case we would be dealing with a violation or deviation from the laws constitutive of a corporeal substance's nature. In fact, according to Descartes we are describing deviations from laws that emanate from God's own immutable nature. And as a result, we would effectively be dealing with miracles when describing illness. This is certainly not the basis on which to build a reliable and scientific medicine. Thus, however tempted we might be to claim that our bodies are letting us down when we fall ill, on the interpretation I am considering this cannot literally be the case. Instead, all we could be claiming with such expressions is that our own expectations or purposes have not been met. In other words, the vocabulary of "natural/un-natural" and "healthy/diseased" is without a foundation in Descartes' mechanical philosophy.

As I have indicated several times now, this conclusion places a great deal of emphasis on Descartes' dualism and what he says about a clock in Meditation Six.

[53] This understanding of Descartes' position is advocated, e.g., in Wilson, *Descartes*, p. 99 and Voss, "Descartes: Heart and Soul," pp. 173–196.

Yet, this interpretation makes it very difficult to understand how Descartes' recognition of a "true error of nature" in the case of the "composite" of mind and body is compatible with his mechanical philosophy and his expectation to found a scientific medicine. Perhaps one might want to insist at this point that the mechanical philosophy can take into account the full human being—the composite of mind and body. But to the extent Descartes addresses himself to the composite of mind and body when acting as a mechanical philosopher, he does so only insofar as he supposes the body of a human being is a machine.[54] And this supposition in the *Traité de l'homme*, *Meditationes*, and elsewhere just lands us back where we were in the previous paragraph. Whether we are simply proceeding as if the human being's body is a machine, or have concluded that it is a machine because of our successful mechanical explanations, in Meditation Six Descartes denies the truth of any claim that attributes a corruptible nature to a machine identified with corporeal substance. This includes the human being's body, when viewed as a machine.

The primary mistake committed in the interpretation described in the last three paragraphs is to read the claim that a "nature" is "extrinsic" as implying that the healthy or unhealthy states of the human being's body, when viewed as a machine, are somehow illusory or un-real, simply in the eye of the beholder or unrelated to the human machine as it is in itself. "Extrinsic denominations" are, however, constitutive of any machine, human or otherwise. To appreciate this claim, what is needed is a better understanding of extrinsic denominations.

Extrinsic denominations and their opposite, intrinsic denominations, make frequent appearances in scholastic philosophy after Aquinas. They are cited in disputes about the analogical predication of properties shared by God and finite beings, the nature of relations and even the semantic as opposed to ontological character of Aristotle's *Categories*.[55] They also figure in disagreements about formal and objective

[54] We must be cautious about over-reading the stipulation in the letter preceding the text of the *Passions* where Descartes describes his efforts to discuss the passions of the soul as a "Physicist [*en Physicien*]." Descartes is not, I think, discussing the soul *per se* in the *Passions*, but only insofar as it can effect and be effected by the body.

[55] As with a great number of scholastic technical terms, reading primary texts often amounts to multiplying meanings, and especially so in the case of analogical predication, where "extrinsic denomination" is first introduced. The difficulties I have encountered in tracing the origin of the intrinsic/extrinsic distinction are not mine alone. Joshua Hochschild notes that in general "it is remarkable that there is so little explicit reflection and explanation of the notions of intrinsic and extrinsic denomination, both in modern scholarship and in the medieval authors. While the distinction has obvious precedents in Aquinas and before, it appears as a technical term only later, and the examples and applications quickly become familiar, but ... the notion of extrinsic denomination is taken for granted and neither fully defined nor explained," Hochschild, "Logic or Metaphysics in Cajetan's Theory of Analogy," p. 54. Similarly, Stephen Menn notes that scholastic "terminology on the kinds of analogy is confusing. Sometimes the scholastics count analogous terms as a special kind of equivocals, sometimes as midway between equivocals and univocals; sometimes it is linguistic items and sometimes their significata that are equivocal or univocal or analogous." Menn, "Metaphysics," p. 170. Dozens of uses of "extrinsic denomination" in late scholasticism are documented in Doyle, "Prolegomena to a Study of Extrinsic Denomination in the Work of Francis Suarez, S.J.," pp. 121–160.

concepts, and Descartes adeptly handles them in this connection in his replies to Caterus.[56] Antoine Arnauld and Pierre Nicole also mention them in *La logique ou l'art de penser*. Specifically identifying denominations of substances as "modes," Arnauld and Nicole advise us to:

> note that some may be called internal because they are conceived in the substance, such as "round" and "square." Others may be called external because they are taken from something that is not in the substance, such as "loved," "seen," and "desired," names derived from the actions of something else. In the Schools these are called *external denominations* [*dénomination externe*].[57]

There are immediately two points to notice. First, to call extrinsic denominations "modes," even if they mean modes taken from something "not in the substance," is to grant them ontological status in a substance-mode ontology. Now perhaps they are relational modes, but still they are modes. This suggests that there might be room for extrinsic denominations in Descartes' own substance-mode ontology. Second, the cited examples of extrinsic denominations suggest that they can appear in a true statement about a substance. So take the denomination "loved," which relates two substances, as is also the case for "seen" and "desired."[58] "Romeo is loved by Juliet," is a true statement about Romeo. It is what the scholastics would call an accidental property of Romeo that Romeo is loved by Juliet, but accidental properties, just like Cartesian modes, have being. They inhere in or depend on a substance for their existence, and they are not unreal for that.

The falsification of "Romeo is loved by Juliet" reveals what is distinctive about an extrinsic denomination. It is, in the words of Arnauld and Nicole, and using the present example, "derived" from one of Juliet's "actions."[59] Put a slightly different way, for "Romeo is loved by Juliet" to be false, Juliet's action of loving Romeo must cease. Romeo can, of course, indirectly change the truth of "Romeo is loved by Juliet," but by changing himself Romeo really is seeking to change Juliet's act of loving. This is no doubt why some are tempted to think of extrinsic denominations like "loved" as less than a real mode of Romeo. Nevertheless, we will only be tempted in this way if "loved" is compared to non-relational modes. When relational-properties are compared to non-existent properties—properties that do not so much as enter into true claims about a

[56] Descartes, *Œuvres*, vol. VII, pp. 102–103, transl. in *The Philosophical Writings of Descartes*, vol. II, pp. 74–75.

[57] Arnauld and Nicole, *Logic or the Art of Thinking*, p. 32.

[58] Self-referential cases are interesting because the extrinsic denomination "loved" must be in the substance that is loved, but I will not be discussing such cases here.

[59] Instead of thinking of "loved" as a relation, this may make it seem as though "loved" belongs in the Aristotelian category of passions and "loves" in the category of actions. Regardless, this only goes to show that identifying extrinsic denominations with relations is not sensitive enough to the broad use made of extrinsic denominations in the seventeenth century and before.

substance—the fact that they exist and attach to substances should be apparent. It may also be worth recalling that relations are included in Aristotle's *Categories* as one of the class of accidental properties.[60]

A further important point about extrinsic denominations has to do with their connection to intrinsic denominations. We saw that "loved" is a relation, but by virtue of being an extrinsic denomination of Romeo it does not follow that the extrinsic denomination's source in Juliet must also be an extrinsic denomination. To the contrary, "Romeo is loved by Juliet" is an example of an extrinsic denomination, and "Juliet loves Romeo" is an example of an intrinsic denomination. The latter is falsified when Juliet falls out of love with Romeo and it is fundamentally a change in Juliet that makes this happen, however unlovable Romeo becomes. The acts of naming or predicating "loved" of Romeo are based on Romeo's relation to a mode of "love" in Juliet. Extrinsic denominations according to Arnauld and Nicole thus require two things: (a) an intrinsic denomination in a substance or subject and (b) a relation to that intrinsic denomination.

These two conditions are precisely what we find articulated, with greater nuance, by scholastics like Cajetan. In Cajetan's commentary to Aquinas' *Summa*, a text that Descartes used during his time at La Fleche, we learn that:

> Denomination is twofold. One is intrinsic and the other extrinsic. A denomination is called intrinsic when the form of the denominative is that which is denominated, say, white, quantity, etc.; whereas a denomination is extrinsic if the form of the denominative is not in the denominated thing .... In two ways a thing may be called such or such after something extrinsic. In one way, if the reason [*ratio*] for the denomination is the very relationship to something extrinsic .... In another way, if the reason for the denomination is not a relationship of similitude of any other, but a form which is the foundation of a relationship of similitude to an extrinsic thing.[61]

---

[60] This point is ably made by Mark Henninger. The following is especially a propos: "Despite the variety of theories, no one held that real relations are completely mind-dependent .... Given the pervasive Aristotelianism, it would be extremely difficult to deny all extra-mental reality to relations, for the scholastics interpreted Aristotle as explicitly teaching that relation is one of the ten categories of extra-mental being" (Henniger, *Relations*, p. 174). Discussions of the *Categories* as understood by a variety of scholastics can be found in Biard and Rosier-Catach (eds.), *La tradition médiévale des catégories*. An overview of the available positions that were being taken closer to Descartes' time can be found in McMahon, "The Categories in some Post-medieval Spanish philosophers," pp. 355–370.

[61] This translation comes from Cajetan, *The Analogy of Names* (I have not consulted the original in its entirety but a portion of the Latin is reproduced in Hochschild, "Logic or Metaphysics in Cajetan's Theory of Analogy," p. 55 and the translation is accurate). One finds something similar in John of Saint Thomas's lengthy discussion in his *Cursus philosophicus*: "Et quantum ad ipsam formam, manifestum est esse aliquid reale, sicut visio, qua paries denominatur visus, realis forma est in oculo; applicatio tamen eius, ut tangit subiectum denominatum, non est aliquid reale, quia nihil in ipso pariete point. Omne autem non reale apprenhensum est quid rationis, et sic ex parte applicationis in forma denominata aliquid rationis est denominatio estrinseca. Dicitur tamen denominatum subiectum ante operationem intellectus, non ratione eius, quod in illo point, sed ratione eius, quod extra illud supponit, quia in se est forma realis, sed non realiter existit in eo, quod denominat. Unde ratione non existentiale sumitur ut ens rationis, ratione auten praexistentiae in alio, a quo respicit rem denominatam, dicitur denominare ante operationem intellectus," Poinsot, *Tractatus de signis*, p. 55. Similar to this is the account found in Rubius's *In universam Aristotelis dialecticam*, cited in Gilson, *Index scholastico-cartésien*, p. 68.

Extrinsic denominations can occur either because of an extrinsic relation or an extrinsic foundation of a relation. In either case, the "form" is just the property denominated and for an extrinsic denomination the form is in whatever is intrinsically denominated. In the case of "loved," the intrinsic denomination is "loves" and the relation supporting "Romeo is loved by Juliet" is the fact that Romeo is the object of Juliet's love. A case of extrinsic denomination is thus a case of predication in which something true is said of Romeo and it is true because Romeo stands in a specific relation to Juliet. Still, Juliet carries the primary burden of justifying the predication's truth because the property denominated of Romeo is one of her intrinsically denominated "forms."

There is one more point to make about extrinsic denominations before returning to Descartes. When Aquinas introduced the idea of extrinsic denomination, though not precisely the phrase, he had other examples besides "loved" in mind. He was specifically thinking of properties like being and goodness as well as Aristotle's other examples of paronyms.[62] Here, is one of Aristotle's examples from the *Metaphysics*:

> Everything which is healthy is related to health, one thing in the sense that it preserves health, another in the sense that it produces it, another in the sense that it is a symptom of health, another because it is capable of it.[63]

Aquinas also utilized the predicate "healthy" in *De veritate* when introducing extrinsic denomination:

> [T]hus, urine is called healthy in respect to the health of the animal. For, the understanding of "healthy" as predicated of urine serves as a sign of the health of the animal. In such cases, what is denominated in respect to something else does not receive its denomination from a form inherent in it, but from something extrinsic, to which it is referred.[64]

The same example was also used by Cajetan and by many scholastics after Aquinas.[65] There common position was that "healthy" was an intrinsic denomination of animals because the form or property of health resides in animals. So long as

[62] Julie Ward calls these examples, meant to describe an alternative to homonymy and synonymy, "systematic homonyms" in her *Aristotle on Homonym*. The standard account of Aristotle on this topic comes from Joseph Owens, who describes these examples in terms of the "focal meaning" assigned to intrinsic denominations, Owens, *The Doctrine of Being in the Aristotelian Metaphysics*, pp. 107–136.

[63] Aristotle, *The Complete Works*, 1003a 35–38.

[64] "… sic urina dicitur sana per respectum ad sanitatem animalis. Ratio enim sani, secundum quod de urina praedicatur, est esse signum sanitatis animalis. Et in talibus, quod denominatur per respectum ad alterum, non denominatur ab aliqua forma sibi inhaerente, sed ab aliquo extrinseco ad quod refertur" (*De veritas*, q. 21, art. 4). Aquinas refers to extrinsic denominations at least two other times to my knowledge, both occurring in the *Summa theologica* I. q. 16, art. 6 and III, q. 60, art. 1.

[65] "Dupliciter enim contingit aliquid dici tale ab aliquot extrinsico. Uno modo, ita quod ratio denominationis sit ipsa relatio ad extrinsecum, ut urina dicitur sana, sola ratione signi ad sanitatem," Cajetan, *Commentaria in summam Theologiae St Thomae* I. 6. 4; cited in Hochschild, "Logic or Metaphysics in Cajetan's Theory of Analogy," p. 55. Contemporaries of Descartes use this example frequently. Rudolphus Goclenius mentions it in his *Lexicon philosophicum* and

other things stand in appropriate relation to the intrinsically denominated "healthy animal," they can be called "healthy" too, such as medicine because it causes health, or urine because it is a sign of health. Medicine and urine are extrinsically denominated "healthy," but only because there exists (a) an intrinsic denomination of health in the animal and (b) these healthy things enjoy a relation to the intrinsically denominated "form" of health.

Returning to Descartes, the most obvious difference in the passage from Meditation Six and the frequently cited example of health among the scholastics is that instead of speaking of animals as denominated "healthy" by an intrinsic denomination, Descartes' initial discussion of illness leads him to discuss the denomination of "nature" and the relations that support predications of "natural/unnatural" as extrinsic denominations. For Descartes the form or mode supporting talk of a corrupt nature is restricted to human beings, to the composite of mind and body. This suggests that *all* other things, and specifically the human body, when viewed as a machine separate from the union, can at most be extrinsically denominated "healthy." Like medicine and urine in the scholastic account, only an appropriate relation to the human being who is intrinsically denominated "healthy" supports calling *anything* else "healthy."

We find additional evidence that this is Descartes' view in a late letter to Henry More. Although Descartes' response to More was motivated by a different topic, in the course of his discussion he drew the following comparison: "Now just as we say that health belongs only to human beings [*soli homini*], though by analogy [*per analogiam*] medicine and a temperate climate and many other things are also called healthy."[66] The repetition of this common scholastic claim, now couched in the terminology of analogy of attribution, obviously parallels the general claim from Meditation Six. It is striking for two reasons. First, it appears 8 years after the publication of the *Meditationes*, suggesting a consistent view on Descartes' part. Second, because Descartes thought more would find the point familiar and uncontroversial, we have reason to believe he thought the same would be the case for readers of the *Meditationes*. Neither More nor anyone else, to my knowledge, took exception to the example.

Eustachius a Santo Paulo uses the example of health at least twice, though he does not identify it as an extrinsic denomination, choosing instead to describe it as an analogy of attribution in the logic section of his work, *Lexicon philosophicum*, p. 507 and *Summa philosophiae quadripartita*, pp. 18 and 19 respectively. Francisco Suarez appears to be an exception to the rule of using the example of health in connection with extrinsic denomination, though it is often cited in Suarez's discussion of final causes. For references to extrinsic denominations in Suarez see Doyle, "Prolegomena to a Study of Extrinsic Denomination in the Work of Francis Suarez, S.J."

[66] Descartes, *Œuvres*, vol. V, p. 270, transl. in *The Philosophical Writings of Descartes*, vol. III, p. 362. This is a case of Descartes speaking about health in terms of an "analogy of attribution" where in the *Meditationes* he referred to health in terms of an "extrinsic denomination" (see the previous note for a reference to health and analogy of attribution in Eustachius a Santo Paulo). To give some indication of the easy shift between "analogy of attribution" and "extrinsic denomination," we need only look as far as Suarez who, on at least one occasion, referred to "denominatio per attributionem" (as cited in Suarez, *Suarez on Individuation*, p. 203).

The interpretation of Meditation Six that I want to defend relies heavily on the points made above, and specifically on the claim that an extrinsic denomination requires a relation to an intrinsic denomination. I believe that having identified an intrinsic denomination like "natural" or "healthy" Descartes proceeds to use the denomination in a principled way to denominate other things as "natural" or "healthy." Any alternative leaves us not with an extrinsic denomination, not even a "mere extrinsic denomination," but an imposition of a name wholly without foundation. But if anything is clear, however, it is that in Meditation Six Descartes is trying to provide extrinsic denominations of "nature" with a foundation. The remaining task is to specify the intrinsic denomination of "nature" and the relation that the human body, when thought of as a machine, bears to the thing intrinsically denominated. Returning to the text of Meditation Six, the first candidate for such a foundation would seem to be the human being's body, but matters are not so simple:

> [W]hen I consider the machine of the human body in relation to the movements which normally occur in it [as judged by what normally occurs in the human being's body that the machine imitates], I may think that [the machine] too is deviating from its nature .... As I have just used it, 'nature' is simply a denomination which depends on my thought; it is quite extrinsic to the things to which it is applied, and depends simply on my comparison between the idea of a sick man and ... the idea of a healthy man .... When we say, then, with respect to the body [viewed as a machine] suffering from dropsy, that it has a disordered nature because it has a dry throat and yet does not need drink, the term 'nature' is here used merely as an extrinsic denomination.[67]

It is not clear what the intrinsic denomination is meant to be in this passage. The denomination that accounts for our assessment of what is natural is said to depend "on my thought," but are we to believe that an idea or the objectively real object of the idea could be intrinsically denominated with a "nature" in the relevant sense? Are my ideas of a sick man and a healthy man mind-dependent? If so, how does this avoid the conclusion that the intrinsic denomination is a human invention, which is the conclusion I have wanted to resist all along? And, how does an idea intrinsically denominated with a "nature" relate to what normally occurs in a human being's body?

To begin to answer these new questions it will help to realize that there is no basis for claiming "healthy," "natural" or "sick" are intrinsic denomination of the human being's body when viewed as a machine. This much the previous interpretation got right for, simply stated, none of the denominations can be intrinsic denominations of the human machine or any other machine thought to be an extended substance governed entirely by the laws of nature. Yet, Descartes does tell us we can find intrinsic denominations of "nature" supporting talk of "errors of nature" in the composite of mind and body. Our being healthy or ill is not a matter of standing in relation to anything else and so these denominations must be intrinsic denominations.[68]

---

[67] Descartes, *Œuvres complètes*, vol. VII, pp. 85–6; transl. in *The Philosophical Writings of Descartes*, vol. II, p. 59.

[68] I will not attempt to say precisely what the nature of the composite or "union" is for Descartes. At the very least, though, it involves two commitments. First, our nature involves interaction between mind and body that is not *wholly described* by Descartes' laws of nature. Not every true

This is the clue to Descartes' position: we human beings have an intrinsic nature susceptible to corruption. When we have erroneous desires, desires that draw us toward what is in fact bad for us, we are ill. Assuming the intrinsic denominations of human beings are the foundation for extrinsic denominations of "nature," and therefore "healthy" and "diseased," the remaining question is this: what relation does our idea of the human being's body, when viewed as a machine, have to the human being's body in the union? In other words, what relation supports the denomination of human machines as "healthy?" Now that we understand the logic of denomination, one answer we can rule out is that the human body in union with the mind and the human machine are identical. They are not identical because the human being has an intrinsic denomination that the human body, when viewed as a machine, does not have. The missing denomination from the machine is the one that the supposition in in the *Traité de l'homme* and in the *Meditationes* specifically deny it: a machine does not have the same nature as a human being, but suppose it does. Abstracting from the human body in union with the mind to the human machine in this way does not eliminate from our idea of the latter the thought of what should normally occur in the former, but it does eliminate the intrinsic denomination of a non-corporeal nature.

It should now be clear that my interpretation shares with the previous one a commitment to machines lacking intrinsically denominated natures subject to corruption. A machine's corruptible nature, insofar as it has one, is an extrinsic denomination coming from a literal act of creation, as in the case of a clock, in the thought experiment described in the *Traité de l'homme* and the *Meditationes*, or in the causal relations that exist between corporeal substances and the human being. A machine is, quoting La Forge once again, "a body composed of several organic parts, which being united, work together to produce several movements, of which they would not be capable if they were separated." La Forge was simply pursuing Descartes' idea that the nature of a machine—what allows a clock to be a good one or a bad one, or what allows the machine body to be a healthy one or a sick one—is ultimately an extrinsic denomination. In other words, La Forge's definition confined itself to what can be denominated intrinsically of a machine, that is, what makes it a corporeal substance.

If it is correct, that Descartes' strategy of supposing or hypothesizing that a machine will have the same nature as a human being's body is what supports the extrinsic denomination I have been discussing, it certainly begins to look as though the denominations are not only extrinsic, but also that they are extraneous or even

claim about us, in other words, can be expressed within the confines of natural philosophy. Second, the nature of the union requires that the union be a subject for the intrinsic denomination of "nature." This follows from the logic of denomination discussed above, yet this second commitment does not in itself answer the vexing question of whether the union must be a substance in Descartes' technical sense. Though I do not believe the union is a substance, to call it a subject of predication remains ambiguous. One possibility might be that the union is none other than the human body, which properly understood is a mode of both mind and body.

arbitrary.[69] Yet, as we saw earlier, extrinsic denominations are used to make true claims and they are based on a relation to an intrinsically denominated thing with the same denomination—e.g. "healthy" or "natural". Thus, Descartes' insistence to More that health only belongs to a human being and his insistence in Meditation Six that the only "true error of nature" takes place when a human being's body is corrupted does not mean he believed it is false to say "this is unhealthy weather" or "the human machine is ill" or even that "that animal is sick." In fact, if this were the case then Descartes' medicine would include a significant collection of false claims about animals and human machines.

To sum up, for a predicate like "healthy" to denominate a given subject extrinsically, "healthy" must denominate another subject intrinsically. And, equally important, there are conditions that constrain the relations between intrinsically and extrinsically denominated objects, such that they are not "just in the mind." It is not "just in the mind" that healthy food is a cause of health in the body of a human being, and it is not "just in the mind" that a machine duplicate of our body is indistinguishable from our body from the standpoint of a mechanical philosophy. In the case Descartes is most interested in, we might say the conditions that prevent extrinsic denominations of "nature" from being extraneous or arbitrary are the shared intrinsic denominations between the human being's body, when viewed as a machine, and the human being's body, when thought of in union with the mind. Both the *Traité de l'homme* and *Meditationes* would have us believe that the only intrinsic denomination separating the human machine and a human being's body is the latter's intrinsic denomination of "nature." The human machine is as close a copy of the union as can be made given the nature of corporeal substance. Were we to denominate a rock "healthy," or any other material object that lacks the extent of shared qualities the human machine has with the body of a human being, there we would be entering into the realm of the extraneous or arbitrary denomination. This, however, is not where Descartes goes in Meditation Six.

10.6 Conclusion

The conclusion I have been arguing for is this: Descartes' introduction of mechanical thinking into natural philosophy preserves "natural/unnatural" and "healthy/diseased" denominations as instances of extrinsic denominations. This is obvious when Descartes relates natural philosophy to medicine, but it applies wherever "natural/unnatural" is applied to a machine and "healthy/diseased" is applied to living things. I now want to end by briefly addressing three questions. First, what is the difference

[69] Cottingham et al. *The Philosophical Writings of Descartes*, translate "extrinsica" as "extraneous" in the Meditation Six passage we have been discussing. As we are about to see, this is a misleading translation.

between a corporeal substance and a machine? Second, how exactly do extrinsically denominated machines provide any help to the physician, who presumably takes an interest only in the health and disease of the genuine human body? And finally, third, what is Descartes telling us about the health of animals?

The second question is the easiest of the three to answer. The primary allure of the mechanical philosophy to seventeenth-century medicine was tied to the rationale it provided for intervention. It gave physicians a new etiology of disease, and Descartes' strategy of utilizing knowledge gained about the human being's body, when viewed as a machine, was his way of introducing this new etiology. His challenge was to re-describe the use of traditional remedies in order to explain how they acted on a machine, which in turn called for a new rationale for their application. For the physicians who worried that the health of a machine could not apply to the health of the human being, Descartes could remind them that they were getting things backwards. The health of the machine is parasitic on the health of the human being, not the other way around. This is why the human machine and the mechanical philosophy mattered to later physicians.

The first question is more difficult to answer. From what we have learned, however, I think we can say this: machines are not just corporeal substances. For Descartes, machines have natures that can be corrupted, whether they are clocks or human machines. This is not true of mere corporeal substances, which never act unnaturally in violation of the laws of nature. The sense in which machines can be said to "have" corruptible natures is the question and, of course, the answer is that they have them by extrinsic denomination. In other words, machines are corporeal substances that are extrinsically denominated with a nature over and above the nature they have as corporeal substances.

The example of the clock from Meditation Six can clarify this difference. Descartes wrote:

> '[N]ature' is simply a denomination which depends on my thought; it is quite extrinsic to the things to which it is applied, and depends simply on my comparison between ... my idea of ... a badly made clock ... and my idea of ... a well made clock.

Is there an intrinsic denomination supporting the extrinsic denomination of "nature" in this passage? Descartes wrote that the denomination "depends on my thought." Yet notice that clocks lack a relation to an intrinsic denomination of a corruptible nature existing outside our thought. This contrasts with the case of the human being's body when viewed as a machine. For the human machine, the intrinsic denomination of a corruptible nature is present in the human being's body and this denomination operates in the background for the human machine. What this means, perhaps unsurprisingly, is that a familiar machine's corruptible nature is imposed or literally created by us. It is our ideas that we impose, as artisans, on the world of corporeal substance. In the case of the human machine, however, there is less choice in the machine's nature since we do not literally make the machine.

Even saying this, however, is not to say that other machines are wholly mind-dependent—wishful thinking does not assemble a clock. The already existing corporeal substance, including the laws of nature, must be respected for us to produce machines with the extrinsically denominated natures and functions that we desire. Descartes was well aware that the more we understand how corporeal substance exists and operates independently of us the better we can make machines. When a clock no longer does what we expected it to do when we made it, it is because its intrinsically denominated nature, as a corporeal substance, leads to unexpected or undesirable effects. But we can say this only when the corporeal substance is compared to the extrinsically denominated nature we assign to the machine in our literal act of creation.[70]

This brings us finally to healthy and unhealthy animals. As I have tried to show, the nature of a machine that justifies talk of "natural/unnatural" is an extrinsic denomination. The question when it comes to animals and really to any natural machine is this: are they more like the human being's body, when viewed as a machine, or are they more like familiar human artifacts that we create using corporeal substance? In his natural philosophical practice Descartes favors the answer that animals are more like the human machine. Recall that by hypothesis the human machine is as near a duplicate of the human being's body as can be imagined. I think we can say something similar about animals and other living things, which are as similar to the human machine as can be imagined. For animals and other living things, the intrinsic denomination of nature to which their extrinsically denominated "nature" relates is the human being's body when in union with the mind. In their case, however, it is through the intermediary of the human machine that they relate to the appropriate intrinsic denomination of "nature."

What I am tempted to call "Cartesian anthropocentrism" manifests in a number of ways in Descartes' mechanical philosophy. It appears in his claim to have written a "prolegomenon" to medicine that concerned itself solely with animals and in his willingness to account for the motion of plants by describing their "heart" and circulatory systems.[71] What bears emphasizing is that we no more impose the extrinsic denomination of health and disease on living machines any more than we impose the extrinsic denomination of health and disease on the human being's body when viewed as a machine. Human being's bodies may be the exception to the rule of extrinsic denominations, but the same cannot be said of either the human machine or the animal machine.

---

[70] There is a special case, however. The universe as whole, i.e. God's machine, never deviates from the nature given to it in God's act of creation. In the case of this one machine its nature is effectively intrinsic and does not enable talk of "unnatural" or "malfunctioning" (see also note 13 above).

[71] Descartes, *Œuvres*, vol. IV, p. 329, transl. in *The Philosophical Writings of Descartes*, vol. III, p. 274, and Descartes, *Œuvres*, vol. II, p. 329 respectively.

Chapter 11
Mechanism and Surgery: Dionis' *Anatomy* (1690)

Jacques Lambert

Pierre Dionis has not yet been the subject of much study. At best, he is most often simply cited as the author of a declaration that historians frequently use to illustrate the state of mind of most scientists upon learning that, out of more than a million spermatozoids, only one or a few would be used for reproduction. Faced with this discovery, Dionis reacted by writing that "it truly was wasted seed."[1] This reaction of astonishment, or even outrage, at such waste from Nature probably deserves the attention of historians.

But a historian of science has other reasons for taking interest in Dionis and his work. I propose to show the limits of mechanism in a work which is neither by a philosopher nor by a scientist, nor by a physician, but by a practitioner of anatomy in a very particular intellectual context. Not only does mechanism take on different realities in different contexts (different times, places, professions, authors) but it can also designate a way of seeing, a method, a research program, a medical or philosophical doctrine. The *Anatomie* of Dionis is only one example of this, rather banal and perhaps even a bit disappointing with respect to other works, but interesting with respect to the difficulties that it shows or hides.

I would like to thank Mark Naimark, who translated this chapter.

[1] "C'était bien de la graine perdue," Dionis, *L'anatomie de l'homme suivant la circulation du sang et les nouvelles découvertes,* p. 302.

J. Lambert (✉)
Département de philosophie, Université Grenoble II, Grenoble, France
e-mail : jacquelambert@wanadoo.fr

D. Garber and S. Roux (eds.), *The Mechanization of Natural Philosophy*, Boston Studies in the Philosophy of Science 300, DOI 10.1007/978-94-007-4345-8_11,

11.1 Pierre Dionis, Surgeon

Pierre Dionis (1643–1718) was neither a philosopher nor a physician, but a surgeon. To understand the author and his work, we need to examine briefly the context of surgery in France toward the end of the seventeenth century and the early eighteenth century. During this period, surgery was beginning to gain recognition as a discipline, though this recognition faced great difficulties, mainly due to the authority of the University and the *Collège des médecins* (College of Physicians), who, seeing surgeons as rivals, wanted to hold onto their power and privileges. The job description of the surgeon was poorly defined: there was little in common between the *chirurgien ordinaire du roi* (the King's Surgeon) at the Royal Court, and those who worked in the countryside. Moreover, the profession is divided between barber-surgeons (*corporation de Saint Damien*) who practiced blood-letting, and *chirurgiens jurés* ("sworn surgeons" of the *corporation de Saint Côme*), who, continuing the work of Ambroise Paré, were trained in anatomy and had a certain general level of culture. The *Faculté* used all its resources, legal and "other", to have the Paris *Parlement* condemn the teaching of surgery, and to prevent, physically if need be, the activity of *chirurgiens jurés*. Because physicians used the services of barber-surgeons for blood-letting, they were able to call on them to oppose *chirurgiens jurés*. Thus Dionis had to struggle against the *Collège des médecins*, and at the same time take on the barbers, from whom he endeavoured to set himself apart. His goal was to show that surgeons were experts in anatomy, that anatomy needed to be recognized as the foundation of medicine, and that, since anatomy was a genuine science with its own principles and method, it must constitute an important part, even the most important part, of the practice of medicine.

We should add that Louis XIV greatly contributed to the intellectual and social recognition of surgeons, upon whom he called personally, preferring them to physicians. He always defended the guild from the attacks of the *Faculté* and the verdicts of the *Parlement*. While he was able to recognize surgeons officially and regulate the exercise of surgery, in particular by outlawing its practice by untrained individuals, such as certain clerics (the order of the Brothers of Charity in particular), he was unable to abolish the requirement that, in order to acquire the title of Doctor, surgeons had to undergo the tests of the *Faculté*. In a verdict from 1660, delivered after a lengthy lawsuit, surgeons were not allowed to create a *collège*, and could only form a *communauté*, which prevented them from teaching.[2] The *Faculté* kept for itself the exclusivity of cadavers of the executed for the

[2] In 1724, Mareschal, *premier chirurgien du roi*, succeeded in creating five positions of demonstrator in the Collège de Saint Côme. The reaction was immediate and spectacular: on January 16, 1725, the physicians, with the Dean leading them, marched in a procession to the *Collège de chirurgie* to take the building by assault, erasing the word *collège*, and attempting physically to remove the demonstrator from his chair. This event was repeated on July 26, and ended with a general melee (Delaunay, *La vie médicale aux* XVIe, XVIIe *et* XVIIIe *siècles*, pp. 325f).

practice of anatomy. The situation is best summarized in these words: "The history of surgical practice in the grand siècle is that of an ongoing confrontation with the medical institutions."[3]

11.2 Pierre Dionis at the *Jardin Du Roi*

It is in these circumstances that Pierre Dionis was named demonstrator of anatomy and surgery at the *Jardin du Roi* (Royal Botanical Garden), a position he held from 1672 to 1680. The better-known successes of Joseph Duverney often lead us to forget the pioneering work of Dionis. Founded by the *médecin ordinaire du roi* in 1626 as a simple garden for growing and acclimatizing plants, it was established as an institution for teaching and research in 1635, before opening in 1640. Three chairs for "demonstrators" were created at first: botany, chemistry, and anatomy. Teaching was in French, and took the form of "demonstrations." The *Faculté* immediately reacted very strongly, and filed suit to close down this rival institution. Like the *Collège de chirurgie*, the *Jardin du Roi* encountered a major obstacle in that the *Faculté* still had the unique privilege of awarding the title of Doctor. But this obstacle was overcome by the nomination of *surintendants* and demonstrators recruited from among the *médecins ordinaires du roi*, who were trained in the rival University of Montpellier. We can cite the example of Joseph Pitton de Tournefort (1656–1708), the famous botanist trained in Montpellier, who became a demonstrator, then *surintendant* from 1702 to his death, or that of Jean Pecquet (1622–1674), the friend of Fouquet and La Fontaine, an anatomist known for his discovery of the cisterna chyli and the thoracic duct (1651). Once again, it took the full authority of the king to overturn the unfavorable verdicts of the *Parlement*. The situation was even more serious for Pierre Dionis, since he was in charge of teaching the circulation of blood. The teaching of this discovery, we recall, was condemned by the *Faculté* because it was contrary to the Galenic doctrine. By playing on the Latin word "circulatores," the *Collège des médecins* associated proponents of circulation with travelling physicians and charlatans! In 1675, in the heart of the period when Dionis was teaching, Nicolas Boileau notably ridiculed this condemnation in his *Arrêt burlesque*.[4]

---

[3] Gélis, "chirurgiens," p. 323.

[4] [Boileau, Bernier and Racine], *Requeste des maîtres ès arts, professeurs et régens de l'Université de Paris, présentée à la Cour souveraine de Parnasse, ensemble l'Arrest intervenu sur ladite requeste contre tous ceux qui prétendent faire, enseigner ou croire de nouvelles découvertes qui ne soient pas dans Aristote*. The *Arrest intervenu sur ladite requeste Requeste* was written by Nicolas Boileau, perhaps at the request of the *Président au Parlement* Lamoignon, and *Requeste des maîtres ès arts* was written by the Gassendist François Bernier. The share due to each has been debated, and the participation of Racine has been suggested. During the following year (1672), the year in which Dionis took up his position at the *Jardin du Roi*, Guy Patin, who represented, with his famous verve and irony, the retrograde ideas of the *Faculté*, died; the year after (1673) Molière's *Imaginary Invalid*, in which the playwright ironically commented on the physicians of the *Faculté*, was performed.

We have a great deal of difficulty today understanding the consequences of William Harvey's discovery. Let us simply say that, while the great anatomical discoveries and novelties introduced since the publication of Vesalius's *De corporis humani fabrica* in 1543 could be construed as mere improvements on ancient anatomy, Harvey's physiology was a radical departure. Harvey did not simply introduce a "novelty," difficult to accept for the Ancients: he attacked the very bases of the Galenic system by questioning the role of the liver in the production of blood (hepatic hematosis) and in nutrition in general. The later physiological discoveries would precipitate the collapse of the entire Galenic edifice that the contemporaries of Dionis among the physicians and professors of the *Faculté* fervently defended by means of harassment, lawsuits, and court judgments.

Anatomy demonstrations were aimed primarily at surgery students. These students held tickets to reserve their seats in case of large numbers of auditors. There was a surgical anatomy course over 10 days during the first semester, followed by the same number of days devoted to surgical operations during the second semester. The anatomy demonstrations were collected and published under the title *Anatomie de l'homme suivant la circulation du sang et les nouvelles découvertes*, published in Paris in 1690. This work enjoyed several reprints (the sixth French edition appeared in 1780) and many translations into various languages, including one in Tartar by the Reverend Father Parrenin at the request of the Emperor of China, Kangxi. As for the surgical operations, they were collected and published in Paris in 1707 under the title *Cours d'opérations de chirurgie démontrées au Jardin Royal.* This book was also reprinted several times (the eighth edition with additions by Georges de la Faye appeared in Paris in 1782) and translated into the principal languages of Europe. It enjoyed exceptional longevity for a field such as surgery, although this book did appear before the true modern era, when such a work would quickly become outdated.[5] To be complete in this presentation of the work of Dionis, we should mention a *Traité général des accouchements*, published in Paris in 1718. As *chirurgien ordinaire* of Marie-Thérèse, the dauphins, and the duchess of Burgundy, Dionis delivered the duchess of Berry, and made his name in obstetrics in a particular period of the history of this discipline.[6]

---

[5] "Dionis brings together in a clear and methodical synthesis the surgical knowledge of his time. It will be the Bible of surgery students of the eighteenth century" (Gélis "Chirurgiens," p. 322). Things only truly change with the dawn of modern surgery, which can be dated to the years 1860–1865. The conjuncture of several factors will radically transform surgical techniques, and consequently, the boldness and the field of surgery itself. At issue are anaesthesia (T. G. Morton, 1846), antiseptics (mainly carbolic acid), asepsis (Lister, 1867), and the use of artery forceps (Koeberlé, 1864; Péan, 1867).

[6] It was in the seventeenth century that the profession of "obstetrician (*accoucheur*)", heretofore reserved for midwives, appeared in France. In 1663, Louise de La Vallière, the mistress of Louis XIV, called upon a talented and well-known surgeon, Julien Clément, for her delivery. The news of this event spread beyond the Court. In 1659, on the occasion of an abortion also carried out at the Court by a midwife, La Constantin, on Mademoiselle de Guerchy, it was discovered that this practice was fairly common. In 1668 the *Traité des maladies des femmes grosses* by François Mauriceau

The final and most important reason that this work should receive scholarly attention takes us to the heart of the subject. Dionis summarized in the simplest and fullest manner possible the mechanistic program of the new anatomy in these words to his students: "You will understand well what each part does when I show you how it is made."[7] This is truly a "maxim" in the sense given in Furetière's *Dictionnaire*: "principle, foundation of an art or science." This "research program" is of course not new to the seventeenth century: more than 30 years earlier, Descartes had become the theoretician of this new philosophy. Although it was nothing new, the maxim remained thoroughly pertinent. In particular, the demonstration of the circulation of blood represented the great anatomo-mechanistic cause that would strike the fatal blow to Galenic hepatocentrism, and in so doing, to the entire traditional medical edifice. It would be up to Gaspare Aselli, Jean Pecquet, Olaf Rudbeck, and Thomas Bartholin to finish it off with the later discoveries of the lymphatic system. Here are these "modern discoveries" announced in the title of the work, after that of the circulatory system. Thomas Bartholin wrote the obituary and the epitaph in his "funeral of the liver": his *Post inventa vasa lymphatica hepatis exsequiae*, published in Copenhagen in 1653, ended with the description of the tombstone inviting passers-by to remember the liver, the most important personage of ancient medicine.[8]

On the subject of the relation between these discoveries and mechanism, it seems necessary at this point to make a distinction among various items.

The discovery of the circulation of blood, along with that of lymphatic circulation, was undoubtedly experienced by contemporaries as the failure of Galenism. But that is not the same as embracing the theses of mechanism. We have an illustration of this in the work of Harvey himself, a first-rate anatomist who was able to recognize his debt to mechanistic models, but who was not truly a mechanist at heart. The position of Dionis is rather original. We first note that he demonstrated the circulation of blood and its consequences, while never referring to the work of Harvey, whose name appears only twice in over 600 pages: first in the Preface, spelled "Harvée," and second in the body of the work under the name, undoubtedly incorrectly transcribed by the printer, of "Horrée."[9] In the latter passage in particular, Dionis insists on saying that he will not take the trouble to prove the circulation of blood. The ligature experiment is both common and infallible.

What is more interesting is the mechanistic interpretation that seems to impose itself upon him following this discovery. Not only does he oppose Anatomists to the Ancients (as Harvey opposed Anatomists to Galenists), but for him, anatomical

was published. The first royal edict rendering the profession of midwife subject to surgeons was issued in 1692. Within a few years, surgeons, who earlier learned from midwives, became their indispensable instructors. It was a student of Julien Clément, Nicolas Puzos (1686–1753), who would become famous in the obstetrics courses.

[7] Dionis, *Anatomie*, p. 14: "Vous connaîtrez assez ce que chaque partie fait quand je vous aurai fait voir comment elle est faite."

[8] Leclerc, *Bibliotheca anatomica*, vol. II, pp. 692–698.

[9] Dionis, *Anatomie*, p. 397.

mechanics seems to demonstrate mechanism. Galenic blood physiology relied on anatomical errors, and worse than that, it presumed an anatomical impossibility:

> The Ancients imagined that it was the liver that made blood, and distributed it to the parts of the body to nourish them, and that the chyle could not be carried elsewhere. And because of this, they thought that it was led by the same veins that they said carried blood from the liver to the intestines. To refute this opinion, it suffices to examine the opposing movements they attribute to chyle and to blood, as they do not appear to believe that two liquors, one rising and one descending, could pass through the same canal; indeed, the circulation of blood found in our time is so contrary to this distribution of blood by the veins, that far from carrying blood to the parts of the body, they have in fact no other use than returning it to the heart.[10]

Dionis accordes much importance to this argument since according to him, it destroys the "*principalitas*" of the liver. The pre-eminence of the liver had been established through an analogy between the primordial function of the liver as well as its volume, its position, its ruddy substance, its large vessels. In remarking that the blood cannot circulate in two opposite directions, he makes evident that contradiction of the physicians who considered that a rational argument could be better proof than a factual argument; to see this contradiction, Dionis says that "all we need to do is examine motions."

On other occasions, the author returns to the anatomical contradictions of the Ancients; such contradictions were inherent to a physiology governed by humors and faculties. We could say that for Dionis, the discovery of the circulation of blood definitively destroyed Galenic anatomy and physiology, and that in so doing, it consecrated mechanism. Of course, it is not at all obvious that mechanism is the only consequence that can be drawn from the anatomical system at stake in Harvey's physiology. Beyond the fact that Harvey's demonstrations were oriented toward practical ends, and did not deal with more philosophical questions, it is clear that we have here an interpretation of Harvey's discovery by surgeons as if it had definitively established mechanism.

11.3 Demonstration and Anatomo-Surgical Mechanism

The theses of medical mechanism were still far from being accepted when Dionis delivered his maxim. Dionis' maxim took on a particular coloration in the surgical context. The reasons that could lead a philosophical historian to pay little attention to a work that is neither that of a physician reasoning on the principles and methods

[10] "Les Anciens se sont imaginés que c'était le foie qui faisait le sang, et qui le distribuait aux parties pour leur nourriture, et que le chile ne pouvait être porté ailleurs; et pour cet effet ils voulaient qu'il y fût conduit par les mêmes veines qu'ils disaient porter le sang du foie aux intestins. Pour détruire cette opinion, il ne faut qu'examiner les mouvements opposés qu'ils donnaient au chile et au sang, n'y ayant pas apparence de croire que deux liqueurs dont l'une, selon eux, montait, et l'autre descendait, pussent passer en même temps par un même canal; d'ailleurs la circulation du sang que l'on a découverte de nos jours, s'est trouvée si opposée à cette distribution du sang par les veines, que bien loin de le porter aux parties, elles n'ont au contraire point d'autre usage que celui de le reporter au cœur" (ibid., pp. 192f).

of his art, nor that of a philosopher arguing as to his vision of Nature, are the very reasons for its originality. Only with the benefit of royal protection could a surgeon teach that which was unacceptable to the *Faculté* and subject to its punishment. But we should also appreciate another aspect of its originality, an intrinsic one. It is remarkable that Dionis is presenting a work of anatomy at a very special moment in its history, when the subject is quickly developing and is beginning truly to organize itself. And it is even more original because the anatomy is presented merely as a preparation for surgical practice. This context gives the work a characteristic content and presentation.

In the Preface Dionis noted that "anatomical demonstrations are so rarely done in most provinces that surgeons there can hardly see one during their entire lifetime."[11] For this reason the author sets out a book to take the place of a course intended for future surgeons. The work is presented, as befits the *Jardin du Roi*, in the form of ten "demonstrations." Each demonstration is introduced by a plate showing the bodily parts studied in the lesson. Similarly, the 10 days corresponding to the various surgical operations are preceded by plates showing the *appareils* (devices) or sets of instruments required for the operation in question. This symmetry is thus not purely formal. To say that the work takes the place of the demonstrations is certainly an exaggeration. It is doubtful that they served the functions of the anatomical demonstrations that they certainly do not replace. The plate on the frontispiece shows an anatomical amphitheatre with its seats filled with auditors, and in the foreground a demonstrator in front of the corpse of a woman lying on a table. On the left and right are two "persons of quality" seated and seen from behind. As can be seen sometimes in works on anatomy from the period, we can see a gap between the details in the text and the lack of precision of the images. As for the surgical plates, as noted above, they show all instruments needed for each surgical operation: bleeding the arm, abdominal sutures, skull fractures, etc. Each instrument is assigned a letter that links to a description in the text. At least we have the impression that the surgical tool tray is full, while the letters written on the parts of the skeleton in the anatomical plates, for example, only roughly indicate parts. The style of the *Anatomie* is oral, concrete, descriptive, clear, and concise. Certain simple terms, such as *désunion* for separation, are still part of the French surgical vocabulary today. To the academic, or even scholastic, terminology for the abdominal muscles found for example in the Latin text of Jean Riolan's *Enchiridion anatomicum* then in use at the *Faculté*, we can contrast the incisive description that derives from the autopsy: "We can consider the oblique and the transversal [muscles] as hands laid one atop the other in different directions, and the recti as two large rods that prevent the others from tensing too much and pressing the intestines against the spine and the lumbar vertebrae."[12]

[11] "[L]es anatomies se font si rarement dans la plupart des provinces qu'à peine les chirurgiens qui s'y trouvent en peuvent-ils voir une en toute leur vie" (ibid., Preface, n.p.).

[12] "L'on peut considérer les obliques et les transversaux comme des mains posées les unes sur les autres en différents sens, et les droits comme deux grosses barres qui empêchent que les autres en agissant ne bandent et ne compriment trop les intestins contre l'épine et les vertèbres des lombes" (ibid., pp. 149f).

Such a description would obviously be out of place in a traditional anatomy treatise written by a physician. But if the aim of the book seems to require that it have a certain style, a certain type of content, and more important, a particular method, it also requires a certain theoretical character. And here is where the originality and the interest of this work is located. By serving during the first semester as the basis for surgical practice to be taught in the following semester, anatomy was being represented as a fundamental theoretical science of which surgical technique was an application. The beginning of the book is explicit on this point, and even if it seems to go without saying today, it must have seemed surprising, or even provocative, for the author's contemporaries:

> You know, gentlemen, that surgery is an operation of the understanding that knows the ills of the human body, and at the same time, an operation of the hand that holds the instruments and the remedies to cure them.[13]

When we recall the historical context, we cannot fail to see a certain audacity in this declaration from a surgeon, publicly giving a definition of surgery that has all the marks of medicine, only adding at the end the idea that it is a mechanical art, usually the only characteristic mentioned. The first part of Dionis' course on surgical operations develops the theme of surgery as a science in its own right, opposing "the opinion of a certain people who treat it simply as a mechanical art. It deserves to be called a science as much as does mathematics, which traces on paper with ruler and compass the figures and demonstrations that the mind imagines."[14] Seeing in it "the most noble, the most certain, and the most necessary of all sciences," Dionis is not only using rhetorical arguments for recognizing the mechanical arts as being part of the liberal arts, arguments used since such quarrels in the fifteenth century. Bringing together surgery and mathematics was yet another play on the word "demonstration." By demonstration, we of course understand any indicating or exhibiting having cognitive value: the botanist of the *Jardin du Roi* would demonstrate with the help of a stick. But demonstrating also and especially meant establishing a necessary judgment through reasoning. The geometer alone demonstrates, as Furetière's *Dictionnaire* reminds us.[15]

To summarize, the word "mechanical" was used for arts executed by means of the hand, as opposed to the "liberal" arts, for which geometry was the classic example. But the separation of theory from practice is just as impossible in geometry as in surgery. In both cases, we are dealing with demonstrations using the hand, but just

---

[13] "Vous savez, Messieurs, que la chirurgie est une opération de l'entendement qui connaît les maux du corps humain; et en même temps une opération de la main, qui y porte les instruments et les remèdes pour les guérir" (ibid., p. 1).

[14] "[L]'opinion de quelques-uns qui la traitent d'art simplement mécanique. Elle ne mérite pas moins le nom de Science que les Mathématiques qui tracent sur le papier avec la règle et le compas, les figures et les démonstrations que l'esprit imagine" (Dionis, *Cours d'opérations de chirurgie démontrées au jardin du Roi*, Preface, p. IX).

[15] Furetière, *Dictionnaire universel*, art. "démonstration," n.p.

as much with demonstrations in the logical and intellectual sense, since "the hand only executes what the mind tells it to." Going farther, Dionis sees in surgery the most noble of sciences, "for what makes for the nobility of a science is the dignity of its object,"[16] an Aristotelian argument well known to the *Faculté*.

Accepting anatomy as the basis for all of the medical profession, and not just for surgery, means giving it the status of a true science that gives principles to the other medical sciences. In this case, the principles are mechanical, as in any mechanistic interpretation of the operation of the body. While the author never makes these principles explicit or defines them, he never ceases to use them throughout the book, which seems fitting for the practical goals of this work. The reader has no difficulties finding shape and motion as the true and unique principles to explain all the activities of the human machine. The scientific value of anatomy ultimately rests on both the rational and logical principles that lead one to believe that the anatomy can be comprehended only by means of geometrico-mechanical principles, established through the observations and experiences offered as evidence for them. We have already referred to the experiment, both banal and obvious, of blood-letting proving the circulation of blood, and in so doing, the mechanism that forms its basis. On two occasions, Dionis recounts interesting personal experiments, one regarding the chyliferous vessels observed on the cadaver of a condemned man he took care to feed 3 or 4 hours before his death, and the other on the thoracic canal leading to the right ventricle, to confirm the new discoveries of lymphatic circulation.[17] It goes without saying that the many decisive discoveries that were taking place in anatomy confirmed that anatomy appeared more and more as the fundamental science for all of medicine, and that the philosophy underlying these discoveries was justified.

The comparison of anatomy with geometry meant much more for contemporaries of Dionis than it does for us. They sometimes spoke of mechanically solving a geometry problem, meaning to solve the problem "by trial and error, without total certainty, by using a compass or other instruments, rather than doing it by reasoning alone, not taking the matter into account with all imaginable certainty and precision."[18] Geometry, the theoretical science of principles *par excellence*, can thus itself be practiced as an art.

It is the assimilation of the body to a machine that makes it possible and legitimate to classify anatomy as a science, a science of principles established on basis of an examination of the sensible parts of the body.[19] The latter, recalls Dionis, "give

[16] Dionis, *Cours d'opérations*, pp. IX f.

[17] Dionis, *Anatomie*, pp. 184 and 442. The relation of the experiment on the chyliferous vessels is important. When practiced on an animal that is not digesting, dissection will not show these vessels. This fact had allowed the *Faculté* to refuse to admit Aselli's discovery.

[18] "[E]n tâtonnant, sans une entière certitude, en se servant du compas ou d'autres instruments, au lieu de le faire par le seul raisonnement, et faisant abstraction de la matière avec toute la certitude et précision imaginables," Furetière, *Dictionnaire universel*, art. "mécaniquement," n.p.

[19] Dionis, *Anatomie*, for example p. 5: "C'était une réponse générale par laquelle [les Anciens] éludaient aussi bien que par leurs qualités occultes, toutes les difficultés qu'on leur proposait; de sorte que les Ecoliers n'étaient pas plus savants après qu'auparavant. Mais aujourd'hui que l'on

reason the opportunity to exercise its judgment on that which is insensible in man."[20] And here is the difficulty, here is all the art. For having failed to take the trouble to examine the things themselves, the Ancients condemned themselves to recounting fables.[21]

But even firmly based on the principle or maxim of anatomical mechanism, can we be sure, even after having taken the trouble to examine the things themselves, that we ourselves are not recounting fables?

11.4 The Principle of Anatomical Mechanism

The principle by which a body can and must be considered as a machine, the principle we can call the principle of anatomical mechanism, has significant consequences for research as well as for the history of physiology.[22] We will discuss four such consequences.

1. First of all, anatomical mechanism legitimates dissection. We note that we can also call it "anatomical analysis." In total opposition to traditional teaching based on a famous text by Aristotle (*Parts of Animals,* 641 a 1–6), which condemns in advance any attempt to explain the living by dissecting cadavers, the new anatomy considers the hand of the cadaver as equivalent to that of the living person, lacking

explique toutes ces mêmes actions par une manière purement mécanique, je vous ferai voir, en vous démontrant chaque partie avec exactitude, que l'action qu'elle fait dépend absolument de sa structure, étant une suite de sa disposition naturelle; en sorte qu'elle ne doit faire autre chose que ce qu'elle fait. [This was a general response with which the Ancients avoided, as they avoided their occult qualities, all the difficulties proposed to them, so that students were no more knowledgeable afterwards than before. But today, when we explain all these same actions in purely mechanical fashion, I will make you see, by demonstrating to you each part with precision, that the action it does depends absolutely on its structure, being a necessary consequence of its natural disposition, so that it may do no other thing than that which it does.]"

[20] "[D]onnent occasion au raisonnement de porter son jugement [sic] sur ce qu'il y a d'insensible dans l'homme," ibid., p. 126.

[21] Ibid., pp. 208 and 371.

[22] André Pichot proposed a distinction between "machinism" and "mechanism," *Histoire de la notion de vie*, pp. 344f. For him, Galen has "machinised" the body by breaking it down into parts, each with a function. Descartes adopted this distinction by substituting the laws of mechanics for the Galenic principles, but could not avoid thinking of finality, a though inherent in the notion of a machine. This remark is interesting. The *méchanè* does designate an organic set of *organa* (organs, instruments). Where Galenism sees an *organic* whole (mainly through faculties), mechanism only sees an *articulated* whole. It seems that the word "finality" could be profitably replaced in the seventeenth century by the word "utility", translating the Latin word *usus*, and referring to the double sense of "use to be made" and "use made." If these two meanings are complementary in Galenism, with a subordination of the second to the first, this is not the case for mechanism. The structure in this case is not the instrument (*organon*) used: it is its cause. This shift is only made possible if the form of the bone is not its hardness but its shape. It is perhaps in this transformation of the notion of form that a radical change takes place.

only its movement. Dissection in this context is no longer merely encouraged: it is obligatory, and it becomes an urgent requirement for deepening our understanding of the living body.

2. Next, anatomical mechanism leads to the possibility of deducing function (we then speak of the "office" or "use" of some anatomical feature) from an examination of the structure. This principle that Claude Bernard would describe, while criticizing it, as "anatomical deduction" is now essential.[23] We have seen that Dionis perfectly summarized this in the expression quoted in the introduction. The exact description of each part of the body would indicate in and of itself its *raison d'être* or its finality. On this point as well, the reversal of tradition is total. Traditionally, the "how" represented only the instrumental part, subordinated to the "why." Far from subordinating the "how" to the "why" modern philosophy banished all considerations of "why". The description acquired an explanatory value since it by itself sufficed for demonstration. This is quite different from the traditional use of description, in which the description or example only served as an illustration once the rational demonstration was established.
3. Anatomical mechanism also leads one to think that a more complete and detailed analysis would necessarily allow one to understand "uses" still awaiting explanation. For example, although the apparent structure of the kidney does not clearly indicate its function as a filter, a more thorough inspection and a more detailed study would easily show it. This thesis coincides with the extraordinary growth of microscopy, and we can observe a relation of mutual reinforcement between them. The idea that an advanced morphological analysis is to be seen as the ultimate explanation sent scientists on an unprecedented exploration of the body, while the spectacular successes of this exploration offered daily proof of the fertility and truth of this idea. This is so whether we think of the geometrico-mechanical structure of muscles presented by Steno (Niels Stensen) in his *De musculis et glandulis observationum specimen* from 1664, or that of the kidneys by Marcello Malpighi in 1666, or that of the salivary glands by Thomas Wharton in the same year in his *Adenographia*.

[23] The critique of what Claude Bernard called "anatomical deduction," that is to say, the idea of being able to deduce function from examination of structure alone, is a recurring theme in the work of the physiologist. This principle of anatomical deduction is presented and affirmed even more clearly in his final works. For example, in *La science expérimentale* (1878), and in particular in the chapter entitled "Le problème de la physiologie générale" where after having developed his arguments in this light (pp. 104–107), the author writes: "En un mot la physiologie n'est point une déduction de l'anatomie [In a word, physiology is not deduced from anatomy]" (p. 107). Claude Bernard sometimes speaks of the "anatomical rut" to characterize the status of subordination of physiology to anatomy in the physiology that preceded Lavoisier and Bichat. For example, in his *Leçons sur les phénomènes de la vie communs aux animaux et aux végétaux*, published in the same year as he published his *La science expérimentale*, he writes: "Trois grands hommes, Lavoisier, Laplace et Bichat, vinrent tirer la science de la vie de l'ornière anatomique où elle menaçait de languir [Three great men, Lavoisier, Laplace, and Bichat, came to pull the science of life out of the anatomical rut where it risked languishing]" (p. 6).

4. Anatomical mechanism finally established as evident the proportionality of a cause with its effect. The laws of statics and dynamics offer innumerable examples of a mechanical-type causal relation (understood through a simple examination of the efficient cause and the material cause) in which an effect always appears proportionate to its cause when we consider the parameters of size, volume, shape, and extension. Thus the laws of mechanics tend to substitute, not without difficulty, for the qualities of analogy.

These four specifications of the principle of anatomical mechanism obviously need to be nuanced and interpreted in accordance with their contexts. Thus, the Galenists are no more Galen than the Aristotelians were Aristotle; the anatomists, including Harvey and Riolan, generally don't speak of "organs" but of parts; for a long time, as a legitimate physician, one opened bodies to look for poison; the use of dissection in the teaching of anatomy had not been established even by the end of the seventeenth century: the Faculté of Paris was not that of Padua! The idea that pertains to anatomical mechanism seems to us to be that anatomy and its mechanical explication are necessary and ideally sufficient.

11.5 The Program of Anatomical Mechanism

With the identification of the body as a machine, the program of anatomical mechanism aimed above all at rejecting a so-called organic spontaneity considered essential. Consequently, all the many entities that the Galenists had imagined to account for this spontaneity are subject to strong criticism. On these grounds, the principal entity attacked by the mechanists is the faculty. At the very beginning of his *Anatomie*, Dionis could not resist the temptation to be ironic, and write in a style that would suit Molière, his contemporary:

> The Ancients used [the word "faculty"] to explain all the actions that take place in the body: they replied that the stomach had a chylific faculty, and the liver a sanguinific one, that bones were formed by an ossific faculty, and cartilage by a cartilaginific one, etc.[24]

The criticism of the faculties is radical, and it constitutes the central argument. It is sometimes directed against a "virtue" or to an "idea", a relic perhaps of the "eidos" of Aristotle, or more certainly of Galen. This is the case when Dionis invites the reader not to worry about "finding an ossific virtue or idea, for otherwise we would have to multiply these virtues."[25] When we examine the texts and the diagram of the machinery, we see that the concept of *disposition* is clearly called on to replace both that of *faculty* and that of *habit* (or *habitus*) that themselves reduce to that of *power* (*energeia*). The rest of the text on supposed faculties is explicit in that respect:

[24] "Les Anciens s'en servaient [du mot "faculté"] pour expliquer toutes les actions qui se font dans le corps; ils répondaient que l'estomac avait une faculté chilifique, et le foie, une sanguifique; que les os se formaient par une faculté ossifique, et les cartilages par une cartilaginifique etc." (Dionis, *Anatomie,* p. 5).

[25] "[À] chercher une idée ou vertu ossifique, autrement il faudrait multiplier ces vertus," ibid., p. 14.

> But today, when we explain all these same actions in purely mechanical terms, I will make you see, by demonstrating to you each part with precision, that the action it does depends absolutely on its structure, being a necessary consequence of its natural disposition, so that it may do nothing other than that which it does.[26]

The use of the term *demonstration* ("démontrant") is essential here, for it holds a double meaning with reference to the machine. It means first that the entire argumentation is contained in the description alone, in the spirit of the lectures at the *Jardin du Roi*, where teaching was done in the form of *démonstrations*. He then incorporates the idea of a necessary explanation, as in the case of a machine whose operation is instantiated in its structure. As for the "disposition", naturally, we must not see in it any imagined fitness carry out any particular project, but the very structure that by itself and itself alone, indicates its possibility for action.

Later, Dionis recalls the famous and passionate quarrel between Cressé and Guillaume Lamy on the subject of the end attributed to the parts of the body. We can see that he sides with Lamy against Cressé, who laid down as a principle the determination of the use. With Cressé, we must not say that a given part is "made for this" but rather that this part "does this." To properly understand a bodily part, one shouldn't say that it was made for such and such uses: "one only needs to examine it" and "work to set out all the small parts that compose it, and you would see that the action it does would be a consequence of its disposition. And hence, that we shouldn't say that the eye was made to see, but that one sees with an eye."[27]

The idea of substantial unity represented a second and important target in the war declared by mechanism against an imaginary spontaneity. Anatomy, which inevitably divides, separates, and breaks things up undermines the activity inherent in this substantial unity. In addition to its dimensions, or that of its vena porta or vena cava, or its arteries, one of the glories of the liver, one of the manifest reasons for its *praestantia* or its *principalitas* was special substance of which it was made up. The etymology of the word *parenchyma* already evoked the representation of the gushing of blood, obviously reinforcing the idea of a co-substantiality of the viscera with the nourishing vital humor that the liver was supposed to produce:

> The Ancients attributed to the liver a very special substance .... The Moderns carefully sought the structure .... The substance of the liver is only a cluster and assembly of an infinite number of small glandular bodies.[28]

---

[26] "Mais aujourd'hui que l'on explique toutes ces mêmes actions par une manière purement mécanique, je vous ferai voir, en vous démontrant chaque partie avec exactitude, que l'action qu'elle fait dépend absolument de sa structure, étant une suite nécessaire de sa disposition naturelle; en sorte qu'elle ne doit faire autre chose que ce qu'elle fait" (ibid., p. 14, see note 20).

[27] "[Il] n'y a qu'à bien l'examiner, ... travailler à développer toutes les particules qui la composent, alors on verrait que l'action qu'elle fait serait une suite de sa disposition et que par conséquent l'on ne devrait point dire que l'œil avait été fait pour voir, mais que l'on voyait avec un œil" (ibid., p. 15).

[28] "Les Anciens attribuaient au foie une substance toute particulière .... Les Modernes ont recherché avec soin la structure .... La substance du foie n'est qu'un amas et un assemblage d'une infinité de petits corps glanduleux" (ibid., p. 190).

Similarly for blood, and later, for the cortical grey matter, which as well is merely an assembly of an infinite number of small glands that filter the animal spirits. It is remarkable that this notion of assemblage, which will be used by the materialists of the eighteenth century, in particular Helvetius, to undermine and reject the idea of a soul conceived as a unique and substantial principle, was used a century earlier by mechanist anatomists to undermine the idea of the Galenic body as a substance.

The consideration of glands leads to the heart of the rejection of spontaneity in mechanist thinking. The new interpretation of glands settles an old debate: is blood the product of an enhancement or enrichment of some pre-existing fluid, or on the contrary, the result of a degradation? The glands, whose exocrine and endocrine functions would be discovered later, are, in the eyes of the mechanist, purely passive structures. They produce nothing, nor do they even restore anything, as do springs, but they are only sieves:

> There is no gland that does not separate out some liquor by its natural disposition, just as a sieve lets pass through its holes the particles that have an appropriate shape.[29]

And what organ is not a sieve in the great hydraulic machine that is our body? The entire digestive system is merely a series of filters. Blood, to remain pure, must be filtered and decanted. In the same way, the breasts separate the milky parts from the mass of blood and send them out by way of the excretory canal.[30] The brain itself is only a sieve. Separating is one kind of "passive operation."

Another kind of notable passivity in total opposition to the idea of spontaneity is the behavior of the sense organs, imagined as pure receptor/selectors:

> Smelling is not doing something, but only being acted upon … tasting is not doing something, but only receiving … seeing is not doing something … vision is itself merely a shaking."[31]

All these manifestations of an essential passivity seem quite consistent with the Christian thought of the philosophers and theologians for whom Nature, far from being creative, is only a creature that derives from God alone everything that it is, everything that it has, everything that it can be.

11.6 Aporia and Traps of Anatomical Mechanism

The other important concept of this anatomical mechanism is obviously that of *use*. This is a key term in the *Anatomie*. It corresponds to that of *disposition* to which we referred earlier. What is at issue can only be *use made* and not at all that of the *use*

[29] "Il n'y a pas une glande qui ne sépare quelque liqueur par sa disposition naturelle de même qu'un crible laisse passer par ses trous des particules qui en ont la figure" (ibid., p. 208).

[30] Ibid., p. 359.

[31] "[F]lairer n'est pas faire quelque chose mais seulement souffrir, … goûter n'est pas faire quelque chose mais seulement recevoir, … voir n'est pas faire une chose, … la vision n'étant elle-même qu'un ébranlement" (ibid., pp. 519, 528, 504).

to be made. But despite all the precautions taken to ward off the dangers generated by the idea of spontaneity, the term "use" seems to fall victim to all the traps and perils inherent in the representation of a final cause. Furthermore, the language seems also to show different degrees of inherent difficulty. While the expression "ce qui fait que" ("that which brings it about that"), commonly employed, is clear enough for indicating a relationship of consequence, as when we say that the esophagus is made up of three membranes, "ce qui fait qu'il se peut dilater aisément" ("which brings it about that it can dilate easily"), the expression *parce que* ("because"), for its part, shows a certain ambiguity. Thus, when speaking of the vagus nerve, Dionis writes that "it is covered with strong membranes *because* it extends a long distance."[32] We must understand that it is the proportionality of the effect to the cause that authorizes here a reversal of the causal relation in his formulation.

We must distinguish the language used in several types of apparent reference to finality. The simplest case is presented through a simple comparison, for example, the association of the spine with an aqueduct, the jaw with a millstone, or the ligaments with ropes. Here Dionis simply uses the adverb *comme* ("like"). The search for an effective way of communicating his ideas can lead to more tolerance for a language more suggestive of final causes. A formulation such as "pour que le sang soit distribué" ("for the blood to be distributed") can be easily reinterpreted in non-finalistic language. With respect to the fatty reserves accumulated in the omentum and the mesentery of hibernating animals, one can write that since they neither move nor deplete their stores, these animals "do not need much food."[33] We understand in turn, and in a symmetrical fashion, that if we take into consideration a particular diet, that *should also* direct our attention to the discovery of reserves. We are in presence of a certain circularity of structure and function. The discourse thus authorizes one to present the first through the second.

The surgeon is naturally familiar with this form of thought that goes back and forth from structure to function and from function to structure. All finalism, and even further, all providentialism set aside, it is reasonable to think that if the vena cava has a wide diameter, it is because a great deal of blood flows through it. To use *pour que* ("so that") in place of *parce que* ("because") cannot, in this case, lead to error. This simply shows that we are dealing with a reversal due to the symmetry of the principle of causal proportionality described above. These types of expressions and representations, which all arise from the chosen method of exposition, are unimportant. But this is not the case for the following examples, which deal with the method of research, something we would today describe as heuristic models. We should distinguish a number of different types.

In an elementary case, we are dealing with an obvious macroscopic operation, and more often than not, a single well-defined function. This is the case for the jaw, whose use clearly is to crush just like a millstone in a mill.

---

[32] "[I]l est revêtu de membranes fortes *parce qu'*il fait un long chemin." Ibid., pp. 431 and 423.

[33] Ibid., p. 142.

Legitimating its use is riskier when one affirms that a sieve can do nothing but filter. Undoubtedly a filter is made to filter, but the question remains as to know where and when it is possible or even necessary to see filters. In the case of the kidney, Dionis rightly observes that everyone agrees in recognizing its use as a filter; in this connection we noted above the remarkable work of Marcello Malpigni published in 1666.[34] Pierre Dionis' is more original in *seeing* sieves and filters in a great number of anatomical parts, in all digestive organs, and even in the brain. We have seen that the absence, evident or presumed, of all spontaneity in the operations of the body leads one to favor this form of what we might call hypothetical reasoning which, refusing any production, any genuine synthesis, can find no explanation for any properly observed modification in state other than through hypothesis. The reasoning breaks down into three phases: the observation of a presumed change necessarily presupposes a "filtration" which in turn presupposes the existence of filters to accomplish it. It is obvious for the mechanist that this model cannot be called into question without damaging mechanism as a whole in the process.

A double sanction both justifies and limits the use of imagination: from below, we might say, the sanction of facts truly observed, for example, a perfectly observed change in the blood; from above, the sanction of rational and well-established principles whose reasonableness cannot contest the evidence of the observations. It is between these two types of evidence, one sensory, and the other purely theoretical, that the talent of the anatomist shows itself, as he is placed in the position, happily, of "imagining" uses.[35] This "fortunate imagination" obviously is exercised within a mechanistic scheme: the mechanistic scheme fixes a necessary condition for the "imagination" to be "fortunate."

Here we find the opposition between a strictly mechanical model and a chemical model, between iatromechanics and iatrochemistry, which to a great extent will fuel the physiological debates of the time, and which around 1715 will provide the opportunity for a remarkable materialization in the famous quarrel over the interpretation one should give the operations of digestion by the stomach. Philippe Hecquet, from the Paris *Faculté*, had to defend the mechanistic thesis of trituration and crushing against Jean Astruc of the Montpellier *Faculté*, who affirmed the intervention of a solvent. Dionis, the surgeon, firmly took the side of the mechanical interpretation; this gives him the opportunity to criticize Duncan, the head of the chimiatric School, on the filtration and precipitation carried out in the liver. In addition, he does not fail to note that chemists admit that yeasts are present everywhere that "filtrations" take place. Dionis defends the thesis according to which the circulation of blood authorizes and even requires a mechanical interpretation, judged by him necessary and sufficient.

[34] "Tout le monde reconnaît l'usage de filtre du rein," ibid., p. 215.

[35] Fontenelle clearly defined this faculty for imagination in the elogium he gave to the great anatomist Daniel Tauvry on his death at age 31 in 1701: "À la grande connaissance qu'il avait de l'Anatomie, il joignait le talent d'imaginer heureusement les usages des structures [to the great knowledge he had of anatomy, he brought the talent of fortunately imagining the uses of structure]" ("Éloge de Tauvry," in *Œuvres de Monsieur de Fontenelle*, vol. V, p. 53).

This thesis, according to him, makes superfluous any recourse to agents other than the mechanical agents appealed to in circulation. Thus, the great impulsion of the heart, the spring of the arteries, the entire circulatory dynamic show that blood is continuously agitated and that "the various molecules in it must be separated" so that "we don't need to have yeasts for precipitations."[36] Let us summarize: the term of "dissolution" generally used to designate a "precipitation" must not confuse us; it can only be a "separation." According to the mechanists, chemists err both in being too cautious and, at the same time, too bold(!). A more attentive examination, a more detailed study, a more thorough and rigorous exploration will always permit us to reveal the true cause of already recognized uses, heretofore hidden and mechanical. This alone can lead to identifying the still unknown uses of things. It is because they did not respect these rules that the Ancients "imagined" chimerical uses. All the discoveries of the Moderns demonstrated and continued to demonstrate each day the prejudices and errors of the Ancients. Dionis never fails to remind us of this.

While this type of argument had been common since Descartes, we should note that Descartes himself falls victim to the same criticism. Reporting the well-known explanation that Descartes had given for the circulation of blood, claiming to correct William Harvey, Dionis declared:

> Here is one of the most beautiful imaginations possible, and it is certain that by assuming this one can explain all phenomena that can be encountered in this domain. We are obliged to this great man for having broken the ice and been the first to explain mechanically the motions of the heart.[37]

The interpretation given by the surgeon is resolutely mechanistic and hard to square with the assumption of yeasts. But whereas Descartes meant to correct Harvey while recognizing his merit for having "broken the ice," Dionis in turn corrected the philosopher, praising him, while excusing him, for having "broken the ice."[38]

---

[36] "[L]es différentes molécules en doivent être séparées ... il n'est pas nécessaire d'admettre des levains pour les precipitations" (Dionis, *Anatomie*, p. 205). Dionis speaks, for example, of the stomach, where "un acide dissolvant qui par ses pointes aigües et tranchantes pénètre et dissout les aliments les plus solides [a dissolving acid, which by it sharp and trenchant points penetrates and dissolves the most solid food]" (ibid., p. 188). Furetière, *Dictionnaire universel*, art. "précipiter" n.p.: "Dissoudre un métal avec de l'eau forte et puis l'en séparer en y versant de l'eau commune ou de l'huile de tartre qui la fait tomber ou précipiter au fond du vaisseau ou bien en faisant évaporer l'eau forte avec le feu. [To dissolve a metal with an acid then separate it out by pouring in ordinary water or tarter oil, which makes it fall or precipitate to the bottom of the vessel, or by evaporating the acid with heat.]"

[37] "Voilà une des plus belles imaginations qu'on puisse avoir et il est certain que par cette supposition l'on peut expliquer tous les phénomènes qui se rencontrent sur cette matière. Nous sommes obligés à ce grand homme d'avoir rompu la glace et d'avoir expliqué le premier par la mécanique les mouvements du cœur" (Dionis, *Anatomie*, p. 380).

[38] "Il ne connaissait pas assez bien la structure du cœur et ses méditations l'occupaient trop pour en avoir une plus grande connaissance. Toujours dirons-nous qu'il a fait tout ce qu'un homme pouvait faire, ne sachant que ce qu'il en savait [He did not know the structure of the heart well enough, and his meditations occupied him too greatly to have a greater understanding. But we will say that he did everything a man could do, knowing only what he knew.]" (ibid.).

Finally, it is easy to note the ambiguity of the word *imagination*, which at times refers to the anticipated correct representation of a use, and at others, a mere fantasy of the intellect. The ambiguity of such a term is obviously not accidental. It condenses the aporia and the dangers of research in this morphological context.

Here we can perceive the problem that results from the encounter of the analogy that justifies the mechanical paradigm in its generality with the observation or the "inspection" of this or that particular structure. We can find an example of this in the problem of the motion of the brain. The necessary existence of animal spirits requires not only an operation of filtering, but an alternating movement, diastolic and systolic, like that of the heart, in order to move these spirits, so as to receive blood and circulate the animal spirits.[39]

There remains a final situation, the most interesting one: the one in which the observer finds himself when he is uncertain and has nothing to propose after a minute examination. And here, we can distinguish four more cases. In the first case, the uses remain to be determined while awaiting a more detailed analysis or a correlation that can send us in a new direction. With regard to the vermiform appendix, for example, Dionis reserves his judgment. He notes only that it is larger in children, "which is extremely embarrassing for anatomists."[40]

In the second case, several possibilities for explanation are offered for the anatomist's judgment. For example, when he reviews the doctrines proposed to explain the problem of generation, Dionis distinguishes four that can be imagined. Of the four, he eliminates two, retaining Jan Swammerdam's thesis of preformation and the ovist theses presented by Nicolaas Hartsoeker. He compares the strengths and weaknesses of each of the two remaining theories in relation to the data of experience and in the extent to which they agree with the principle of mechanics. In accordance with this, he opts for the ovist doctrine. Ovism seems to him more compatible both with the principles of mechanism and with the existence of monsters that cannot have been wanted by God, a thesis that the preformationism would have to allow for.[41]

The third case might be thought of in terms of abduction. It is true that when Dionis attempts to understand in which part of the brain thought takes place, he imagines that there are good reasons to think that the most important things take place under the corpus callosum, in the corona radiata,[42] as if the idea of a long, subtle, and difficult elaboration required for the representation of the formation of ideas requires a passage in a complex and fine structure to be perfected. We have here a good example of what we indicated above with respect to the analogical relation between the great mechanical principles and the particular situations presented through observation.

[39] Ibid., pp. 462f.

[40] "[C]e qui embarrasse extrêmement les anatomistes" (ibid., p. 175).

[41] Ibid., pp. 348f.

[42] Ibid., p. 465. The question will be dealt with in this sense by Thomas Willis in an authoritative work in the subject, published in London in 1664. In his *Cerebri Anatome cui accessit nervorum descriptio et usus*, the great anatomist leaned toward seeing the seat of the soul in the corona radiata, in which impressions are made.

The fourth case can be seen as intermediate between the two preceding ones. Faced with a structure whose use he does not know, but which presents a remarkable complexity, the anatomist, knowing nothing of its use, can nonetheless "imagine" that this the importance of the use in proportion to the complexity of the part. With regard to the choroids plexus, placed under the corpus callosum, Dionis writes that "it is so artistically made than one may believe that is has considerable uses. This is why many people have attempted to discover them."[43]

11.7 Conclusion: Yes to Mechanism, but Which One?

One might expect that the originality of a surgeon's mechanistic thought might lie above all in the absence of developments and arguments often encountered in the writings of philosophers and theologians. All the more when what is at issue is a course for future surgeons. But in the case of Dionis, on the contrary, we are struck by the importance of the philosophical (much more than the theological) grounding on which he presents his anatomy. While it is true that such arguments do not offer an occasion for high-level speculative developments, they are nonetheless present, and often well presented. A simple reminder of the context allows us to realize this.

The work of Dionis can be situated in the historical context of the strong growth of anatomy in Europe. The anatomical science of modern times, inaugurated in the middle of the sixteenth century, continued to undergo a period of extraordinary progress in the last years of the seventeenth century, which confirmed anatomy as an authentic scientific discipline. Nevertheless, there were very serious obstacles to its recognition and to the role given it in the profession and in the medical disciplines. This observation is particularly true for France, where the Paris *Faculté* showed a firm conservatism in doctrine, and great intransigence with respect to practice and teaching for all those who wanted to teach "novelties". To this we must add the changes in surgery, which was in the midst of its own attempts at disciplinary recognition and which was also facing its own bans from the *Faculté*.

In such circumstances, the promotion and development of anatomy was a priority for the anatomist as well as for the surgeon. Dionis advances the growth and the primacy of anatomy and surgery by situating them in the general context of the Quarrel of the Ancients and the Moderns. The anatomy of the Moderns, which he constantly opposed to that of the Ancients, was to serve as the indispensable basis for all of medicine. As a consequence, it is anatomy that gives principles to the other fields of medicine. The study of the structures of the body is no longer complementary

---

[43] "[I]l est si artistement fait que l'on a sujet de croire qu'il a des usages considérables. C'est pourquoi plusieurs se sont efforcés de les découvrir" (ibid., p. 468). Let us note that Steno had shown in this complex structure an infinite number of microscopic glands that could serve as filters. Let us recall that in the seventeenth century, *plusieurs* meant "much" or "many."

or subordinate. It must, on the contrary, enlighten physiology, and in so doing, all of medicine, by describing the organs and their operation. We have seen the formula by which Dionis summarized the primacy of structure and the mechanistic vision. For him, these two ideas necessarily form a whole.

For Dionis, the mechanical conception of bodies seems to represent both the entire philosophy and the entire program of the Moderns in the field of medicine. While we can understand the ambition of the program, there remains the question of the limits of the doctrine that inspires it. The omnipresence of mechanism seems proportionate to its concrete character. By exposing a surgical anatomy, one adopts the most mechanistic point of view possible in the medical sciences. Surgical practice appears as applied mechanics within another applied mechanics: that of Nature. Indeed, surgery is first of all the sciences and arts of "operations." These are classically reduced to four: separating, uniting, removing, adding. Each of these responds to a symmetrical operation in an "art" directed at a Nature that has defects: one brings together that which Nature has separated, one clears that which she has blocked. The "competition" between the two arts, on which the anatomist Leonardo da Vinci based his philosophy, always accompanies the surgeon's thinking. The practice of surgery relies on anatomical knowledge, which itself is the application of the laws of mechanics. And the operations of Nature follow, with the help of geometry, these same rules of mechanics. In the well-established doctrinal context of mechanism at the end of the seventeenth century, considerations of figure and motion, mechanics and geometry, statics, hydraulics, and pneumatics, the recognition of the mechanical arts (that is to say the hand) in the liberal arts, etc., offered a natural framework and setting for a sort of "spontaneous mechanism" for the surgeon.

These reminders, as well the evocation of certain discussions on the frontier of philosophy, what we would call epistemology, must thus be understood in the context of a surgeon, and of a surgeon's desire to demonstrate the scientific and "liberal" quality of his domain. Without entering into the philosophical or even "epistemological" debates that have no place in a course on surgery, it is nonetheless possible to see in this work the difficulties and the limits inherent to the mechanistic thinking. On the "physiological" level, it would be interesting to return to the critique of the chemical interpretation. We think, for example, that Dionis' discretion, indeed his silence about the work and the person of Harvey is not unrelated to the critique of iatrochemistry. Even while accepting the mechanistic thesis of a chemistry in which all the operations are explicable by figures and motion, as is the case here, there remain difficult passages in this mechanical anatomy.

The traps of language and final causes seem to us more obvious and more important. The critique of final causes was one of the essential elements of mechanistic philosophy, as well as the theme mechanistic philosophers perhaps stressed the most. It is unnecessary to remind of the benefits of their elimination in the study of the living body. To say that an artery is large because it must carry a great deal of blood is a formula that allows one to appreciate the role of the cause while observing the scale of the effect. It is quite different when we say that apples ripen in order to be eaten by man. Between these two situations we find examples in which it is not so obvious how to assess them. When Dionis criticizes the well-known considerations the Ancients used to justify the higher position of the eyes and the brain, it is to replace them with a mechanistic consideration by which the reservoir must be placed high to be able properly to irrigate the rest of the body.[44]

But don't we encounter final causes again when it comes to justifying the absence of the *rete mirabile* in man? Justified in animals for slowing the flow of the blood going from the heart to the brain, the *rete mirabile* is useless in man, because of his upright position: "This is why Nature did not give him one," writes our imprudent mechanist.[45] The aporia and the traps seem more pernicious when we come to the critique of spontaneity. They readily appear in the example noted above of a "reservoir brain". We understand that the rejection of Galenic faculties forbids all recourse to imaginary explanations. But why "must" the brain be a reservoir? The "use made" and not the "use to be made" avoids any psychological projection, but the difficulties in representing it only sets back our understanding of an "intelligence" of the living. The anatomist who limits himself to observing the result and the actual use of a body part is in the same situation as the philosopher who observes a fully created nature. In the end, the author is logically driven to the same conclusion, very common in the mechanistic theologies and philosophies of the day.

Let us conclude with the end of Fontenelle's *Éloge* of Méry: "Astronomy, Anatomy, are the two sciences where the characters of the Sovereign Being are the most notably marked. One announces its immensity by that of the celestial spaces, the other its infinite intelligence by the mechanics of animals. We can even believe that anatomy has an advantage: intelligence proves even more than immensity."[46]

---

[44] Ibid., p. 31.

[45] "C'est pourquoi la nature ne lui en a pas donné" (ibid., p. 466). The *rete mirabile* was a structure, not recognized in modern anatomy that was supposed to filter blood between the carotid and the brain.

[46] "L'Astronomie, l'Anatomie sont en effet les deux sciences où sont le plus sensiblement marqués les caractères du Souverain Etre; l'une annonce son immensité par celle des espaces célestes, l'autre son intelligence infinie par la mécanique des animaux. On peut même croire que l'anatomie a quelque avantage; l'intelligence prouve encore plus que l'immensité" (*Œuvres*, vol. VI, p. 181).

Chapter 12
Du Clos and the Mechanization of Chemical Philosophy

Rémi Franckowiak

12.1 Du Clos, Boyle and Fontenelle

> That year, Mr. Du Clos continued the examination that he had begun of Mr. Boyle's *Essays of Chemistry*. This English scholar had undertaken to explain all the chemical phenomena by way of corpuscular philosophy, that is, through the motion and the configurations of small bodies alone. Mr. Du Clos, as great a chemist as Mr. Boyle, but being perhaps more chemistry minded, did not think this science could or even needed to be reduced to such clear principles as shapes and motions, and he readily accepted a certain specious obscurity, which is quite well established .... [C]hemistry, by visible operations, resolves bodies into certain coarse and tangible principles, salts, sulphur, etc. But physics, through delicate speculations, acts on these principles, as chemistry does on bodies, and resolves them into other even simpler principles, to small moving bodies with an endless number of shapes: here is the main difference between physics and chemistry, and almost the same as that which lay between Mr. Boyle and Mr. Du Clos. The spirit of chemistry is more confused, more veiled; resembles more the mixed bodies, where the principles are mixed up with one another, while the spirit of physics is more distinct, clearer; finally it identifies the first origins, and the other does not go through to completion.[1]

[1] Fontenelle, *Histoire de l'Académie Royale des Sciences*, vol. I, pp. 79–81: "M. du Clos continua cette année l'examen qu'il avoit commencé des Essais de Chimie de M. Boyle. Ce savant Anglois avoit entrepris de rendre raison de tous les Phénomenes Chimiques par la Philosophie corpusculaire, c'est-à-dire, par les seuls mouvemens & les seules configurations des petits corps. M. du Clos, grand Chimiste, aussi-bien que M. Boyle, mais ayant peut-être un tour d'esprit plus Chimiste, ne trouvoit pas qu'il fût nécessaire, ni même possible, de reduire cette Science à des principes aussi clairs que les figures & les mouvemens, & il s'accomodoit sans peine d'une certaine obscurité spécieuse qui s'y est assés établie. [...] La Chimie par des operations visibles résout les corps en certains principes grossiers & palpables, sels, souffres, &c. Mais la Phisique par des spéculations délicates agit sur ces principes, comme la Chimie a fait sur les corps, elle les résout eux-mêmes en d'autres principes encore plus simples, en petits corps mus & figurés d'une infinité de façons: voilà la principale différence de la Phisique & de la Chimie, & presque la même qui étoit entre

R. Franckowiak (✉)
Université Lille I, Villeneuve d'Ascq, France
e-mail: remi.franckowiak@univ.lille1.fr

D. Garber and S. Roux (eds.), *The Mechanization of Natural Philosophy*, Boston Studies in the Philosophy of Science 300, DOI 10.1007/978-94-007-4345-8_12,

It is in these terms that in 1733, Bernard le Bouvier de Fontenelle, the permanent secretary of the Royal Academy of Science in Paris, summarized the ten papers that Samuel Cottereau Du Clos had read about 60 years earlier before the *Compagnie*. These ten papers form his examination of Robert Boyle's 1667 *Tentamina Chimica*, the Latin translation of his 1661 *Certain Physiological Essays*. In his text, Fontenelle seems to give priority in the knowledge of bodies to the methodological reasoning of the mechanical philosophy, to the detriment of the experimental reasoning of chemistry, by establishing an opposition between the heterogeneous nature of the substances resulting from the resolution of mixed bodies by fire, and the more fundamental nature of the corpuscles resulting from their resolution by the mind. He reduces the difference between the two domains to an opposition between confusion and clarity in the practice of chemistry and physics. However, chemistry and physics both proceed in a similar way, using the same method: analysis. But the former, by "visible operations," cannot in fact hope for an analysis of bodies as fine as the latter is able to obtain through "delicate speculations." In other words, the physicist in his study can fully achieve what the chemist in his laboratory only partially can do. Since Du Clos' cast of mind was closer to that of a chemist, he himself was satisfied with the rough principles he obtained. So, Fontenelle argues, Boyle's physics then intervened to carry the chemist's analysis through to completion. Fontenelle's comments are indeed radical because he denies chemists any possibility of access to knowledge of the true causes of the phenomena they produce in their laboratory by the use of their instruments alone. In his opinion, they are unable to go beyond the level of tangible bodies. The practice of chemistry nevertheless cannot be criticized in itself insofar as it provides the matter on which the resolving power of the mind of physics, higher than the power of the alembic, can be exerted.

In fact, Fontenelle was opposing two extreme caricatures of physical science here: the hard-working but not particularly effective chemist, on the one hand, and on the other, the speculative and powerful physicist. These two caricatures correspond neither to Du Clos nor to Boyle; indeed, the two strenuously resisted these exaggerated characterizations. Du Clos' criticism of Boyle's book was not so much directed against the mechanical philosophy as such, but rather against the place Boyle assigned it for the comprehension of chemical phenomena, and its degree of probability compared with the observations and experiments that chemists carried out. For his part, Boyle wanted to establish a natural philosophy placed in a position intermediate between a strict mechanical philosophy (considering only motion and shapes) and simple chemical practices. The purpose of the publication of his essay on the operation of reintegrating saltpetre in the *Tentamina Chimica* was precisely to provide an illustration and an experimental justification for the

---

M. Boyle & M. du Clos. L'esprit de la Chimie est plus confus, plus envelopé; il ressemble plus au mixtes où les principes sont embarassés les uns avec les autres, l'esprit de Phisique est plus net, plus dégagé; enfin il remonte jusqu'aux premieres origines, & l'autre ne va pas jusqu'au bout." Fontenelle is the real author of this report, and not Jean-Baptiste Du Hamel, as some claim.

mechanist doctrine.[2] As for Du Clos, far from "rejecting the mechanistic approach to chemistry completely,"[3] he aimed to establish the most "reasonable and convincing" causes of the chemical phenomena examined by Boyle from experimental work followed closely by the members of the *Académie*. However, this work is limited; it could not identify "the ultimate origins" of bodies, as Fontenelle complained: in other words, it could not go through "to the conclusion" of the analysis. But at the same time this work highlights the boundaries that Boyle, who claimed to rely only on experiments, should not have crossed.

The aim of this chapter is not to analyze Du Clos' examination of Boyle's work as such, but to explore Du Clos' philosophy of chemistry. Du Clos is, to our knowledge, the first French chemist to combine in natural philosophy demonstration using the laws of motion with demonstration using the qualities of chemical principles, and in that way bringing to bear two different and complementary orders of reasons. For him, the mechanical considerations represent a first approach, a stage towards the knowledge of "the truth of things" (*la vérité des choses*).[4]

12.2 Chemistry of Principles Versus Boyle's Mechanical Philosophy

Du Clos (1598–1685) is a forgotten and yet important figure in the history of chemistry.[5] He is known, above all, through Fontenelle's report, as a foil for Boyle's mechanical philosophy and as a confused chemical practitioner. *Médecin ordinaire du roi* in Paris, Du Clos joined the section of physics at the *Académie royale des sciences* as soon as it was created in 1666, and opened the first session. He was then 68 years old, read about 50 papers, left two published works and two manuscripts, and was one of the three best remunerated academicians. His examination of Boyle's book took place at a special moment in the history of chemistry. Du Clos indeed appeared at the *Académie* only a few years after the recognition of Paracelsian chemistry as an essential science in physics. But this recognition led to the result that the chemical principles, Mercury/Sulphur/Salt or Spirit/Oil/Salt/Earth/Water, powerful rivals of hylemorphism's principles because they could be demonstrated in the laboratory, were now studied for their own sakes, and no longer through their confrontation

[2] This same text was also the subject of some criticisms from Spinoza; see Macherey, "Spinoza, lecteur et critique de Boyle," pp. 733f.

[3] As Kim wrongly claims in *Affinity, that Elusive Dream*, p. 66.

[4] On the different kinds of corpuscular theories of matter in seventeenth-century chemistry, see Clericuzio, *Elements, Principles and Corpuscles*. However, Clericuzio does not deal with the mechanical dimension of Du Clos' chemical explanations.

[5] On Du Clos, see Todericiu, "Sur la vraie biographie de Samuel Duclos (Du Clos) Cotreau," pp. 64f.; Stroup, "Censure ou querelles scientifiques," pp. 435–452; and Franckowiak, "Samuel Cottereau Du Clos."

with the Aristotelian elements. This, in turn, highlighted their conceptual weakness.[6] Chemistry came to lose its theoretical weight. In his *Sceptical Chymist*, where he highlighted in particular the important difficulties of the chemistry of principles, the discipline appeared to Boyle as a simple set of practices or experimental know-how, to be explained by the mechanical affections of tiny insensible bodies.[7] Boyle indeed preferred to reduce chemistry, or the chemistry of the "vulgar chemists" as he called it, to a simple craftsman's activity. Chemical operations were to have their rationale within the framework of his "mechanical hypothesis" and not to be explained by a certain number of would-be homogeneous chemical principles that supposedly constituted all mixed bodies. According to Boyle, bodies were not composed of certain elementary substances, be they three or five in number; rather, they were composed of a texture of shaped corpuscles in motion by which their qualities were explained. Visible effects could only come from a mechanical change. Nevertheless, Boyle was not a defender of experimental chemistry any more than he was an opponent of the Paracelsian doctrine based on the mechanical philosophy. His aim was rather to provide the practice of chemistry with philosophical principles and to give natural philosophy an experimental basis, in other words, to create a "Chemico-Physical" intermediary that was no longer chemistry—because only observations and experiments were to be preserved—nor strict mechanical physics.[8] Thus Boyle wanted "vulgar chemists" to stop producing their own discourse on chemical operations, break with their tradition, and submit to the mechanical hypothesis.[9] Here, chemistry takes on the role of an instrument for natural philosophy.[10]

---

[6] See Franckowiak, "Du Clos, un chimiste post-*Sceptical Chymist*," pp. 361–377.

[7] See Boyle, *The Sceptical Chymist*.

[8] See Principe, *The Aspiring Adept*, p. 208; Clericuzio, "Carneades and the Chemists," pp. 79–90; Clericuzio, *Elements, Principles and Corpuscles*; and Franckowiak, "La chimie du 17e siècle."

[9] The more elevated sort of chemistry, which aims at the preparation of so-called "Philosophical Mercury" in particular, on the other hand, retains the status of real science, a well-established reasoned practice within a strong tradition. On Boyle's relation to this tradition, see Principe, *The Aspiring Adept*, pp. 153 and ff. Moreover, as early as 1660, one year before the publication of the *Sceptical Chymist*, Samuel Sorbière expressed the wish that chemists would allow other more learned people to speak about their experiments; see his letter of July 13, 1660, in *Relations, Lettres et Discours de Mr Sorbière sur diverses matières curieuses*, pp. 167–168. It is also important to note that the image Sorbière gives of the hard-working chemist is close to the etymology of the name of Philoponus, the character of the vulgar chemist in the *Sceptical Chymist*.

[10] In Boyle's experimentalism, the practice of Chemistry is considered as a means to produce an experimental piece of evidence of the action of the human mind in matter; see Hamou, "Descartes, Newton et l'intelligibilité de la nature," pp. 146f. A study more precisely focused on early modern chemistry in France points out that Boyle's influence among eighteenth-century French chemists was very weak. Moreover, his *Sceptical Chymist* and his thoughts on chemistry in general had little influence in France, unlike Du Clos'; see Franckowiak, "La chimie du XVIIe siècle." Boyle's virulent denunciation of chemistry without solid principles is undoubtedly his real contribution to chemistry, or at least to French chemistry. A remark by Christian Huygens illustrates this point. In 1692,

12.3 Du Clos' Examination of the Second Part of Boyle's *Tentamina Chimica*

Du Clos could be regarded as a post-*Sceptical Chymist*, that is, a chemist who continued, after the publication of the *Sceptical Chymist* in 1661, to explain chemical phenomena by the presence of a small number of chemical principles, while taking into account Boyle's main arguments against the principles of a chemistry self-sufficient in practice and whose theory persisted as a tradition.[11] His project at the *Académie* was to rebuild chemistry within the framework of a reappraisal of the chemical principles so as to resist its reduction to mechanical principles, and to keep it as a real and specific science within natural philosophy. If only the practice of chemistry deserved to be preserved, Du Clos tried, session by session, firmly to establish true chemical principles through clear reference to their experimental reality. According to him, these principles were fairly close to Jean-Baptist van Helmont's doctrine, from which he also adopted the title of "modern chemist" for himself.[12] The requirement of experimental confirmation that can be observed in all Du Clos' papers is important insofar as it constitutes a new way of setting forth theory in chemistry. It is also undoubtedly the only way to obtain coherence between chemical practice and chemical doctrine. On this basis, and following the *Sceptical Chymist*'s admonitions, Du Clos thus examined the arguments put forward in Boyle's *Tentamina Chimica*: chemical theory is just a hypothesis in the same way as mechanism is, that is, both are to be debated on the basis of experimental evidence. Boyle had systematically confronted the doctrines that chemists advanced with their practice. Du Clos was to do the same with Boyle's. He also confronted it with his own experiments, carried out in front of his peers at the *Académie*, who constituted the public to be convinced. He spoke about "probable" cause justified by experiments and observations. His method was simple: Du Clos "observed," he "saw," he "noticed," he "made experiments"; and, at times, "the *Compagnie* wished to see" and in turn, Du Clos "showed," "thus the *Compagnie* recognized that" and is

just a few weeks after Boyle's death, Huygens answered a letter from Leibniz in which he expressed his disappointment at the sterility of Boyle's chemical experimental work, whose mechanism, according to him, was not new at all (see Leibniz to Huygens, 8 January 1692, in Huygens, *Œuvres complètes,* vol. X, pp. 228f.). See as well Huygens to Leibniz, 4 February 1692, in ibid., p. 239: "Mr. Boyle est mort, comme vous sçaurez desja sans doute. Il paroit assez etrange qu'il n'ait rien basti sur tant d'experiences dont ses livres sont pleins ; mais la chose est difficile, et je ne l'ay jamais cru capable d'une aussi grande application qu'il faut pour establir des principes vraisemblables. Il a bien fait cependant en contredisant à ceux des Chymistes."

[11] Franckowiak, "Du Clos, un chimiste post-*Sceptical Chymist*," pp. 361–377.

[12] Van Helmont, *Ortus medicinae*. See Du Clos, "Project d'exercitations physiques," *Procès Verbaux de séances, Registre de physique*, Dec. 31, 1666, t. 1, p. 1; and "De la recherche des principes des mixtes naturels," in ibid., pp. 4–16. All the references here to Du Clos' works at the *Académie* come from the manuscript *Procès Verbaux de séances*.

"satisfied," and so Du Clos concluded that it "just seems to him that." On the other hand, if Boyle "had seen," if he had more experimental knowledge, he could have similarly concluded that ....[13]

On September 1, 1668, Du Clos announced that he had recently read Boyle's *Tentamina Chimica* in which "he noted several things worth examining in the *Assemblée*"[14]; which he did in ten sessions until January 26, 1669.[15] Du Clos completely agreed with Boyle when the latter stressed the importance of accumulating observations and well-established experiments in order to found the physical sciences. On the January 5, 1669, Du Clos examined the second part of the *Tentamina Chimica*, the part with which we are dealing. After presenting his conception of the nature of saltpeter, he granted Boyle "[t]hat the effervescence comes indeed from the motion, but Mr. Boyle does not assign the cause of this tumultuous motion which he maybe could not find in the shape and position of the particles, in the way he thinks he has found that of the fluidity of bodies."[16] Du Clos thought that he could identity the cause of this tumult. The cause of the colors that saltpeter acquires through fire "is not really well explained by the new position of the particles. He considered it would have been better to impute it to the exaltation of some sulfur contained in this matter or acquired from coal."[17]

> Mr. Boyle did not find the cause in accordance with the corpuscular philosophy, since he said nothing about it. It is true that he ingenuously declared elsewhere, notably on page 23, that the assumptions of this doctrine are not enough to explain everything.[18]

Du Clos wrote, still in relation to his own experiments, that "Mr. Boyle did not find the solution to this question in the corpuscular philosophy; but this is easily and quickly found in the philosophy of the chemists who recognize that the Earth has

---

[13] On such experiments made in front of his peers in the *Académie*, see among others the papers of Nov. 3 and 24, 1668, Jan. 5, 1669, and Jan. 26, 1669 (ff. 26v–27r). At the beginning of November 1668, the Academy had recently set up a laboratory. We may thus suppose that the Academicians' desire to see for themselves what Du Clos spoke of also corresponded to a legitimate curiosity about this new show (the change of colors, the change of physical state, etc.). In other words, what is at issue here is chemistry staged in the laboratory as a theater. Later on, this theatrical element disappeared. Boyle too often used such expressions as "we have observed" and offers generous details of his experiments. However, he only practiced chemistry for a short time, only until the end of 1660s. See Newman and Principe, *Alchemy tried in the Fire*, pp. 15–30.

[14] Du Clos, *Procès Verbaux de séances*, t. 4, Sept. 1, 1668, f. 196v. Du Clos' reading of the *Tentamina Chimica* is certainly selective but respects Boyle's text.

[15] *Tentamina Chimica* was not the first of Boyle's writings Du Clos examined. From the beginning of 1667, Du Clos had commented several times on Boyle's writings about "alchemical" subjects, which he considered insufficiently grounded in a real knowledge of the chemical literature or chemical substances. See Franckowiak, "Du Clos, un chimiste post-*Sceptical Chymist*."

[16] Du Clos, *Procès Verbaux de séances*, *Registre de mathématiques*, t. 5, vol. 1, Jan. 5, 1669, f. 4v. "C'est bien le mouvement qui fait l'effervescence, mais Mr. Boyle, poursuit-il, n'assigne pas la cause de ce mouvement tumultueux, que peut estre il n'a pû trouver dans la figure et disposition des particules, comme il pense y avoir trouvé celles de la fluidité des corps."

[17] Ibid., f. 5r.

[18] Ibid., f. 5v.

much symbol [= attractive affinity] with sulfurs."[19] Du Clos then wondered: "But what acceptable and convincing arguments can be put forward to support the corpuscular doctrine?"[20] He did not believe the explanation of the "imaginary" position of the particles was "convincing" or "reasonable" because it is not "manifest" or "apparent." And Du Clos did not fail to quote Boyle when he characterized his writings on the fluidity and solidity of bodies as "an imperfect draft" "so that he could excuse," according to Du Clos, "what is less well explained and further from the atomic or corpuscular doctrine which Mr. Boyle did not find as easy to establish through chemical experiments as he had intended."[21] This was indeed the weakness in Boyle's doctrine that led Du Clos as a result to grant little credibility to the thinking that was so far from experiment and so difficult to see in the laboratory. Actually, Du Clos was not calling into question the quality of Boyle's experiments at all, but was simply rejecting their interpretation. In other words, Du Clos seems to have been using Carneades' remark in the *Sceptical Chymist* against Boyle himself: "I told You already … that there is a great Difference betwixt the being able to make Experiments, and the being able to give Philosophical Account of them."[22] While putting his "opinion"[23] forward, he was asking for more solid evidence: "Who can know if these differences come from the change of texture in particles, which cannot be seen,"[24] he wrote in his last communication. The reversal of roles is striking: Boyle became the vulgar chemist and Du Clos the sceptical chemist.[25]

12.4 Integration of Mechanical Explanations into Du Clos' Chemistry

In his examination of *Tentamina Chimica*, though, Du Clos does not at all refuse to integrate mechanical considerations into chemistry. He claims that bodies can indeed be perceived—up to a certain point—as composed of "particles" in motion that "unite"[26] and "bind closely."[27] Thus, for example, the reason why a few acidic

[19] Du Clos, *Procès Verbaux de séances*, Jan. 12, 1669, ff. 7r–7v.

[20] Ibid., f. 9r.

[21] Du Clos, *Procès Verbaux de séances*, Jan. 19, 1669, f. 14r.

[22] Boyle, *Sceptical Chymist*, p. 208.

[23] Du Clos, *Procès Verbaux de séances*, Jan. 26, 1669, f. 21v.

[24] Ibid., ff. 24v–25r.

[25] Du Clos did not exercise his skepticism against Boyle alone. In a communication read in April 1667 concerning the examination of Le Givre's book on mineral waters, whose conclusions appeared hasty and faulty to him, he writes (Du Clos, "Examen du livre des Eaux Minerales du Sieur Le Givre", *Procès Verbaux de séances*, March 12, 1667, pp. 57–70): "Voilà toujours l'auteur dans les suppositions, et le lecteur sans preuve, qui le satisfasse …. [T]outes les expériences et observations que cet auteur a rapportées ne prouvent rien de ce qu'il pretend."

[26] Du Clos, *Procès Verbaux de séances*, Nov. 24, 1668, f. 308v.

[27] Ibid., f. 309r.

liquids that have been excessively rectified, that is, repeatedly distilled, dissolve metals badly is because of the lack of tenuousness of salt particles in the liquids; they are not able to penetrate the metal pores, stick to them, and weigh them down by giving them acrimony.[28] Nevertheless, Du Clos confines these kinds of considerations to the level of sensible bodies alone. Mechanical arguments are strictly limited to phenomena concerning the coarse part of the world. In Du Clos' chemical philosophy, there are two domains of reality, as in the Paracelsian doctrine.[29] So every single body has two parts: on the one hand a tangible and visible part, and on the other hand an invisible and intangible part. The former is an extended and divisible part, which concerns the superficial interactions of matter, and only here are mechanical explanations at issue. It represents a kind of "bark" for the latter part, that is, a spiritual matter full of forces and qualities. In fact, Du Clos holds that there are three "true" principles for every single natural body (principles not yet proved experimentally): (1) an active and incorporeal principle called "Nature," an informative cause which takes place in (2) the passive and concrete principle called "Body" (*Corps*) (that is, elementary water joined with the elements air and earth), through the necessary mediation of (3) a last and intermediate principle, the "Spirit," neither completely incorporeal nor absolutely corporeal (that is, it has some extent, but it is indivisible, it establishes the link between the two other principles). These come together to produce natural mixed bodies. But only the passive corporeal principle is of relevance to mechanical explanations, according to Du Clos; it is obvious that the incorporeal stands outside of mechanism, and relates to that which is typically considered in the domain of the chemical. The Spirit, as for it, is both chemical and mechanical.

Thus for Du Clos, it is only passive and corporeal bodies that relate to mechanical interpretations, which highlight their physical properties. The incorporeal, which produces their chemical properties, is obviously irrelevant to mechanism. The generation of saltpeter nicely illustrates the way in which Du Clos breaks up a natural phenomenon into physical considerations and chemical ones. Its natural formation on the walls of a cellar, for example, may thus be understood in a mechanical way. For Du Clos, natural saltpeter results from the condensation of the air in a sulfurized salt, that is, in an alkali, which attracted it, reduced its agitation and retained it.[30] Moisture can help this formation to relax "the compaction of the parts" of sulfurized salt, and thus facilitate the entry of air.[31] Fulmination occurring when saltpeter is brought into contact with burning coals is also due to the expulsion of this condensed air.[32] On the other hand, an alkali is acrid because of the igneous spirit

[28] Ibid., ff. 302r–302v.

[29] See Koyré, *Paracelse*, pp. 28–30; Bianchi, "The Visible and the Invisible," pp. 17–50.

[30] Boyle also admits that fixed nitre (= salt of tartar) attracts the air (he talks about "its aptness to attract the air"); see Boyle, *Certain Physiological Essays*, section X, in *Works*, vol. II.

[31] See Du Clos, *Procès Verbaux de séances*, Jan. 5, 1669.

[32] See also Du Clos, "Observations du Salpestre, de sa generation et de sa vertu fulminante," *Procès Verbaux de séances*, Jan. 22, 1667, pp. 16–22.

principle it contains; aerial spirit principle is, the cause of acidity in acidic substances.[33] Also, the artificial generation of saltpeter, that is, its "reintegration" by "the meeting of the volatile and fixed parts," that is, the coming back together of the liquid "acid spirit" and the "bitter fixed salt" (both of which result from the decomposition of saltpeter), corresponds to the mechanical condensation of air "which brings about the acidity" of the spirit of the saltpeter retained, connected with the fixed salt. Thus acridity and acidity in the end completely arise from chemical properties that cannot be explained mechanically, unlike the completely mechanical mode of reunification of the parts of saltpeter which should not be understood as resulting from a remote attraction at all. Indeed, the "suitability" (*convenance*), or the "symbol" (*symbole*) between the spirit and the salt of saltpeter sets the particles into motion, which adds to and produces effervescence in the mixed body,[34] in the same way as the example of the phenomenon of air transported to a source of heat, which is the driving force for this motion:

> Because heat excites motion in air surrounding the fiery matter, this moving air changes place, and since this place cannot be empty, another portion of air succeeds it there. And as long as the motion lasts this succession of air continues, thus air continuously passing on top of this fiery Matter .... It is not so much fire which truly attracts air toward it, but it is rather by rarefying, and pushing the closest air that this pushed air is followed by other air in order to prevent the vacuum.[35]

Du Clos is also opposed to Boyle's proposal concerning particles moving on their own and always moving.[36] For him they are at rest, apart from the moment of action initiated by the tendency to bind to passive and corporeal parts of the substances in question.[37]

Actually, according to Du Clos, in comparison with mechanical explanations, chemical explanations provide easy solutions to some problems, as the experiments on saltpeter show.[38] Overall, these problems concern tangible qualities (odor, color, savor, coldness, heat, etc.), and modification of quality in particular. For these phenomena, we should prefer chemical explanations, through the presence of a rarefied sulfur in the mixed bodies, for example, or through the "manifest suitability"

---

[33] See Du Clos, *Procès Verbaux de séances*, Dec. 31, 1666, pp. 11f.

[34] Du Clos, *Procès Verbaux de séances*, Jan. 5, 1669, f. 4v. Boyle, *Certain Physiological Essays*, also admits a conflict between the two opposed substances.

[35] See Du Clos, "Expériences de l'augmentation du poids de certaines matières en les calcinant à la chaleur du Soleil, ou du feu ordinaire," April 1667, pp. 49–51: "Car la chaleur excite du Mouvement en l'air qui environne la matiere embrasée, cet air meu change de place, et ce lieu ne pouvant estre vuide, il y succede une autre portion d'air. Et tant que le mouvement dure cette succession d'air continüe, l'air passant ainsi continuellement sur cette Matiere embrasée .... Ce n'est pas tant que le feu attire veritablement l'air à soy; mais c'est qu'en rarefiant, et poussant celuy qui luy est le plus proche, cet air poussé est suivy d'un autre air, qui luy succede pour empescher le vuide."

[36] Boyle's explanation is not very clear; this motion might be initiated by a current of subtle matter. See Boyle, *Certain Physiological Essays,* section XXVI, in *Works*, vol. II, p. XXX.

[37] See Du Clos, *Procès Verbaux de séances*, Jan. 19, 1669, and Jan. 26, 1669.

[38] Ibid., Jan. 12, 1669.

(*convenance manifeste*) to a body for another, and so on. Thus, if the "crackling" of the spirit of saltpeter poured over fixed saltpeter is to be explained, "the analogy with the bouncing of balls colliding on the billiard table," as Boyle suggests, would be wholly appropriate, according to Du Clos.[39] It might indeed be so; the explanation is "probable." But, this crackling is accompanied by a "moderation of acrimony" which must be "more precisely attributed" to a chemical reason (e.g. to "the opposition of qualities" [*contrariété des qualités*]) rather than to "the shape and combination of particles."[40] In the same way, as regards metals dissolved in corrosive spirits, there must be "in addition to the tenuousness of their particles produced by fire … some quality specific to the dissolution of metals."[41] The fluidity a metal acquires with its dissolution by a corrosive liquid appears to Du Clos as just a "discontinuity" of their particles, which by coming back together can recover their first solidity. But the total fluidity of a body corresponds to the radical resolution of this body by the alkahest—Van Helmont's universal solvent by which the prime and aqueous matter can be reached—"must proceed from a cause other than the discontinuity of the attenuated particles" (*discontinuation des particules atténuées*).[42] So in chemistry the mechanical approach is only justified as a first and superficial approach to laboratory phenomena. But for Du Clos, especially when the tangible qualities of substances are in question, there is probably a more fundamental level, which is completely chemical.[43] In other words, contrary to a certain tradition, Du Clos would affirm that the only true science of matter is chemistry.

A few months before his examination of Boyle's book appeared, Du Clos had also announced in a paper on a certain category of extractive liquids, that the goal of chemistry is to partake in the "knowledge of nature and the quality of mixed bodies." And so to this end, Du Clos distinguished between chemical and mechanical instruments in two papers on the means of analyses in chemistry (fire, air, *menstrua*) dated June 1668.[44] Both kinds of instrument—chemical and mechanical—are really used in the laboratory and not just in the mind, as Fontenelle maintained in physics, but their roles are reversed, since chemical instruments are the only ones able to get beyond a crude analysis of matter and to attain intimate or even radical knowledge.

[39] See Boyle, *Certain Physiological Essays,* Section XXIV, in *Works*, vol. II, p. XXX.

[40] Du Clos, *Procès Verbaux de séances*, Jan. 12, 1669, f. 8v.

[41] Ibid., f. 9v.

[42] Du Clos, *Procès Verbaux de séances*, Jan. 19, 1669, f. 16v.

[43] Du Clos, *Procès Verbaux de séances*, Jan. 12, 1669, ff. 11r–11v: "There is hardly appearance as I already said, that it is the various position and shape of particles of saltpeter which make the difference of qualities which are noticed there, and which make it sometimes acrid, sometimes acid; sometimes sulfurized, sometimes mercurial, sometimes overheating sometimes refreshing &c. These differences could, it seems to me, to be better related to various matters which are in Saltpeter and to their alternating predominance; because some [matters] are manifestly sulfurized or igneous; the other ones aerial, as one can observed with the confection of saltpeter and with its resolution."

[44] Du Clos, *Procès Verbaux de séances*, June 16, 1668, ff. 58r–63r, and June 23, 1668, ff. 63v–66r.

The operation of distillation is thus presented completely mechanically here. According to Du Clos, in the division of parts of mixed bodies, fire acts mainly through its heat, stimulating

> [M]otion in [kinds of] mobile matter, in accordance with their degree of mobility…; those which are easiest to be moved are more agitated and rarefied sooner by the heat, rising up first and leaving behind those which are less mobile but which can follow, once they are moved by a stronger heat.[45]

Therefore only kinds of matter with the same degree of mobility are set in motion, "agitated" equally by the same degree of heat, which "makes them rise" and separate from those which show a greater fixity at this degree of heat. It seems insufficient for Du Clos simply to advance the claim that the function of heat is simply to put together similar parts and to separate those of a different nature.[46] Its role is limited to initiating motion in the body subject to distillation: initially to its most mobile parts, then progressively to those that are less mobile. The former are agitated, untied, rarefied (that is, they occupy more space) to move towards the top of the alembic. Remaining at the bottom, only some of the fixed parts will become fused. For Du Clos, this mechanism is the reason why the operation of distillation reveals in turn spirit, phlegm, oil, that is, three of the five principles of the "vulgar chemists" and three of the first chemical mixed bodies for him; "heat being graduated in proportion to the quality of these kinds of matter," to leave in the lower part of the cucurbit a bulk made up of a little oil and salt "bound" to earth which they had "penetrated." [47] Depending on the kinds of matter, what the heat of fire cannot divide will be divided "by the blaze and combustion" of fire. How is this motion initiated by fire in the matter? It is thanks to the impulsion which fire gives to parts of matter, as Du Clos writes:

> This heat and this blaze provide an impulsion to the parts of the warmed or burning subject and this impulsion is followed by their rarefaction. The impulsion and the rarefaction cause these agitated and rarefied parts to rise. This rise happens more quickly close to fire, where the impulsion is stronger, though the rarefaction is less there, as one sees in the smoke which rises up from burning wood, and goes up through the flue of chimney, since it rises up more quickly in the flue, where it is less rarefied than when it is going out and it is rarefying in the air, where it spreads and disperses. Thus, it is not only the rarefaction which makes the smoke rise, but also the impulsion which it receives from the fire, an impulsion stronger close to the fire which gives it.[48]

[45] Ibid., June 16, 1668, f. 58r.

[46] Boyle seems to have held such a view, along with a number of "vulgar chemists"; see Boyle, *Certain Physiological Essays*, part II, section XXVI, in *Works*, vol. II, p. 105.

[47] Boyle seems to have held this view as well, though in his *Sceptical Chymist*, he prefers to speak of *prima mista, mista primaria*, or "secundary principals, or mixed bodies of peculiar sort"; see Boyle, *Sceptical Chymist*, pp. 215 and 273.

[48] Du Clos, *Procès Verbaux de séances*, June 16, 1668, ff. 62r–62v. "Cette chaleur et cet embrasement font impulsion aux parties du subject eschauffé ou embrasé et ceste Impulsion est suivie de leur rarefaction. L'Impulsion et la rarefaction causent l'eslevation des dictes parties agitées et rarefiées. Cette Eslevation est plus prompte proche du feu, où l'Impulsion est plus forte, quoy que

Consequently, fixed matter is that which "is not disposed to receive enough impulse by fire" to rise up; it is either "absolutely motionless confronted with fire," or is endowed with a small degree of mobility, sufficient to melt it, such as metals, or to swell like certain metal calxes. The matter which is most fixed is, according to Du Clos, that which is most terrestrial; and conversely, the wettest matter is that which is least fixed by fire. It should be noted that the role of fire in the phenomenon of distillation—the main operation in chemistry—was of course known, but generally chemists considered volatility, as an intrinsic quality of some mixed bodies, to be the most important factor in explaining it. For Du Clos, though, distillation is linked to mechanical considerations.

Kinds of matter made up of immobile and incombustible parts, endowed with a certain compactness, would be submitted to the action of three sorts of *menstrua* (solvents), two of which act chemically.[49] The activity of *menstrua* called "corrosive" (*aqua fortis* for example) just give rise to the mechanical separation of their "integral parts" (*parties intégrantes*). On the other hand, *menstrua* known as "extractive" are of a different nature. Their activity corresponds to the extraction of some "constituent parts" (*parties constitutives*) of a body which "is symbolic" to the *menstruum* (that is, which shows affinity for it) through their union to it. *Menstrua* of this kind are mixed (that is, made up of sulfurized (acrid) salt and mercurial (acid) salt), and are, according to Du Clos, "specific to chemical analyses true solvents … which we should stock up on for use in projects proposed both for the search for principles of natural mixed bodies and for analytical observations which may help us to understand the nature and qualities of these mixed

la rarefaction y soit moindre, comme on le voit en la fumée qui s'esleve du Boys embrasé, et qui monte par le Tuyau d'une cheminée; car elle monte plus viste dans ledict tuyau, ou elle est moins rarefiée que quand elle en est sortie et qu'elle se rarefie d'avantage dans l'air, où elle s'estend au large et se dissipe. Ce n'est donc pas la seule rarefaction qui fait monter la fumée, mais aussi l'impulsion qu'elle reçoit du feu, laquelle Impulsion est plus forte proche du feu qui la donne." This mechanical theory of the impulsion of fire was quickly put into practice in a proposal Du Clos made for the desalination of sea water in a paper of July 1668. According to him, the salinity of sea water is an accident, and can for this reason be separated. One of the means selected to remove the salt is the following (Du Clos, ibid., f. 120v): "It would be necessary to make the water rise up through sand or from ground, because the salt, which is coarser than water, could not easily go up with it, and the water, which is more subtle and more mobile than salt, less resisting to the Impulsion of what forces it to rise up, would go up more easily than the salt through the ground or the sand, and thus it would separate from it and would become soft." Clay could just as well be used for this separation, but it is not, according to Du Clos, "porous enough to give passage to water" (f. 120v). Du Clos presents still another means—chemical this time—by precipitation of the salt of sea water by means of the attraction of salt by Glauber's "specular stone" (ff. 121v–122r).

[49] Ibid., *Procès Verbaux de séances*, June 23, 1668, ff. 63v–66r. According to Du Clos fire is no longer the instrument for the resolution of mixed bodies into their true principles, as the "vulgar chemists" used to think (see the first two of Du Clos' papers at the Académie). It is also for this reason that Du Clos suggested collecting a history of plants through chemical means, and thus not only through an analysis by fire, giving rise to a controversy with Denis Dodart; see Alice Stroup, *A Company of Scientists*.

bodies."[50] Finally, the *menstrua* known as "resolutive" (*résolutifs*) come completely under the concept of the chemical. They "radically" resolve bodies, and they thus have to "be symbolic with respect to the subjects on which they are used"; they are most perfectly represented by the *alkahest,* about which Du Clos read several papers.[51]

As one can see, Du Clos' chemistry recognizes both mechanical and chemical explanations. For Du Clos they seem quite complementary, and in this sense his thought is very original, not only in chemistry but in natural philosophy too.[52] Corresponding to the mechanical impulsion, which initiates the motion necessary to part of the matter in the retort for it to rise up during the distillation, there is a chemical "symbolic" attraction between two substances. In Du Clos' tripartite conception of natural bodies, developed in his 1677 *Dissertation sur les mixtes naturels*, the substance that unites both aspects in a single essence would be the Spirit principle. This principle is an intermediate means between the mobile corporeal passive body—that is, the informable "mobile subject" (*sujet mobile*)—and the incorporeal nature, the guiding principle of motions which it determines.[53] According to its intermediate nature, the spirit is an extended but indivisible substance, both mobile and in motion. It is the "luminous, very subtle and very active substance, to illuminate, penetrate and move corporeal matter."[54] This light, pushed by luminous stars, like the sun, continuously spreads everywhere while rarefying, but can be condensed, concentrated by convex mirrors or glass. In any case, it is the impulse of these rays of the sun (which, turning on its center, communicates motion to them), which is,

---

[50] Du Clos, *Procès Verbaux de séances*, June 23, 1668, ff. 65r–65v: "Les liqueurs mixtes qui consistent de sels sulphurez, et de sels mercuriels, joincts ensemble, sont seules capables de dissoudre l'union des parties constitutives principales, tant mercurielles que sulphurées, chacun de ces sels agissant sur la partie qui luy est symbolique et la separant de l'autre, laquelle estant pareillement resoute par un sel symbolique ne fait plus de resistance. Ces Menstrües mixtes sont les vrayes (f. 65v) dissolvants propres aux analyses chymiques, et ceux dont nous devons faire provision pour nous servir aux desseins proposez tant de la Recherche des Principes des mixtes naturels, que des observations analytiques, qui peuvent aider à la connoissance de la nature et des qualitez de ces mixtes. Les menstrües meslez sont ou universels, ou particuliers. Les universels doivent estre tirez des sels les moins specifiez, mais de nature mixte et temperée. Tel est le sel commun, duquel Paracelse a faict son grand dissolvant, qu'il nomme sel circulé."

[51] Du Clos, *op. cit. in* n. 16, for the 11, 18 and 25 Aug. 1668, ff. 127v–175r. About the *alkahest*, see Joly, "L'alkahest, dissolvant universel," pp. 305–344; about its medical use, see Porto, "*Summus atque felicissimus salium*," pp. 1–29.

[52] This use of the two kinds of explanation in chemistry will be found very clearly though in a somewhat different way in Wilhelm Homberg at the beginning of the eighteenth century; see Franckowiak and Peterschmitt, "La chimie de Homberg," pp. 65–90. The application of mechanism in chemistry required great concessions on behalf of mechanism. Among Cartesian philosophers who had the ambition to extend mechanism to chemical phenomena, one can observe the abandonment of its metaphysical dimension and use of chemical explanations; see Peterschmitt, "The Cartesian Chemistry," pp. 193–202.

[53] See Du Clos, *Dissertation sur les principes des mixtes naturels, faite en l'an 1677.*

[54] Ibid., pp. 4, 7 and 20.

according to Du Clos, the cause of the circular motion of the planets.[55] And since it is said that the activity of solar light is comparable to fire,[56] one may thus consider the fact that the matter escaping by the impulsion of the heat of the fire escapes by the mechanical impulsion of the spirit principle that Du Clos called "igneous spirit" (*Esprit igné*) precisely because it "stimulates by the motion of heat in the bodies it penetrates."[57] The motion it initiates in the air in contact with the calcined bodies also takes part in the mechanism of their increase in weight through the addition onto their surface of sulfured "igneous particles" contained in the air. This addition of particles or "igneous exhalations," poorly described but certainly similar to fire, is produced thanks to affinity, thanks to the "Symbol" (*Symbole*), that connects terrestrial sulfurs with these calcined matters.[58] This "symbolic" (*symbolique*) attraction also occurs by means of the Spirit:

> [A]nd it is also through this spirit that the particular natures of mixed bodies have some power to act on other mixed bodies without corporeal contact, and even at very considerable distances from one body to the other, by the sole mediation of the spirit with which they are connected, and by the extent to which particular natures of these mixed bodies can have a certain sphere of activity out of the bodies that they inform.[59]

Thus, spirit is involved in this chemical way in the attraction and the fixation of the air by calcined saltpeter and in the union by affinity of the saline parts of a contrary nature in the reintegration of saltpeter.

12.5 Conclusion

Du Clos sees, touches, feels, tastes, observes, and what he has seen is visible to everyone, what he has felt is tangible to everyone. Therefore he valorizes experimental demonstration in his examination of Boyle's book. His method—actually similar to Boyle's—recognizes that experiment has the ability to bring about conviction. Even if experiment does not force one to acknowledge the truth of Du Clos' hypothesis in a definitive way, it gets us to incline toward it as the most reasonable

---

[55] Ibid., pp. 7 and 27.

[56] Ibid., p. 28.

[57] Ibid., p. 8.

[58] See Du Clos, *Procès Verbaux de séances*, April 1667, pp. 40–52. One notes that this communication was read at the *Académie Royale des Sciences* more than 100 years before Lavoisier's works on those phenomena.

[59] Du Clos, *Dissertation sur les principes des mixtes naturels, faite en l'an 1677*, p. 15: "[C]'est aussi par cet esprit que les natures particulières des Mixtes ont de l'extension pour agir sur d'autres Mixtes sans attouchement corporel, & même en distances tres-notable (sic) d'un corps à l'autre, par la seule médiation de l'Esprit qu'ils participent, & par l'étenduë duquel les natures particulières de ces Mixtes peuvent avoir une certaine sphère d'activité hors des corps qu'elles informent."

explanation, that is, that which is most in accord with experience. This does not make Du Clos the radical opponent of the mechanical philosophy Fontenelle wanted to make him out to be in his report, though. No more than Boyle, neither is he a physicist who takes body, through a purely rational analysis, all the way to the size, shape, and motion of its parts. Mechanical considerations introduced into his chemistry have limits those imposed by the senses, which allow us to approach a minimal knowledge of things. Du Clos writes: "The sense perceiving the sensible thing knows well that it is .... What the senses announce to us about the mixed bodies ... is obvious. They teach us about the existence of these mixed bodies and some modes of their Being."[60] The expression of the obviousness, of that which is immediately verifiable by everyone, or indeed of the truth in chemistry, just seems to be mechanism, which only concerns the corporeal and superficial part—thus the sensible part—of natural things in the world; that is, this passive and divisible extension, made up of dense, solid, very small, shaped and mobile elementary "corpuscles" of water, earth and air, air being more subtle than the others and compressible to allow the bodies to move.[61] But senses "do not reveal all the causes" of natural phenomena which cannot occur without motion, and which are underpinned, according to Du Clos, by other principles that are, in contrast, "active and able to move." He claims: "The assumption of these principles is good and true; but their determination is not easy for the understanding which can consider them in different ways."[62] Therefore he suggests establishing the probable cause of natural motions in corporeal bodies; he assigns this cause to an intermediate half-chemical, half-mechanical principle. This active principle is the igneous spirit, it is the solar fire "which appears to our eyes through its light, and which is felt through the heat that its motion stimulates in bodies" in order to alter and generate mixed bodies.[63] This spirit, which is in fact the means of action used by Nature—that is, the guiding principle, the true moving, incorporeal and informative force—is also what gives natural mixed bodies an extra-corporeal activity and a chemical identity (symbol and qualities). Nevertheless, Du Clos' three principles presented at this stage are only, according to him, "conjectures" which he hopes are well founded; he "postulates them as probable."[64] In other words, the mechanical explanations are posited to explain the evident corporeal and passive behavior of mixed bodies, but they are insufficient from the point of view of knowledge of the "truth of the things."[65] However, it is the

[60] Ibid., p. 1: "Le sens apercevant la chose sensible, connoît bien qu'elle est.... Ce que les sens nous annoncent des Mixtes ... est évident. Ils nous apprennent l'existence de ces Mixtes, & quelques modifications de leur Etre."

[61] Ibid., pp. 3–6.

[62] Ibid., p. 2: "La supposition de ces Principes est bonne & vraie ; mais leur détermination n'est pas facile à l'Entendement, qui les peut considérer diversement."

[63] Ibid., p. 7.

[64] Ibid., p. 24.

[65] Du Clos' expression, ibid., p. 1.

consideration of the corporeal, passive and sensible body, which allows knowledge of the incorporeal, active and solely intelligible to be reached. According to Du Clos, the mechanical, corporeal body can carry our understanding to the recognition of an immediate and chemical cause, and in the end, to the absolute and first cause: God.[66]

Thus Fontenelle was not entirely wrong to characterize Du Clos' chemistry as a crude science. He is right, because Du Clos' chemistry is really a "chemical physics," to use an expression coined by Nicaise Le Febvre, a chemist who profited from Du Clos' teaching, a science which aims for knowledge of causes of natural phenomena and natural principles of bodies by experimental practice.[67] But he is also wrong, because for Du Clos, the particles of matter, subjected to the laws of motion, still represent a crude level of matter, which must be surpassed by chemistry; the true causes of phenomena are to be sought on a more fundamental level. The alkahest is for this reason the chemical instrument *par excellence*, conceptual as well as concrete, whose activity is mechanically incomprehensible. Its interest lies in the promise it gives to attain the knowledge of true principles of bodies through the radical resolution it offers, a resolution that does not at all involve a simple separation of bodies into their smallest pieces. The alkahest is an extreme example of the gap between the mechanical affections and the chemical, but also between Boyle's scientific practice and that of Du Clos'. The former is connected with "physiology"—it is the name under which he gathers his reflections in the work on which Du Clos commented—where the study of matter and its constitution is fully included in a philosophy of nature, that is, in a project that aims to give an account of nature in general.[68] The latter is connected with physics, in the section of physics in the *Académie*, which is more precisely centered on matter, indeed on particular substances. For Du Clos, the way of gaining access to the knowledge of substances is no longer based on a conception of the world which had earlier underpinned the theory of principles in chemistry (in particular, that resulting from the three or five Paracelsian principles) which committed one to the pursuit of a cosmogony. Du Clos' chemistry is neither a simple, "vulgar" chemistry nor is it a "resolutive physics," so it cannot be reduced to a Cartesian style of mechanism.[69] Du Clos presented himself as a "Platonist," which he certainly would have interpreted as the claim that the corporeal part of the world, which is, according to him, completely lacking in its own significant qualities, received the impression of the "ideas," of the characters of the

[66] Ibid., pp. 10 and 19–23.

[67] The term is from Le Febvre's *Traité de la Chymie* (1660). The expression will be used again in 1702 by Wilhelm Homberg who will continue the inversion of the status of chemical physics and mechanical physics; only chemistry according to him is true, whereas mechanism, the explanation by shape and motion of the parts of the bodies while "probable" is also contestable; see Franckowiak and Peterschmitt, "La chimie de Homberg."

[68] The work in question is Boyle's *Certain Physiological Essays*.

[69] Using Barlet's expression in *Le vray et méthodique cours de la Physique résolutive, vulgairement dite Chymie.*

Nature principle.[70] Those which it displays comes to it through the mediation of the alternative Spirit, widespread in the corporeal part of the world, and which constitute, along with the Nature, the two true chemical principles, properly speaking, which imprint the receptacle which receives them with the qualities it has. Thus the qualities of a substance are the expression of an inner chemical activity that grounds its essence. It is the imposition of an incorporeal and higher ground of sensible qualities, enclosed deep within a perfectly passive manipulable Body, movable by active principles, having as its only proper motion that which heaviness imposes on it, which gives Du Clos this "more chemical cast of mind" which Fontenelle attributes to him to distinguish him from Boyle. The fact remains that Du Clos may be the first after Boyle in the history of chemistry to endow chemistry with mechanical explanations, something, in addition, completely consistent with the chemistry of principles, characterized at that time as "modern," the science of substances, the physics of qualities.[71]

[70] See Stroup, "Censure ou querelles scientifiques," pp. 435–452.

[71] In the quarrel of Ancients and Moderns, chemists belong among the moderns, of course. Moreover, following Van Helmont, from the 1660s, chemists such as Boyle, Du Clos and Le Febvre were in the habit of calling themselves "modern chemists," as opposed to "vulgar Chemists," whose practice was centered on a simple application of the Paracelsian principles to the preparation of pharmaceutical remedies.

ERRATUM

The Mechanization of Natural Philosophy

Edited by Daniel Garber and Sophie Roux

Erratum to: DOI 10.1007/978-94-007-4345-8

The book *The Mechanization of Natural Philosophy* by Daniel Garber and Sophie Roux (eds.), which is volume 300 in the BSPS series, was originally published in error with volume number 282.

Bibliography

Manuscripts

Archives of the Académie des sciences, Paris. 1666–1685. *Procès Verbaux de séances de l'Académie Royale des Sciences*, 11 vols.
Blasius of Parma, *Expositio in libros meteorologicorum*. Vaticana, ms. Vat. Lat. 2160.
Buridan, J., *Expositio libri meteororum*. Vaticana, ms. Vat. Lat. 2162.
Fabri, H., *Œuvres du Père Honoré Fabri*. Bibliothèque Municipale de Lyon, ms. 701–704.
Gregorie, J., *Hydrostatica*. Aberdeen University Library, ms. 2206/7.
Pendasio, F., *Lectiones in quartum librum meteorologicorum*, Biblioteca Ambrosiana, ms. D. 396 inf.
Persio, A., *De natura ignis*. Biblioteca Corsiniana dell'Accademia dei Lincei, Rome, Archivio Linceo 6.
Pomponazzi, P., *In libros meteororum*. Biblioteca Ambrosiana, ms. R. 96 sup.

Primary Literature

Albertus, M. 1890. *Opera Omnia*, vol. IV, ed. A. Borgnet. Paris: L. Vivès.
Alexander, H. (ed.). 1955. *The Leibniz-Clarke correspondence, with extracts from Newton's Principia and Opticks edited with an introduction and notes*. Manchester: Manchester University Press.
Aristotle. 1934. *Physics*. Trans. P.H. Wicksteed and F.M. Cornford, 2 vols., Loeb Classical Library. Cambridge, MA/London: Harvard University Press.
Aristotle. 1982. *De generatione et corruptione*. Trans. with notes by C.J.F. Williams. Oxford: Clarendon Press.
Aristotle. 1984. *The complete works of Aristotle. The revised Oxford translation*, 2 vols., ed. J. Barnes. Princeton: Princeton University Press.
Arnauld, A. 1671. *Plusieurs raisons pour empêcher la censure ou la condamnation de la philosophie de Descartes Fragments philosophiques*, ed. V. Cousin, 303–317.
Arnauld, A. 1999. *Examen du traité de l'essence des corps* (1680). Paris: Fayard.
Arnauld, A., and P. Nicole. 1981. *La logique ou l'art de penser* (1662), ed. P. Clair and F. Girbal. Paris: Vrin.
Arnauld, A., and P. Nicole. 1996. *Logic or the art of thinking*, ed. and trans. J.V. Buroker. Cambridge: Cambridge University Press.
Arriaga, R. 1632. *Cursus philosophicus*. Antwerp: B. Moretus.

S. Roux and D. Garber (eds.), *The Mechanization of Natural Philosophy*, Boston Studies in the Philosophy of Science 300, DOI 10.1007/978-94-007-4345-8,

Averroes. 1562–1574. *Opera*, 12 vols. Venizia: Giunta. Repr. 5 vols., Frankfurt a. M.: Minerva, 1962.

Averroes. 1958. *Middle commentary and epitome on Aristotle's* De generatione et corruptione. Trans. S. Kurland. Cambridge, MA: Mediaeval Academy of America.

Babin, F. 1679. *Journal ou relation fidèle de tout ce qui s'est passé dans l'université d'Angers au sujet de la philosophie de Des Chartes en l'execution des ordres du Roy pendant les années 1675, 1676, 1677 et 1678*. Angers.

Bacon, F. 1627. *Sylva sylvarum or a natural history in ten centuries*. London: W. Lee.

Bacon, F. 1663. *Opuscula varia posthuma, philosophica, civilia, et theologica... Cura et fide Guilielmi Rawley*. Amsterdam: J. Ravenstein.

Bacon, F. 1676. *The Novum Organum of Sir Francis Bacon of Verulam, Viscount St. Albans. Epitomiz'd for a Clearer Understanding of his Natural History. Translated and Taken out of the Latine by M. D. B. D.* London: T.S Lee.

Bacon, F. 1679. *Baconiana. Or certain genuine remains of Sr. Francis Bacon*, ed. T. Tenison. London: J.D. for R. Chiswell.

Bacon, F. 1857–1874. *Works*, ed. J. Spedding, R.L. Ellis, and D.D. Heath. London: Longman.

Bacon, F. 1996a. *The major works*, ed. B. Vickers. Oxford: Oxford University Press.

Bacon, F. 1996b. *The Oxford Francis Bacon*, 15 vols., ed. G. Rees and L. Jardine. Oxford: Clarendon Press.

Barlet, A. 1653. *Le vray et méthodique cours de la physique résolutive, vulgairement dite chymie*. Paris: N. Barlet.

Barrow, I. 1734. *The usefulness of mathematical learning explained and demonstrated: Being mathematical lectures read in the public schools at the University of Cambridge*. Trans. John Kirkby. London: Austen.

Basson, S. 1621. *Philosophiae naturalis adversus Aristotelem libri XII*. Genève: P. de la Rouvière.

Bayle, P. 1684a. *Recueil de quelques pièces curieuses concernant la philosophie de M. Descartes*. Amsterdam: Mortier.

Bayle, P. 1684b. *Nouvelles de la République des lettres*. Amsterdam: Mortier.

Beeckman, I. 1939–1953. *Journal tenu par Isaac Beeckman de 1604 à 1634*, 4 vols., ed. C. de Waard. The Hague: M. Nijhoff.

Bernier, F. 1684. Éclaircissement sur le livre de M. de la Ville. *Recueil de quelques pièces curieuses concernant la philosophie de M. Descartes*, ed. P. Bayle, 45–91. Amsterdam: Chez H. Desbordes.

Boate, A., and G. Boate. 1641. *Philosophia naturalis reformata, id est, Philosophae aristotelicae accurate examination ac solida confutation et novae ac verioris introductio*. Dublin: ex officina typographica Societatis Bibliopolarum.

Boileau, N., N. Bernier, and J. Racine. 1992. *Requeste des maistres es arts, professeurs, & regens de l'Université de Paris presentée à la Cour souveraine de Parnasse: en semble l'Arrest intervenu sur ladite Requeste* (1671), *Corpus*, 20–21, 231–239.

Bossuet, J.-B. 1909–1929. *Correspondance de Bossuet*, 15 vols., ed. Ch. Urbain and E. Levesque. Paris: Hachette.

Boyle, R. 1661. *The sceptical chymist*. London: J. Cadwell for J. Crooke.

Boyle, R. 1996. *A free enquiry into the vulgarly received notion of nature*, ed. E.B. Davis and M. Hunter. Cambridge: Cambridge University Press.

Boyle, R. 1999–2000. *The works of Robert Boyle*, ed. M. Hunter and E.B. Davis. London: Pickering and Chatto.

Boyle, R. 2001. *Correspondence*, ed. M. Hunter, A. Clericuzio, and L.M. Principe. London: Pickering and Chatto.

Cabeo, N. 1646. *Commentaria in libros meteorologicorum*. Rome: F. Corbelletti.

Cally, P. 1674. *Institutio philosophiae*. Caen: Poisson.

Cally, P. 1675. *Universae philosophiae institutio*, 2 vols. Caen: J. Cavelier.

Carlyle, A. 1973. *Anecdotes and characters of the times*, ed. J. Kinsley. London/New York/Toronto: Oxford University Press.

Carpenter, N. [pseudonym: N.C. Cosmopolitanus] 1621. *Philosophia libera triplici exercitationum decade proposita*. Frankfurt: Hulsian.

Castelli, B. 1669. *Alcuni opuscoli filosofici di Benedetto Castelli*. Bologna: G. Monti.
Charles-Joseph de Troyes [Père P.C.J.D.T.P.C.]. 1682. *La philosophie de Monsieur Descartes contraire à la foi catholique avec la réfutation d'un imprimé fait depuis peu pour sa défense*. Paris: G. Caillou.
Charleton, W. 1654. *Physiologia Epicuro-Gassendo-Charltoniana: Or a fabrick of science natural, upon the hypothesis of atoms, founded by Epicurus, repaired by Petrus Gassendus, augmented by Walter Charleton*. London: T. Newcomb for T. Heath.
Charpentier, F. 1676. *Deffence de la langue françoise, pour l'inscription de l'Arc de triomphe*. Paris: C. Barbin.
Chassins, G. 1614. *De natura sive de mundo*. Lyon: S. Rigaud.
Collegium Conimbricenses. 1606. *Commentarii Collegii Conimbricensis in quatuor libros de coelo*. Venice: J. Vincentius and J. Amadinus.
Collegium Conimbricenses. 1608. *In libros meteorologicos commentaria*. Lyon: H. Cardon.
Comenius, J.A. 1967. *Opera omnia*. Prague: Academia Scientiarum Bohemoslovaca.
Conway, A. 1982. *The principles of the most ancient and modern philosophy*. Dordrecht: Kluwer.
Cudworth, R. 1995. *The true intellectual system of the universe*. London: Tegg. repr. Bristol: Thoemmes.
Daniel, G. 1691. *Voyage du monde de Descartes*. Paris: Vve Bénard.
Daniel, G. 1696. *Suite du voyage du monde de Descartes ou Nouvelles difficultés proposées à l'auteur du Voyage du monde de Descartes, touchant la connaissance des bêtes*. Amsterdam: P. Mortier.
de Cordemoy, G. 1668. *Discours physique de la parole*. Paris: F. Lambert.
de Cordemoy, G. 1968. *Six discours sur la distinction et l'union du corps et de l'âme*. *Œuvres philosophiques*, ed. P. Clair and F. Girbal. Paris: Presses universitaires de France.
de Launay, G. 1667. *Les essais physiques*. Paris: L'auteur et C. Barbin.
de Launay, G. 1668. *Dissertation de la philosophie*. Paris: J. Cusson.
de Launay, G. 1672. *Les essais métaphysiques du sieur de Launay*. Paris: J. Langlois.
Denis, J.-B. 1672. *Recueil des mémoires et conférences qui ont esté presentées à Monseigneur le Dauphin*. Paris: F. Leonard.
Derodon, D. 1662. *Disputatio de libertate et atomis*. Nîmes. (But also Genève: J.A. and S. de Tournes)
Derodon, D. 1664. *Philosophia contracta*. Genève: P. Chouët.
Descartes, R. 1664. *L'homme de René Descartes et un Traitté de la formation du foetus du mesme autheur avec les remarques de Louys La Forge, Docteur en Medecine, demeurant à La Fleche, sur le Traitté de l'homme de René Descartes et sur les figures par luy inventées*. Paris: C. Angot.
Descartes, R. 1897–1913. *Œuvres de Descartes*, 13 vols., ed. C. Adam and P. Tannery. Paris. Nouv. présentation, Paris: Vrin, 1964–1974.
Descartes, R. 1985–1991. *The philosophical writings of Descartes*, ed. and trans. J. Cottingham, R. Stoothoff, D. Murdoch, and A. Kenny. Cambridge: Cambridge University Press.
Descartes, R. 2000. *Écrits physiologiques et médicaux*, ed. and trans. V. Aucante. Paris: Presses universitaires de France.
Descartes, R. 2001. *Discourse on method, optics, geometry, and meteorology*. Trans. P. Olscamp. Indianapolis: Hackett.
Desgabets, R. 1675. *La critique de la critique de la Recherche de la vérité*. Paris: J. du Puis.
Deusing, A. 1661. *Disquisitio physico-mathematica gemina de vacuo et de attractione*. Amsterdam: P. van der Berge.
Dionis, P. 1701. *L'anatomie de l'homme suivant la circulation du sang et les nouvelles découvertes*, 3rd ed. Lyon: Thened.
Dionis, P. 1718. *Traité général des accouchements, qui instruit de tout ce qu'il faut faire pour être habile accoucheur*. Paris: C.-M. d'Houry.

Dionis, P. 1782. *Cours d'opérations de chirurgie démontrées au Jardin du Roi*, 8e édition, augmentée par Georges de la Faye, 2 vols. Paris: Vve d'Houry.

Du Clos, S. 1680. *Dissertation sur les principes des mixtes naturels, faite en l'an 1677*. Amsterdam: D. Elzevier.

Duhamel, J.-B. 1660. *De meteoris et fossilibus*. Paris: P. Lamy.

Duhamel, J. 1692. *Réflexions critiques sur le système cartésien de la philosophie de M. Régis*. Paris: Courterot.

Duhamel, J. 1705. *Philosophia universalis, sive commentarius in universam Aristotelis philosophiam ad usum scholarum comparatam quaedam recentiorum philosophorum ac praesertim Cartesii propositiones damnatae et prohibitae*, 5 vols. Paris: Thisboust et Esclassen.

Du Roure, J. 1653. *La physique expliquée suivant les sentiments des anciens et nouveaux philosophes; et principalement de Descartes*. Paris: Chez l'auteur.

Eschinardi, F. 1658. *Microcosmus physico-mathematicus*. Rome: H. Corbelletti.

Eschinardi, F. 1666a. *Dialogus opticus*. Rome: H. Corbelletti.

Eschinardi, F. 1666b. *Centuria problematum opticorum, seu Dialogi Optici*. Rome: H. Corbelletti.

Espagnier, J.d'. *La philosophie naturelle restablie en sa pureté*. Trans. J. Bachou. Paris: E. Pepingué.

Eustachius a Sancto Paulo. 1638. *Summa philosophiae quadripartita. De rebus dialectis, moralibus, physicis, & metaphysicis*. Cologne: P. Albertus.

Fabri, H. 1646. *Tractatus physicus de motu locali, in quo effectus omnes, qui ad impetum, motum naturalem, violentum, et mixtum pertinent, explicantur. Auctore P. Mousnerio cuncta excerpta ex praelectionibus R.P. Honorati Fabri*. Lyon: Champion.

Fabri, H. 1648. *Metaphysica demonstrativa, sive scientia rationum universalium, auctore Petro Mousnerio doctore medico. Cuncta excerpta ex praelectionibus R.P Honorati Fabri*. Lyon: Champion.

Fabri, H. 1667. *Synopsis optica*. Lyon: H. Boissat and G. Remeus.

Fabri, H. 1669. *Dialogi de lumine*. Lyon: A. Moulin.

Fabri, H. 1669–1671. *Physica, id est, scientia rerum corporearum, in decem tractatus distributa*, 4 vols. Lyon: L. Anisson.

Fabri, H. 1674. *Epistolae tres de sua hypothesi philosophica*. Mainz: J.P. Zubrodt.

Ferguson, J. 1760. *Lectures on select subjects in mechanics, hydrostatics, pneumatics, and optics*. London: A. Millar.

Ficino, M. 1576. *Opera omnia*, 2 vols. Basel: H. Petrus.

Foucher, S. 1675. *Critique de la Recherche de la vérité, où l'on examine en méme-tems une partie des principes de Mr Descartes*. Paris: M. Coustelier.

Frey, J.-C. 2003. *Cribrum philosophorum (1646)*, ed. D. Garber and R. Ariew. Lecce: Conte Editore.

Fromondus, L. 1631. *Labyrinthus sive de compositione continui liber unus*. Antwerp: B. Moretus.

Fromondus, L. 1639. *Libri sex meteorologicorum*. Oxford: W. Turner.

Gadroys, C. 1671. *Discours sur les influences des astres selon les principes de M. Descartes ou l'on fait voir qu'il sort continuellement des Astres une matiere par le moien de laquelle on explique les choses que les anciens ont attribuées aux Influences occultes*. Paris: J.-B. Coignard.

Galen. 1951. *Galen's hygiene*. Trans. R.M. Green. Springfield: C. Thomas.

Galen. 1997. *Selected works*, ed. and trans. P.N. Singer. Oxford: Oxford University Press.

Galen. 2006. *On diseases and symptoms*, ed. and trans. I. Johnston. Cambridge: Cambridge University Press.

Galilei, G. 1890–1909. *Le Opere di Galileo Galilei, Edizione Nazionale*, 20 vols., ed. A. Favaro. Florence: Barbera.

Galilei, G. 1957. *Discoveries and opinions of Galileo*, ed. and trans. S. Drake. Garden City: Doubleday.

Galilei, G. 1960. *The assayer*. Trans. S. Drake. *The controversy on the comets of 1618*, S. Drake and C.D. O'Malley (eds.). Philadelphia: University of Pennsylvania Press.

Galileo, G. 1967. *Dialogue concerning the two chief world systems*, 2nd ed., ed. and trans. S. Drake. Berkeley: University of California Press.

Galilei, G. 1974. *Two new sciences*. Trans. S. Drake. Madison: University of Wisconsin Press.

Galilei, G., and C. Scheiner 2010. *On sunspots*, ed. and trans. E. Reeves and A. van Helden. Chicago: University of Chicago Press.
Gassendi, P. 1658. *Opera omnia in sex tomos divisa*. Lyon: Anisson and Devenet.
Gautruche, P. 1661. *Philosophiae ac mathematicae totius institutio*. Wien: J. Blaeu.
Glanvill, J. 1676. *Essays on several important subjects*. London: J. Baker and H. Mortlock.
Glisson, F. 1665. *Anatomia hepatis*. Amsterdam: J. Jansson à Waesberge and E. Wayerstraten.
Glisson, F. 1672. *De natura substantiae energetica*. London: Flesher.
Glisson, F. 1677. *Tractatus de ventriculo et intestinis*. Amsterdam: J. Junior.
Goclenius, R. 1613. *Lexicon philosophicum*. Frankfurt. Repr. Hildesheim: G. Olms, 1964, Typis viduae Matthiae Beckeri.
Godart, P. 1666. *Totius philosophiae summa: in brevitate foecundissima, ordine luculentissima, & suavitate latinatis ornatissima*. Paris: A. Alliot.
Godart, P. 1677. *Dissertatio in qua egregie ludicra Cartesii revelantur*. Paris: J. de la Caille.
Gómez Pereira, A. 1554. *Antoniana margarita opus nempe physicis, medicis ac theologis*. Medina del Campo: G. de Millis.
Gómez Pereira, A. 1558. *Nova veraque medicina experimentis et evidentibus rationibus comprobata*. Medina del Campo: F. del Canto.
Gregory, D. 1695. *Catoptricae et dioptricae sphaericae elementa*. Oxford: ex Theatro Sheldoniano.
Gregory, D. 1702. *Astronomiæ, physicæ et geometriæ elementa*. Oxford: ex Theatro Sheldoniano.
Gregory, D. 1715. *Dr Gregory's elements of catoptrics and dioptrics*. Trans. W. Brown. London: E. Curll, J. Pemberton.
Grimaldi, F. 1665. *Physico-mathesis de lumine*. Bologna: Heirs of V. Benatius.
Hakewill, G. 1627. *An apologie of the power and providence of god in the government of the world*. Oxford: J. Litchfield and W. Turner. [Later editions give the title as *An Apologie or Declaration…*]
Hale, M. 1674. *Difficiles nugae, or, observations touching the Torricellian experiment, and the various solutions of the same, especially touching the weight and elasticity of the air*. London: W. Godbid for W. Shrewsbury.
Hale, M. 1675. *An essay touching the gravitation, or non-gravitation of fluid bodies, and the reasons thereof*. London: W. Godbid for W. Shrewsbury.
Hartlib, S. 1970. Considerations tending to the happy accomplishment of Englands Reformation. *Samuel Hartlib and the advancement of learning*, ed. C. Webster, 119–139. Cambridge: Cambridge University Press.
Hill, N. 1601. *De philosophia epicuraea, democritiana, theophrastica*. Paris: R. Thiery.
Hobbes, T. 1644. *Tractatus opticus*. *Universae geometriae mixtaeque mathematicae synopsis*, ed. M. Mersenne, 471–591.
Hobbes, T. 1655. *Elementorum philosophiae sectio prima de corpore [De corpore]*. London: A. Crook.
Hobbes, T. 1656. *Elements of philosophy first section concerning body*. London: A. Crook.
Huet, P.D. 1689. *Censura philosophiae cartesianae*. Paris: Horthemels.
Huet, P.D. [M. G. de l'A]. 1996. *Nouveaux mémoires pour servir à l'histoire du cartésianisme* (1692). Rezé: Séquences.
Huygens, C. 1888–1950. *Œuvres complètes de Christiaan Huygens, publiées par la Société hollandaise des sciences*, 22 vols. The Hague: Martinus Nijhoff.
Keill, J. 1702. *Introductio ad veram physicam, seu Lectiones physicae*. Oxford: ex Theatro Sheldoniano, T. Bennet.
Keill, J. 1720. *An introduction to natural philosophy, or philosophical lectures read in the University of Oxford Anno 1700*. London: Innys and Osborn.
Kircher, A. 1645. *Ars magna lucis et umbrae*. Rome: L. Grignan for H. Scheus.
Kircher, A. 1654. *Magnes, sive de arte magnetica*. Rome: V. Mascardi, B. Deversin and Z. Masotti.
Lamy, B. 1966. *Entretiens sur les sciences* (1683), ed. P. Clair and F. Girbal. Paris: Presses universitaires de France.
La Forge, L. de 1974. *Traité de l'esprit de l'homme* (1666). *Œuvres philosophiques*, ed. P. Clair. Paris: Presses universitaires de France.

La Grange, J. de 1675. *Les principes contre les nouveaux philosophes Descartes, Rohault, Gassendi, le père Maignan, etc*. Paris: G. Josse.

Le Bovier de Fontenelle, B. 1686. *Entretiens sur la pluralité des mondes*. Paris: Vve C. Blageart.

Le Bovier de Fontenelle, B. 1731–1733. *Histoire de l'Académie Royale des Sciences, depuis son établissement en 1666 jusqu'en 1686*, 4 vols. Paris: G. Martin, J.-B. Coignard fils, H. L.Guérin.

Le Bovier de Fontenelle, B. 1767. *Œuvres*, 11 vols. Paris: Saillant and Desaint.

Le Bovier de Fontenelle, B. 1989–2001. *Œuvres complètes*, 9 vols., ed. A. Niderst. Paris: Fayard.

Leclerc, Daniel. 1685. *Bibliotheca Anatomica, sive Recens in Anatomia inventorum thesaurus*. Genève: J.A.Chouet.

Le Febvre, N. 1660. *Traité de la chymie*. Paris: T. Jolly.

Le Grand, A. 1675. *Institutio philosophiae, secundum principia Renati Descartes, nova methodo adornata et explicata ad usum juventatis academicae*. London: J. Martyn.

Le Grand, A. 1679. *Apologia pro Renato Descartes, contra Samuelem Parkerum*. London: M. Clark.

Le Laboureur, L. 1669. *Avantages de la langue françoise sur la langue latine*. Paris: G. de Luyne.

Le Valois [La Ville], L. de. 1680. *Sentimens de M. Descartes touchant l'essence et les proprietez des corps, opposez à la doctrine de l'Eglise, et conforme aux erreurs de Calvin, sur le sujet de l'Eucharistie*. Paris: É. Michallet.

Leibniz, G.W. 1875–1890. *Die philosophischen Schriften*, 7 vols., ed. C.I. Gerhardt. Berlin: Weidmann.

Leibniz, G.W. 1989. *Philosophical essays*, ed. and trans. R. Ariew and D. Garber. Indianapolis/ Cambridge: Hackett.

Leibniz, G.W. 1993. *Leibniz-Thomasius. Correspondance 1663–1672*, ed. R. Bodéüs. Paris: Vrin.

Magnen, J.C. 1646. *Democritus reviviscens, sive de atomis*. Pavia: Magrium.

Maignan, E. 1648. *Perspectiva horaria, sive de horographia, gnomonica libri*. Rome: P. Rubei.

Maignan, E. 1653. *Cursus philosophicus concinnatus ex notissimis cuique principiis: ac praesertim quoad res physicas, instauratis ex lege naturae sensatis experimentis passim comprobata*. Toulouse: Bosc.

Maignan, E. 1661–1672. *Philosophia sacra, sive, Entis tum supernaturalis, tum increate*. Toulouse: A. Colomerius.

Malebranche, N. 1958–1970. *Œuvres complètes de Malebranche*, 18 vols., ed. A. Robinet, 21 t. Paris: Vrin.

Malebranche, N. 1979. *De la recherche de la vérité. Malebranche. Œuvres*, vol. I, ed. G. Rodis-Lewis with the collaboration of G. Malbreil. Paris: Gallimard.

Malpighi, M. 1669. *De viscerum structura exercitatio anatomica*. Amsterdam: P. Le Grand.

Mas, D. 1599. *Commentariorum in universam philosophiam Aristotelis. Una cum quaestionibus quae a gravissimis philosophis agitantur…* Valencia: P.P. Mey.

Mathers, P. [pseud.] 1672. *The great and new art of weighing vanity, or, a discovery of the ignorance and arrogance of the great and new artist, in his pseudo-philosophical writings*. Glasgow: R. Sanders.

Mauriceau, F. 1668. *Traité des maladies des femmes grosses et accouchées. Avec la bonne et véritable Méthode de les bien aider en leurs accouchemens naturels, et les moyens de remédier à tous ceux qui sont contre-nature, et aux indispositions des enfans nouveau-nés*. Paris: Henault, d'Houry, de Ninville and Coignard.

Mersenne, M. 1623. *Quaestiones in Genesim*. Paris: S. Cramoisy.

Mersenne, M. 1624. *L'impiété des déistes, athées et libertins de ce temps*. Paris: P. Bilaine.

Mersenne, M. 1625. *La vérité des sciences contre les sceptiques ou pyrrhoniens*. Paris: T. du Bray.

Mersenne, M. 1636. *Harmonie universelle contenant la theorie et la pratique de la musique*, 2 vols. Paris: S. Cramoisy.

Mersenne, M. 1644. *Universae geometriae mixtaeque mathematicae synopsis*. Paris: A. Bertier.

Mersenne, M. 1945–1988. *Correspondance du P. Marin Mersenne, religieux minime*, commencée par Mme P. Tannery, publiée et annotée par C. De Waard et continuée par M.B. Rochot et A. Beaulieu, 17 vols. Paris: C.N.R.S.

Mersenne, M. 1985. *Questions inouyes* (1634). Paris: Fayard.
More, H. 1659. *The immortality of the soul*. London: W. Morden.
More, H. 1995. *Manual of metaphysics: A translation of the Enchiridium metaphysicum (1679)*, ed. and trans. A. Jacob. Hildesheim/New York: Olms.
Müller, P. 1619. *De cometa anni 1618 commentatis physico-mathematica specialis et generalis*. Leipzig: G. Liger.
Naudé, G. 1644. *Advis pour dresser une bibliothèque* (1627). Paris: Rolet Le Duc.
Naudé, G. 1669. *Apologie pour tous les grands hommes qui ont esté accusez de magie* (1625). Paris: Eschart.
Newton, I. 1952. *Opticks*, with a foreword by A. Einstein, an introduction by E. Whittaker, a preface by I.B. Cohen. New York: Dover Publications.
Newton, I. 1962. *Unpublished scientific papers of Isaac Newton*, chosen, ed. and trans. A.R. Hall and M. Boas Hall. Cambridge: Cambridge University Press.
Newton, I. 1978. *Papers and letters on natural philosophy*, 2nd ed, ed. I.B. Cohen and R.E. Schofield. Cambridge, MA: Harvard University Press.
Newton, I. 1983. *Certain philosophical questions: Newton's Trinity notebook*, ed. J.E. McGuire and M. Tamny. Cambridge: Cambridge University Press.
Newton, I. 1999. *The Principia. Mathematical principles of natural philosophy*. Trans. I.B. Cohen and A. Whitman, with a supplement by I.B. Cohen. Berkeley: University of California Press.
Nifo, A. 1552. *Expositio super octo Aristotelis Stagiritae libros de physico auditu*. Venice: Giunta.
Noël, E. 1648. *Physica vetus et nova*. Paris: S. and G. Cramoisy.
Oldenburg, H. 1965–1986. *The correspondence of Henry Oldenburg*, ed. A.R. Hall and M.B. Hall, 13 vols. Madison/London: University of Wisconsin Press/Mansell.
Olesa i Rovira, P.B.d' 1536. *Summa totius philosophiae et medicinae*. Valencia: Salvaynach y Monpesat.
Olympiodorus, Alexandrinus. 1900. *Aristotelis meteora commentaria*, ed. Wilhelm Stüve. Berlin: Reimer.
Pantheo, G.A. 1618. *Ars et theoria transmutationis*. Venice: J. Tacuin.
Pardies, I.-G. 1672. *Discours de la connaissance des bêtes*. Paris: S. Mabre-Cramoisy.
Pascal, B. 1963. *Œuvres complètes*, ed. L. Lafuma. Paris: Seuil.
Patrick, S. 1662. *A brief account of the new sect of latitude-men, together with some reflections upon the new philosophy*. London: [s.n.].
Pemble, W. 1629. *De formarum origine ... tractatus*. London: R. Young and I. Bartlet.
Persio, A. 1999. *Trattato dell'ingegno dell'huomo*, Venice, 1576. *Supplementi di Bruniana et Campanelliana. Testi*, n. II, ed. L. Artese.
Petty, W. 1674. *The discourse made before the Royal Society the 26. of November, 1674, concerning the use of duplicate proportion*. London: Martyn.
Poinsot, J. 1883. *Cursus philosophicus thomisticus. Tomus tertius philosophia naturalis*. Paris: L. Vivès.
Poinsot, J. 1985. *Tractatus de signis: The Semiotic of John Poinsot*, ed. and trans. J.N. Deely. Berkeley: University of California Press.
Poisson, N.J. 1670. *Commentaire ou remarques sur la méthode de René Descartes*. Vendôme: S. Hip. Repr. New York/London: Garland Publishing, 1987.
Poullain de la Barre, F. 1674. *De l'éducation des dames pour la conduite de l'esprit dans les sciences et dans les mœurs*. Paris: J. du Puis.
Power, H. 1664. *Experimental philosophy*. London: T. Roycroft for J. Martin and J. Allestry.
Ramus, P. 1569. *Scholarum mathematicarum libri unus et triginta*. Basel: Eusebius.
Rapin, R. 1725a. *La comparaison de Platon et d'Aristote avec les sentiments des Pères sur leur doctrine* (1671). *Les œuvres du Père Rapin*, 2 vols., vol. 1., 271–432. Den Haag: P. Gosse.
Rapin, R. 1725b. *Réflexions sur la philosophie ancienne et moderne et sur l'usage qu'on doit faire pour la religion* (1684). *Les œuvres du Père Rapin*, 2 vols., vol. II, 287–392. Den Haag: P. Gosse.
Rapin, R. 1725c. *Réflexions sur la physique* (1684). *Les œuvres du Père Rapin*, 2 vols., vol. II, 432–463. Den Haag: P. Gosse.

Régis, P.-S. 1690. *Système de philosophie concernant la logique, la physique et la morale*, 3 vols. Paris: Anisson, Posuel and Rigaud.

Régis, P.-S. 1691. *Réponse au livre qui a pour titre* P. Danielis Huetii Censura philosophiae cartesianae*: servant d'éclaircissement à toutes les parties de la philosophie, surtout à la métaphysique*. Paris: J. Cusson.

Resta, F. 1644. *Meteorologia de igneis aereis aqueisque corporibus*. Rome: Moneta.

Rocco, A. 1633. *Esercitazioni filosofiche*. Venizia: F. Baba.

Rochon, A. [anonymous]. 1672. *Lettre d'un philosophe à un cartésien de ses amis*. Paris: T. Jolly.

Rohault, J. 1681. *Traité de physique* (1671), 2 vols., 4th ed. Paris: C. Galbit.

Rohault, J. 1682. *Œuvres posthumes*. Paris: G. Desprez.

Rohault, J. 1978. *Entretiens de philosophie* (1672). *Jacques Rohault (1618–1672). Bio-bibliographie. Avec l'édition critique des* Entretiens sur la philosophie, ed. P. Clair. Paris: CNRS.

Ross, A. 1652. *Arcana microcosmi or, the hid secrets of man's body discovered; in an anatomical duel between Aristotle and Galen concerning the parts thereof*. London: T. Newcomb.

Schott, G. 1657. *Magiae universalis naturae, & artis. Partem priman opticam, sive thaumaturgum opticum*. Würtzburg: J.G. Schönwetter.

Schott, G. 1662. *Physica curiosa, sive mirabilia naturae et artis libris XII*. Nürnberg: J.A. Endter.

Sennert, D. 1624. *Epitome naturalis scientiae*. Wittenberg: C. Heiden.

'sGravesande, W.J. 1721–1726. *Mathematical elements of natural philosophy*, 2 vols., 2nd ed. London: Senex, Innys, Osborn and Longman.

Sinclair, G. 1661. *Tyrocinia mathematica, sive, Juniorum ad matheses addicendas introductio in quatuor tractatus videlicet arithmeticum, sphaericum, geographicum & echometricum divisa*. Glasgow: n.p.

Sinclair, G. 1669a. *Ars nova et magna gravitatis et levitatis*. Rotterdam: A. Leers.

Sinclair, G. 1669b. *Vindication of the preface of the book intituled ... Ars nova et magna gravitatis et levitatis*. Edinburgh: n.p.

Sinclair, G. 1672. *Hydrostaticks, or the weight, force, and pressure of fluid bodies, made evident by physical and sensible experiments. Together with ... a short history of coal, etc*. Edinburgh: G. Swintoun, J. Glen and T. Brown.

Sinclair, G. 1683. *Natural philosophy improven by new experiments*. Edinburgh: G. Shaw.

Sinclair, G. 1685. *Satan's Invisible world discovered, or, a choice collection of modern relations proving evidently against the saducees and atheists of this present age, that there are devils, spirits, witches, and apparitions, from authentick records, attestations of famous witnesses and undoubted verity*. Edinburgh: J. Reid.

Sinclair, G. 1688. *The principles of astronomy and navigation*. Edinburgh: The Heirs of A. Anderson.

Sorbière, S. 1660. *Relations, Lettres et Discours de Mr Sorbière sur diverses matières curieuses*. Paris: R. de Ninville.

Sperling, J. 1648. *Antiparasceve. Pro traduce*. Wittenberg: J. Berger.

Sprat, T. 1667. *The history of the Royal Society of London for the improving of natural knowledge*. London: J. Martyn. Repr. by J.I. Cope and H.W. Jones. St. Louis/London: Washington University Press/Routledge/Kegan Paul, 1966.

Steno, N. 1664. *De musculis et glandulis observationum specimen*. København. Repr. in *Bibliotheca anatomica*, vol. II, ed. D. Leclerc, 527–563.

Steno, N. 1671. *Dissertatio de cerebri anatome*. Leiden. Repr. in *Bibliotheca Anatomica*, vol. II, ed. D. Leclerc, 326–334.

Stewart, D. 1817 and 1822. *Dissertation first exhibiting the progress of metaphysical, ethical and political philosophy since the revival of letters*. Boston: Wells and Lilly.

Stevin, S. 1605a. *Wisconstige gedachtenissen*. Leiden: I. Bouvvensz.

Stevin, S. 1605b. *Hypomnemata mathematica*. Leiden: J. Patius.

Stevin, S. 1605c. *Mémoires mathématiques*. Leiden: I.P. Iacobsz.

Stevin, S. 1634. *Les œuvres mathématiques de Simon Stevin*. Leiden: B. and A. Elzevier.

Suarez, F. 1982. *Suarez on individuation*. Trans. J.J.E. Gracia. Milwaukee: Marquette University Press.

Tauvry, D. 1693. *Nouvelle anatomie raisonnée où l'on explique les usages de la structure du corps de l'homme et de quelques autres animaux suivant les lois des mécaniques*. Paris: É. Michallet.

Telesio, B. 1967–1976. *De rerum natura iuxta propria principia*, ed. L. De Franco. Florence/ Cosenza: La Nuova Italia.

Theophrastus. 1929. *Metaphysics*, ed. and trans. D. Ross and F.H. Forbes. Oxford: Clarendon Press.

Thomas Cajetan (Tommaso de Vio) 1953. *The analogy of names and the concept of being*. Trans. E.A. Bushinski. Pittsburgh: Duquesne University Press.

Toletus. 1579. *Commentaria ... in octo libros Aristotelis de physica auscultatione*. Cologne: apud hæredes Arnoldi Birckmanni.

Traber, Z. 1675. *Nervus opticus sive tractatus theoricus, in tres libros opticam, catoptricam, dioptricam distributus*. Vienna: J.C. Cosmerovin.

Valles F. 1556. *Controversiarum medicarum et philosophicarum libri decem*. Alcalá: J. de Brocar.

Valles F. 1562. *Octo librorum Aristotelis de physica doctrina versio recens et commentaría*. Alcalá: J. de Brocar.

Valles F. 1563. *Controversiarum naturalium ad tyrones pars prima, continens eas quae spectant ad octo libros Aristotelis de physica doctrina*. Alcalá: A. de Angulo.

Valles F. 1588. *In quartum librum Meteorologicorum Aristotelis commentaria, iam secundo edita*. Turíno: H.N. Bevilacqua.

Valles F. 1592. *De iis, quae scripta sunt physice in libris sacris, sive de sacra philosophia, liber singularis*, 2nd ed. Lyon: S.a Porta.

van Helmont, J.B. 1648. *Ortus medicinae*. Amsterdam: L. Elzevier.

van Hogelande, C. 1646. *Cogitationes quibus Dei existentia, item animae spiritalitas, et possibilis cum corpore unio demonstrantur. Nec non brevis historia oeconomiae corporis animalis proponitur, atque mechanicae explicatur*. Amsterdam: L. Elzevier.

Victor, A. 1667. *Philosophia cartesiana*, vol. 1–5. Paris: P. Promé, and Saumur: F. Ernou, 1671 (vol. 6).

Vincent, J. 1658–1671. *Cursus philosophicus in quo totius scholae quaestiones fere omnes, aqua perspicacitate, ac doctrino in utramque partem propugnantur*. Toulouse: J. Boude.

Vincent, J. 1677. *Discussio peripatetica in qua philosophiae cartesianae principia ... dilicude examinantur*. Toulouse: Colome and Posuel.

Vitruvius. 1999. *Ten books on architecture*. Trans. I. Rowland. Cambridge: Cambridge University Press.

von Arnim, H.F.A. (ed.). 1921–1924. *Stoicorum veterum fragmenta*. Leipzig: Teubner.

Wallis, J. 1670. *Mechanica, sive de motu tractatus geomtricus*. London: G. Godbid.

Wallis, J. 1675. *A discourse of gravity and gravitation, grounded on experimental observations*. London: J. Martyn.

Ward, S., and J. Wilkins. 1654. *Vindiciae Academiarum containing, Some briefe Animadversions upon Mr. Websters Book, Stiled The Examination of Academies*. Oxford: L. Lichfield for T. Robinson. Repr. in *Science and education in the seventeenth Century. The Webster-Ward debate*, ed. A.G. Debus. London/ New York: Macdonald and American Elzevier, 1970.

Webster, J. 1653. *Academiarum examen, or the Examination of Academies*. London: G. Calvert. Repr. in *Science and education in the seventeenth century. The webster-ward debate*, ed. A.G. Debus. London/New York: Macdonald and American Elzevier, 1970.

Wharton, T. 1656. *Adenographia, sive glandularum totius corporis descriptio*. London: J.G. Impens. Repr. in *Bibliotheca Anatomica*, vol. II, ed. D. Leclerc, 726–747.

Willis, T. 1664. *Cerebri Anatome, cui accessit nervorum descriptio et usus, studio*. London: J. Flesher. Repr. in *Bibliotheca Anatomica*, vol. II, ed. D. Leclerc, 241–294.

Wodrow, R. 1937. *Early letters of Robert Wodrow 1698–1709*, ed. L.W. Sharp. Edinburgh: University of Edinburgh Press.

Zucchi, N. 1652. *Optica philosophica experimentis et ratione constituta*. Lyon: G. Barbier.

Secondary Sources

Alexander, P. 1985. *Ideas, qualities and corpuscles: Locke and Boyle on the external world.* Cambridge: Cambridge University Press.

Alexandrescu, V. (ed.). 2009. *Branching off: The early moderns in quest for the unity of knowledge.* Bucharest: Zeta Books.

Angelelli, I., and P. Pérez-Ilzarbe (eds.). 2000. *Medieval and Renaissance logic in Spain.* Hildesheim: G. Olms Verlag.

Anstey, P. 2000. *The philosophy of Robert Boyle.* London: Routledge.

Anstey, P. 2002. Locke, Bacon and natural history. *Early Science and Medicine* 7: 65–92.

Anstey, P., and M. Hunter. 2008. Robert Boyle's *Designe about natural history. Early Science and Medicine* 13: 83–126.

Anstey, P., and D. Jalobeanu (eds.). 2010. *Vanishing matter and the laws of motion: Descartes and beyond.* London: Routledge.

Ariew, R. 1992. Bernier et les doctrines gassendistes et cartésiennes de l'espace: Réponses au problème de l'explication de l'eucharistie. *Corpus* 20–21: 155–170.

Ariew, R. 1994a. Damned if you do: Cartesians and censorship, 1663–1706. *Perspectives on Science* 2(3): 255–274.

Ariew, R. 1994b. Quelques condamnations du cartésianisme. *Archives de philosophie* 57, cah. 1, Bulletin cartésien 22: 1–6.

Ariew, R. 1999. *Descartes and the late scholastics.* Ithaca/London: Cornell University Press.

Ariew, R. 2010. The new matter theory and its epistemology: Descartes (and late scholastics) on hypotheses and moral certainty. *Vanishing matter and the laws of motion*, ed. P. Anstey and D. Jalobeanu, 31–48. New York: Routledge.

Ariew, R., and A. Gabbey. 1998. The scholastic background. *The Cambridge history of seventeenth-century philosophy*, vol. I, ed. D. Garber and R. Ayers, 425–453. Cambridge/New York: Cambridge University Press.

Armogathe, J.-R. 1977. *Theologia cartesiana. L'explication physique de l'eucharistie chez Descartes et Dom Robert Desgabets.* The Hague: Nijhoff.

Armogathe, J.-R. 2000. The rainbow: A privileged epistemological model. *Descartes' natural philosophy*, ed. S. Gaukroger, J. Schuster, and J. Sutton, 249–257. London/New York: Routledge.

Armogathe, J.-R., and V. Carraud. 2001. La première condamnation des *Œuvres* de Descartes, d'après des documents inédits des Archives du Saint-Office. *Nouvelles de la République des Lettres* 2: 103–137.

Artigas, A. 2001. Un nuovo documento sul caso Galilei: EE 291. *Acta Philosophica* 10: 199–214.

Aucante, V. 2006. *La philosophie médicale de Descartes.* Paris: Presses universitaires de France.

Azouvi, F. 2002. *Descartes et la France. Histoire d'une passion nationale.* Paris: Fayard.

Baffioni, C. 1981. *Il IV libro dei "Meteorologica" di Aristotele.* Naples: Bibliopolis.

Baird, A.W.S. 1970. Pascal's idea of nature. *Isis* 61: 297–320.

Baldini, U., G. Zanier et al. 1977. *Ricerche sull'atomismo del Seicento.* Florence: La Nuova Italia Editrice.

Barfoot, M. 1989. Hume and the culture of science in the early 18th century. *Studies in the philosophy of the Scottish Enlightenment*, ed. M.A. Stewart, 151–190. Oxford: Clarendon.

Bennett, J.A. 1986. The mechanics' philosophy and the mechanical philosophy. *History of Science* 24: 1–28.

Bernard, C. 1878a. *Leçons sur les phénomènes de la vie communs aux animaux et aux végétaux.* Paris: J.B. Baillière et fils. Nouv. prés., avec une préface de G. Canguilhem. Paris: Vrin, 1966.

Bernard, C. 1878b. *La science expérimentale.* Paris: J.B. Baillière et fils.

Bernès, A.-C. (ed.). 1988. *Libert Froidmont et les résistances aux révolutions scientifiques.* Haccourt: Association des Vieilles Familles de Haccourt.

Bernia Pardo, J. 1975. *La diferencia entre el animal y el hombre en la* Antoniana Margarita *de Gómez Pereira.* Ph.D. thesis, University of Valencia.

Berryman, S. 2002. Galen and the mechanical philosophy. *Apeiron* 35: 235–253.

Besomi, O., and M. Camerota. 2000. *Galileo e il Parnaso Tychonico. Un capitolo inedito deldibattito sulle comete tra finzione letteraria e trattazione scientifica*. Florence: Olschki.

Bianchi, M.L. 1994. The visible and the invisible. From alchemy to Paracelsus. *Alchemy and chemistry in the 16th and 17th centuries*, ed. P. Rattansi and A. Clericuzio, 17–50. Dordrecht/ Boston: Kluwer Academic Publishers.

Biard, J., and I. Rosier-Catach (eds.). 2003. *La tradition médiévale des catégories (xii*[e]*-xv*[e] *siècles)*. Louvain: Peeters.

Biener, Z. 2004. Galileo's first new science: The science of matter. *Perspectives on Science* 12: 262–287.

Birch, T. 1756–1757. *History of the Royal Society*, 4 vols. London: A. Millar.

Bitbol-Hespériès, A. 1990. *Le principe de vie chez Descartes*. Paris: Vrin.

Bitbol-Hespéries, A. 1993. Descartes et Regius: leur pensée médicale. *Descartes et Regius*, ed. T. Verbeek. Amsterdam: Rodopi, 47–68.

Bitbol-Hespériès, A. 2000. Cartesian physiology. *Descartes' natural philosophy*, ed. S. Gaukroger, J. Schuster, and J. Sutton, 349–382.

Blackwell, C., and S. Kusukawa. 1999. *Philosophy in the sixteenth and seventeenth centuries*. Aldershot/Brookfield/Singapore/Sydney: Ashgate.

Blay, M. 2001. Force, continuity, and the mathematization of motion at the end of the seventeenth century. *Isaac Newton's natural philosophy*, ed. J.Z. Buchwald and I.B. Cohen, 225–248. Cambridge, MA: MIT Press.

Bloch, O. 1971. *La philosophie de Gassendi. Nominalisme, matérialisme et métaphysique*. The Hague: Nijhoff.

Bluche, F. (ed.). 1990. *Dictionnaire du grand siècle*. Paris: Fayard.

Boas-Hall, M. 1952. The establishment of the mechanical philosophy. *Osiris* 10: 412–541.

Bouillier, F. 1868. *Histoire de la philosophie cartésienne*, 2 vols., 3rd ed. Paris. Repr. Hildesheim/ New York: Georg Olms, 1972.

Bower, A. 1817–1830. *The history of the University of Edinburgh*, 3 vols. Edinburgh: A. Smellie.

Brockliss, L. 1995. Pierre Gautruche et l'enseignement de la philosophie de la nature dans les collèges jésuites français vers 1650. *Les Jésuites à la Renaissance*, ed. L. Giard, 187–219.

Brockliss, L. 1981. Aristotle, Descartes and the new science: Natural philosophy at the University of Paris 1600–1740. *Annals of Science* 38: 33–69.

Brockliss, L. 1987. *French higher education in the seventeenth and eighteenth centuries: A cultural history*. Oxford: Oxford University Press.

Brockliss, L. 1995. Descartes, Gassendi, and the reception of the mechanical philosophy at the French Collèges de Plein Exercice, 1640–1730. *Perspectives on Science* 3(4): 450–479.

Broughton, J., and J. Carriero (eds.). 2008. *A companion to Descartes*. Oxford: Blackwell.

Brockliss, L., and C. Jones (eds.). 1997. *The medical world of early modern France*. Oxford: Clarendon.

Bucciantini, M. 1995. *Contro Galileo. Alle origini dell'affaire*. Florence: Olschki.

Buchwald, J.Z., and I.B. Cohen (eds.). 2001. *Isaac Newton's natural philosophy*. Cambridge, MA: MIT Press.

Camerota, M. 2004. *Galileo Galilei e la cultura scientifica della Controriforma*. Rome: Salerno Editrice.

Campogalliani, P. 1993–1994. Mattoni al sole. Benedetto Castelli, la luce e il calore. *Atti dell'Istituto Veneto di Scienze, Lettere e Arti* 152: 15–31.

Canguilhem, G. 1998. *La connaissance de la vie*. Paris: Vrin.

Capaldi, N. 1975. *David Hume, the Newtonian philosopher*. Boston: Twayne.

Cassirer, E. 1998–2008. *Gesammelte Werke. Hamburger Ausgabe*, 26 vols., ed. B. Recki. Hamburg: F. Meiner Verlag.

Cassirer, E. 1998–2008. *Leibniz' system in seinen wissenschaftlichen Grundlagen. E. Cassirer, Gesammelte Werke*. Hamburger Ausgabe, 26 vols., ed. B. Recki et al., vol. I. Hamburg: F. Meiner Verlag.

Chalmers, A. 1993. The lack of excellency of Boyle's mechanical philosophy. *Studies in History and Philosophy of Science* 24: 541–564.

Clagett, M. 1941. *Giovanni Marliani and late medieval physics*. New York: Columbia University Press.

Clarke, D.M. 1989. *Occult powers and hypotheses. Cartesian natural philosophy under Louis XIV*. Oxford: Oxford University Press.

Clarke, D.M. 2011. Hypotheses. *The Oxford handbook of philosophy in early modern Europe*, ed. D.M. Clarke and C. Wilson, 249–271. Oxford University Press: Oxford.

Clarke, D.M., and C. Wilson (eds.). 2011. *The Oxford handbook of philosophy in early modern Europe*. Oxford: Oxford University Press.

Clericuzio, A. 1990. A redefinition of Boyle's chemistry and corpuscular philosophy. *Annals of Science* 47: 561–589.

Clericuzio, A. 1994. Carneades and the chemists: A study of the *Sceptical chymist* and its impact on seventeenth-century chemistry. *Robert Boyle reconsidered*, ed. M. Hunter, 79–90. Cambridge/New York: Cambridge University Press.

Clericuzio, A. 1998. The mechanical philosophy and the spring of air. New light on Robert Boyle and Robert Hooke. *Nuncius* 13: 69–76.

Clericuzio, A. 2000. *Elements, principles and corpuscles: A study of atomism and chemistry in the seventeenth century*. Dordrecht: Kluwer.

Cohen, I.B. 2002. Newton's concept of force and mass, with notes on the laws of motion. *The Cambridge companion to Newton*, ed. I.B. Cohen and G.E. Smith, 57–84. Cambridge: Cambridge University Press.

Cohen, I.B., and G.E. Smith (eds.). 2002. *The Cambridge companion to Newton*. Cambridge: Cambridge University Press.

Conant, J.B. (ed.). 1957. *Harvard case histories in experimental science*, 2 vols. Cambridge, MA: Harvard University Press.

Copenhaver, B.P. 1984. Scholastic philosophy and Renaissance magic in the *De vita* of Marsilio Ficino. *Renaissance Quarterly* 37: 523–554.

Costabel, P. 1982. *Démarches originales de Descartes savant*. Paris: Vrin.

Costabel, P. 1982. La courbure et son apparition chez Descartes. *Démarches originales de Descartes savant*, 159–166.

Costabel, P. La réfraction de la lumière et la *Dioptrique* de Descartes. *Démarches originales de Descartes savant*, 63–76.

Cottingham, J. 1992. *The Cambridge companion to Descartes*. Cambridge: Cambridge University Press.

Cousin, V. 1866. *Fragments philosophiques pour servir à l'histoire de la philosophie*, 2nd ed. Paris. Repr. Genève: Slatkine, 1970.

Crombie, A.C. 1996. Infinite power and the laws of nature: A medieval speculation. *Science, art and nature in medieval and modern thought*, 67–87. London/Rio Grande: The Hambledon Press.

Daiber, H. 1992. *The meteorology of Theophrastus in Syriac and Arabic translation. Theophrastus: His psychological, doxagraphical, and scientific writings*, ed. W. Fortenbaugh and D. Gutas, 166–293, English trans. of treatise, 261–271.

Dalzel, A. 1862. *History of the University of Edinburgh from its foundation*, 2 vols. Edinburgh: Edmonston and Douglas.

Dear, P. 1990. Miracles, experiments, and the ordinary course of nature. *Isis* 81: 663–683.

Dear, P. 2001. *Revolutionizing the sciences*. Princeton: Princeton University Press.

Dear, P. 2006. *The intelligibility of nature*. Chicago: University of Chicago Press.

de Buzon, F. 2005. Mathématiques et dialectique: Descartes ramiste? *Les Études philosophiques* 4: 455–467.

De Ceglia, F.P. 1999. *De natantibus. Una disputa ai confini tra filosofia e matemática nella Toscana medicea (1611–1616)*. Bari: Laterza.

De Franco, L. 1992. La teoria della luce di Bernardino Telesio. *Bernardino Telesio e la cultura napoletana*, ed. R. Sirri and M. Torrini, 53–77. Naples: Guida.

Deitz, L. 1999. Space, light, and soul in Francesco Patrizi's *Nova de Universis Philosophia* (1591), *Natural Particulars*, ed. A. Grafton and N. Siraisi, 139–169. Cambridge, MA: MIT Press.

Delaunay, P. 1935. *La vie médicale aux xvie, xviie èt xviiie siècles*. Paris: Hippocrate.
Demeulenaere-Douyère, C., and É. Brian (eds.). 2002. *Règlement, usages et science dans la France de l'absolutisme*. Paris: Editions Tec & Doc.
Des Chene, D. 1996. *Physiologia. Natural philosophy in late Aristotelian and Cartesian thought*. Ithaca: Cornell University Press.
Des Chene, D. 2000a. *Life's form. Late Aristotelian conceptions of the soul*. Ithaca: Cornell University Press.
Des Chene, D. 2000b. Life and health in Cartesian natural philosophy. *Descartes' natural philosophy*, ed. S. Gaukroger, J. Schuster, and J. Sutton, 723–735. London/New York: Routledge.
Des Chene, D. 2001. *Spirits and clocks. Machine and organism in Descartes*. Ithaca: Cornell University Press.
Dikjsterhuis, E.J. 1950. *De Mechanisering van het Wereldbeeld*. Amsterdam: Meulenhoff.
Dikjsterhuis, E.J. 1961. *The mechanization of the world picture*. Oxford: Oxford University Press.
Donaldson, G. (ed.). 1983. *Four centuries: Edinburgh University life 1583–1983*. Edinburgh: University of Edinburgh Press.
Doyle, J.P. 1984. Prolegomena to a study of extrinsic denomination in the work of Francis Suarez, S.J. *Vivarium* 22(2): 121–160.
Drake, S., and I.E. Drabkin (eds.). 1969. *Mechanics in sixteenth-century Italy: Selections from tartaglia, Benedetti, Guido ubaldo, & Galileo*. Madison: University of Wisconsin Press.
Dronke, P. (ed.). 1988. *A history of twelfth-century western philosophy*. Cambridge/New York/New Rochelle/Melbourne/Sydney: Cambridge University Press.
Eagles, C. 1977. The mathematical work of David Gregory. Ph.D. thesis, University of Edinburgh.
Edwards, M. 2008. Digressing with Aristotle: Hieronymus Dandinus' *De corpore animato* (1601) and the expansion of late Aristotelian philosophy. *Early Science and Medicine* 13(2): 127–170.
Emerson, R.L. 1977. Scottish universities in the eighteenth century, 1690–1800. *Studies on Voltaire and the Eighteenth Century* 167: 453–474.
Emerson, R.L. 1979. The Philosophical Society of Edinburgh 1737–1747. *British Journal for the History of Science* 12: 154–191.
Emerson, R.L. 1986. Natural philosophy and the problem of the Scottish Enlightenment. *Studies on Voltaire and the Eighteenth Century* 242: 243–291.
Emerson, R.L. 1988. Science and the origins and concerns of the Scottish Enlightenment. *History of Science* 26: 333–366.
Emerson, R.L. 1990. Science and moral philosophy in the Scottish Enlightenment. *Studies in the philosophy of the Scottish Enlightenment*, ed. M.A. Stewart, 11–36. Oxford: Oxford University Press.
Emerson, R.L., and P. Wood. 2002. Science and Enlightenment in Glasgow, 1690–1802. In *Science and medicine in the Scottish Enlightenment*, ed. C.W.J. Withers and P. Wood, 79–142. East Linton: Tuckwell Press.
Emerton, N.E. 1984. *The scientific reinterpretation of form*. Ithaca/London: Cornell University Press.
Fattori, M. (ed.). 1984. *Francis Bacon. Terminologia e fortuna nel XVII secolo*. Rome: Edizioni dell' Ateneo.
Feingold, M. 2001. Mathematicians and naturalists: Sir Isaac Newton and the Royal Society. *Isaac Newton's natural philosophy*, ed. J. Buchwald and I.B. Cohen, 77–102. Cambridge, MA: MIT Press.
Feingold, M. (ed.). 2003. *The new science and Jesuit science*. Dordrecht: Kluwer.
Feingold, M., and V. Navarro (eds.). 2005. *Universities and science in the early modern period*. Dordrecht: Springer.
Festa, E. 2000. Repères pour une analyse historique des notions de continuum et atomisme au XVIIe siècle. In *Atomismo e continuo nel XVII secolo*, ed. E. Festa and R. Gatto, 3–32. Naples: Vivarium.
Festa, E., and R. Gatto. 2000. *Atomismo e continuo nel XVII secolo. Atti del Convegno su Atomisme et continuum au xviie siècle (Napoli, 28–30 aprile 1997)*. Naples: Vivarium.
Festa, E., V. Jullien, and M. Torrini (eds.). 1999. *Géométrie, atomisme et vide dans l'école de Galilée*. Fontenay aux Roses: ENS Editions.

Field, J.V., and F.A.J.L. James (eds.). 1993. *Renaissance and revolution: Humanists, scholars, craftsmen and natural philosophers in early modern Europe*. Cambridge: Cambridge University Press.

Force, J.E. 1987. Hume's interest in Newton and science. *Hume Studies* 13: 166–216.

Fortenbaugh, W., and D. Gutas (eds.). 1992. *Theophrastus: His psychological, doxagraphical, and scientific writing*. New Brunswick: Transaction Books.

Fortenbaugh, W., and R. Sharples (eds.). 1988. *Theophrastean studies*, vol. III. New Brunswick: Transaction Books.

Fortenbaugh, W., et al. (eds.). 1985. *Theophrastus of Eresus: On his life and works*. New Brunswick: Transaction Books.

Fouilloux, É., and B. Hours (eds.). 2005. *Les Jésuites à Lyon. Actes du colloque (Lyon, septembre 2002)*. Lyon: Presses de l'ENS-LSH.

Franckowiak, R. 2008a. La chimie du XVIIe siècle: une question de principes. *Methodos* 8. To be consulted at http://methodos.revues.org/1823

Franckowiak, R. 2008b. Samuel Cottereau [ou Cotreau] Du Clos [ou Duclos]. *Dictionary of 17th-century French philosophers*, ed. L. Foisneau. New York/London: Thoemmes Press.

Franckowiak, R. 2009. Du Clos, un chimiste post-*Sceptical chymist. La philosophie naturelle de Robert Boyle*, ed. C. Ramond and M. Dennehy, 361–377. Paris: Vrin.

Franckowiak, R., and L. Peterschmitt. 2005. La chimie de Homberg: une vérité certaine dans une physique contestable. *Early Science and Medicine* 10(1): 65–90.

Freeland, C. 1990. Scientific explanation and empirical data in Aristotle's *Meteorology*. *Oxford Studies in Ancient Philosophy* 8: 67–102.

French, R. 2000. Where the philosopher finishes, the physician begins: Medicine and the arts course in thirteenth-century oxford. *Dynamis* 20: 75–106.

Furley, D. 1989a. *Cosmic problems*. Cambridge: Cambridge University Press.

Furley, D. 1989b. The mechanics of *Meteorologica* IV: A prolegomenon to biology. *Cosmic problems*, 132–148. Cambridge: Cambridge University Press.

Furley, D. 1989c. The rainfall example in *Physics* II.8. *Cosmic problems*, 115–120. Cambridge: Cambridge University Press.

Gabbey, A. 1982. Philosophia cartesiana triumphata: Henry More (1646–1671). *Problems of Cartesianism*, ed. T.M. Lennon, J.M. Nicholas, and J.W. Davis, 171–250. Kingston: McGill-Queen's University Press.

Gabbey, A. 1993. Between *ars* and *philosophia natualis*: Reflections on the historiography of early modern mechanics. *Renaissance and revolution*, ed. J.V. Field and F.A.J.L. James, 133–145. Cambridge: Cambridge University Press

Gabbey, A. 2001. Mechanical philosophies and their explanations. *Late medieval and early modern corpuscular matter theories*, ed. C. Lüthy, J.E. Murdoch, and W.R. Newman, 441–466. Leiden/Boston: Brill.

Gabbey, A. 2004. What was 'mechanical' about 'the mechanical philosophy'? *The reception of the Galilean science of motion in seventeenth-century Europe*, ed. C.R. Palmerino and J.M.M.H. Thijssen, 11–24. Dordrecht/Boston: Kluwer Academic Publishers.

Galluzzi, P. 2011. *Tra atomi e indivisibili: La materia ambigua di Galileo*. Florence: Olschki.

Garber, D. 1988. Descartes, the Aristotelians, and the revolution that did not happen in 1637. *The Monist* 71: 471–486.

Garber, D. 1992. *Descartes' metaphysical physics*. Chicago: University of Chicago Press.

Garber, D. 2001a. Descartes and experiment in the *Discourse* and *Essays*. *Descartes embodied*, ed. D. Garber, 85–110. Cambridge: Cambridge University Press.

Garber, D. 2001b. *Descartes embodied*. Cambridge: Cambridge University Press.

Garber, D. 2002. Descartes, mechanics, and the mechanical philosophy. *Midwest Studies in Philosophy* 26(1): 185–204.

Garber, D. 2004. On the frontlines of the Scientific Revolution: How Mersenne learned to love Galileo. *Perspectives on Science* 12: 135–163.

Garber, D., and R. Ayers (eds.). 1998. *The Cambridge history of seventeenth-century philosophy*, 2 vols. Cambridge: Cambridge University Press.

Garin, E. 1989. Il termine 'spiritus' in alcune discussioni fra Quattrocento e Cinquecento. *Umanisti, artisti, scienziati. Studi sul Rinascimento italiano*, ed. E. Garin, 295–303. Rome: Editori Riuniti.

Gaukroger, S. 1995. *Descartes: An intellectual biography*. Oxford: Clarendon.

Gaukroger, S. 2002. *Descartes' system of natural philosophy*. Cambridge: Cambridge University Press.

Gaukroger, S. 2006. *The emergence of a scientific culture: Science and the shaping of modernity 1210–1685*. Oxford: Oxford University Press.

Gaukroger, S., and J. Schuster. 2002. The hydrostatic paradox and the origins of Cartesian dynamics. *Studies in History and Philosophy of Science* 33(3): 535–572.

Gaukroger, S., J. Schuster, and J. Sutton (eds.). 2000. *Descartes' natural philosophy*. New York: Routledge.

Gélis, J. 1990. Chirurgiens. *Dictionnaire du grand siècle*, ed. F. Bluche. Paris: Fayard.

Gemelli, B. 2002. *Isaac Beeckman, atomista e lettore critico di Lucrezio*. Florence: Olschki.

Gentile, S., and S. Toussaint. 2006. *Marsilio Ficino: fonti, testi, fortuna. Atti del convegno internazionale, Firenze, 1–3 ottobre 1999*. Rome: Edizioni di Storia e Letteratura.

Georges-Berthier, A. 1914. Le mécanisme cartésien et la physiologie au XVIIe siècle. *Isis* 2: 37–89.

Giard, L. 1997. Sur la compagnie de Jésus et ses collèges vers 1600. *René Descartes (1596–1650): célébrations nationales du quadricentenaire de sa naissance*, ed. Prytanée National Militaire, 199–225. La Flèche (France): Prytanée national militaire.

Giard, L. (ed.). 1995. *Les Jésuites à la Renaissance. Système éducatif et production des savoirs*. Paris: Presses universitaires de France.

Giglioni, G. 2002. The Genesis of Francis Glisson's philosophy of life. Ph.D. thesis, Johns Hopkins University.

Giglioni, G. 2003. The darkness of matter and the light of nature: Notion of matter in Bacon and Comenius and their theological implications. *Acta Comeniana* 17: 9–31.

Giglioni, G. 2005. The hidden life of matter: Techniques for prolonging life in the writings of Francis Bacon. *Francis Bacon and the refiguring of early modern thought*, ed. J.R. Solomon and C. Gimelli Martin, 129–144. Aldershot/Hants/Burlington: Ashgate.

Giglioni, G. 2008. What ever happened to Francis Glisson? Albrecht Haller and the fate of eighteenth-century irritability. *Science in Context* 21: 1–29.

Giglioni, G. 2009. Reading nature without making a book of it: Francis Bacon's *Novum Organum*. *Mélivres/Misbooks,* ed. P. Hummel, 55–70. Paris: Philologicum.

Giglioni, G. 2010a. Fantasy islands: *Utopia*, *the Tempest*, and *New Atlantis* as places of controlled credulousness. *World-building and the early modern imagination*, ed. A.B. Kavey, 91–117. New York: Palgrave Macmillan.

Giglioni, G. 2010b. Mastering the appetites of matter: Francis Bacon's *Sylva Sylvarum*. *The body as object and instrument of knowledge*, ed. C.T. Wolfe and O. Gal, 149–167. Dordrecht/New York: Springer.

Giglioni, G. 2010c. The material foundations of Francis Bacon's Utopia. *Studii de Ştiinţă şi Cultură* 6: 9–17.

Giglioni, G. 2011. *Francesco Bacone*. Rome: Carocci.

Giglioni, G. 2012. *Historia* and *Materia*: The philosophical implications of Francis Bacon's natural history. *Early Science and Medicine* 17: 62–86.

Gill, M.L. 1997. Material necessity and meteorology IV 12. *Aristotelische Biologie: Intentionen, Methoden, Ergebnisse*, ed. W. Kullmann and S. Föllinger, 145–161. Stuttgart: Steiner.

Gilson, É. 1979. *Index scolastico-cartésien*. Paris: Vrin.

Gilson, É. 1984a. Météores cartésiens et météores scolastiques. *Études sur le rôle de la pensée médiévale dans la formation du système cartésien* 102–137. Paris: Vrin.

Gilson, É. 1984b. *Études sur le rôle de la pensée médiévale dans la formation du système cartésien*, 5th ed. Paris: Vrin.

Giudice, F. 1999. *Luce e visione. Thomas Hobbes e la scienza dell'ottica*. Florence: Olschki.

Gómez, S. 2008. The mechanization of light in Galilean science. *Galilaeana. Journal of Galilean Studies* 5: 207–244.

Gómez, S. 2012. Telesio y el debate sobre la naturaleza de la luz en el Renacimiento italiano. *Telesio y la nueva imagen del mundo en el Renacimiento*, ed. M.A. Granada. Madrid: Siruela.

Gómez López, S. 2001. Galileo y la naturaleza de la luz. *Largo Campo di filosofare*, ed. J. Montesinos and C. Solís, 403–419. La Orotava: Fundación Canaria Orotava de Historia de la Ciencia.

Grabiner, J.V. 2002. Maclaurin and Newton: The Newtonian style and the authority of mathematics. *Science and medicine in the Scottish Enlightenment*, ed. C.W.J. Withers and P. Wood, 143–171. East Linton: Tuckwell Press.

Grafton, A., and N. Siraisi (eds.). 1999. *Natural particulars. Nature and the disciplines in Renaissance Europe*. Cambridge, MA: The MIT Pres.

Grant, A. 1884. *The story of the University of Edinburgh during its first three hundred years*, 2 vols. London: Longmans and Green.

Grendler, P.F. 2002. *The universities of the Italian Renaissance*. Baltimore: Johns Hopkins University Press.

Gueroult, M. 1984. *Descartes' philosophy according to the order of reasons*, ed. and trans. R. Ariew. Minneapolis: University of Minnesota Press.

Halkett, S., and J. Laing, 1882–1888. *A dictionary of the anonymous and pseudonymous literature of Great Britain*, 4 vols. Edinburgh: W. Paterson.

Hall, T.S. 1970. Descartes' physiological method: Position, principles, examples. *Journal of the History of Biology* 3(1): 53–79.

Hamilton, G.K. 1974. Three worlds of light: The philosophy of light in Marsilio Ficino, Thomas Vaughan, and Henry Vaughan. Ph.D. thesis. Rochester: University of Rochester.

Hamou, P. 2002. Descartes, Newton et l'intelligibilité de la nature. *Les philosophes et la science*, ed. P. Wagner, 110–165. Paris: Gallimard.

Harrington, T.M. 1982. *Pascal philosophe*. Paris: SEDES-C.D.U.

Hatfield, G. 1992. Descartes' physiology and its relation to his psychology. *The Cambridge companion to Descartes*, ed. J. Cottingham, 335–370. Cambridge/New York: Cambridge University Press.

Hatfield, G. 2008. Animals. *A companion to Descartes*, ed. J. Broughton and J. Carriero, 404–426. Malden: Blackwell.

Haugeland, J. 1998. *Having thought: Essays in the metaphysics of mind*. Cambridge: Harvard University Press.

Heidelberger, M., and G. Schiemann (eds.). 2009. *The significance of the hypothetical in the natural sciences*. New York: Walter de Gruyter.

Henninger, M.G. 1989. *Relations: Medieval theories 1250–1325*. Oxford: Clarendon.

Henry, J. 1986. Occult qualities and the experimental philosophy: Active principles in Pre-Newtonian matter theory. *History of Science* 24: 335–381.

Henry, J. 1989. Robert Hooke, the incongruous mechanist. In *Robert Hooke,* ed. M. Hunter and S. Schaffer, 149–180. Woodbridge: The Boydell Press.

Henry, J. 1994. Robert Boyle and cosmical qualities. *Robert Boyle reconsidered*, ed. M. Hunter, 119–138. Cambridge/New York: Cambridge University Press.

Henry, J. 2001. Void space, mathematical realism and Francesco Patrizi da Cherso's use of atomic arguments. *Late medieval and early modern corpuscular theories*, ed. C. Lüthy, J.E. Murdoch, and W.R. Newman, 133–161. Leiden/Boston: Brill.

Henry, J., and S. Hutton (eds.). 1990. *New perspectives on Renaissance thought: Essays in the history of science, education and philosophy: In memory of Charles B Schmitt*. London: Duckworth.

Heyd, M. 1982. *Between orthodoxy and enlightenment: Jean-Robert Chouet and the introduction of Cartesian science in the academy of Geneva*. The Hague: Nijhoff.

Hochschild, J.P. 2001. Logic or metaphysics in Cajetan's theory of analogy: Can extrinsic denomination be a semantic property? *Proceedings of the Society for Medieval Logic and Metaphysics* 1: 45–69.

Hoquet, T. 2005. *Buffon: Histoire naturelle et philosophie*. Paris: Honoré Champion.

Hummel, P. 2009. *Mélivres/Misbooks. Études sur l'envers et les travers du livre*. Paris: Philologicum.

Hunter, M. 1989a. Latitudinarianism and the 'Ideology' of the early Royal Society: Thomas Sprat's *History of the Royal Society* (1667) *reconsidered. Establishing the new science*, ed. M. Hunter, 45–71. Woodbridge: The Boydell Press.

Hunter, M. (ed.). 1989b. *Establishing the new science. The experience of the early Royal Society*. Woodbridge: The Boydell Press.

Hunter, M. 1992. *Science and society in Restoration England*. Aldershot: Gregg Revivals.

Hunter, M. (ed.). 1994. *Robert Boyle reconsidered*. Cambridge: Cambridge University Press.

Hunter, M. 2007. Robert Boyle and the early Royal Society: A reciprocal exchange in the making of Baconian science. *British Journal for the History of Science* 40: 1–23.

Hunter, M., and E.B. Davis. 1996. The making of Robert Boyle's *Free enquiry into the vulgarly Receiv'd notion of nature* (1686). *Early Science and Medicine* 1: 204–271.

Hunter, M., and S. Schaffer (eds.). 1989. *Robert Hooke. New studies*. Woodbridge: The Boydell Press.

Hunter, M., and P.B. Wood. 1989. Towards Solomon's House: Rival strategies for reforming the early Royal Society. *Establishing the new science. The experience of the early Royal Society*, ed. M. Hunter, 185–244. Woodbridge: Boydell Press.

Hutchison, K. 1982. What happened to occult qualities in the Scientific Revolution. *Isis* 73: 233–253.

Hutchison, K. 1991a. Dormitive virtues, scholastic qualities and the new philosophies. *History of Science* 29: 245–278.

Hutchison, K. 1991b. Individualism, causal location and the eclipse of scholastic philosophy. *Social Studies of Science* 21: 321–350.

Jacob, J.R., and M.C. Jacob. 1980. The Anglican origins of modern science: The metaphysical foundations of the Whig constitution. *Isis* 71: 251–267.

Jacob, J.R., and M.C. Jacob. 1992. The saints embalmed. Scientists, Latitudinarians, and society: A review essay. *Albion* 24: 435–442.

Jacquart, D. 1988. Aristotelian thought in Salerno. *A history of twelfth-century western philosophy*, ed. P. Dronke, 407–428. Cambridge/New York: Cambridge University Press.

Jacquart, D. 2001. Minima in twelfth-century medical texts from Salerno. *Late medieval and early modern corpuscular matter theories*, ed. C. Lüthy, J.E. Murdoch, and W.R. Newman, 39–57. Leiden/Boston: Brill.

Jalobeanu, D. 2008. Bacon's brotherhood and its classical sources: Producing and communicating knowledge in the project of the Great Instauration. *Philosophies of Technology: Francis Bacon and His Contemporaries. Intersections. Yearbook for Early Modern Studies* 11(1): 197–229.

Jalobeanu, D. 2009. The fascination of Solomon's House in seventeenth century England: Baconianism revisited. *Branching off*, ed. V. Alexandrescu, 225–255. Bucharest: Zeta Books.

Janiak, A. 2000. Space, atoms and mathematical divisibility in Newton. *Studies in History and Philosophy of Science* 31: 203–230.

Janiak, A. 2008. *Newton as philosopher*. Cambridge: Cambridge University Press.

Joachim, H. 1904. Aristotle's conception of chemical combination. *Journal of Philology* 29: 72–86.

Joly, B. 1996. L'alkahest, dissolvant universel ou quand la théorie rend pensable une pratique impossible. *Revue d'histoire des sciences* 49(2–3): 305–344.

Jones, P. (ed.). 1988. *Philosophy and science in the Scottish Enlightenment*. Edinburgh: J. Donald.

Joy, L. 1987. *Gassendi the atomist: Advocate of history in an age of science*. Cambridge: Cambridge University Press.

Kahn, D. 2002. La condamanation des thèses d'Antoine de Villon et Étienne de Clave contre Aristote, Paracelse et les 'cabalistes'. *Revue d'histoire des sciences* 55(2): 143–198.

Kargon, R.H. 1966. *Atomism in England from Hariot to Newton*. Oxford: Clarendon.

Kavey, A.B. (ed.). 2010. *World-building and the early modern imagination*. New York: Palgrave.

Kim, M.G. 2003. *Affinity, that elusive dream: A genealogy of chemical revolution*. Cambridge, MA: MIT Press.

Knight, H., and M. Hunter. 2007. Robert Boyle's *Memoirs for the natural history of human blood* (1684): Print, manuscript and the impact of Baconianism in seventeenth-century medical science. *Medical History* 51: 145–164.
Kolesnik, D. (ed.). Forthcoming. *Qu'est-ce qu'être cartésien?* Lyon: Éditions de l'ENS de Lyon.
Koyré, A. 1966. *Études galiléennes*. Hermann: Paris.
Koyré, A. 1968a. Pascal savant. *Metaphysics and measurement*, 131–156. Cambridge, MA: Harvard University Press.
Koyré, A. 1968b. *Metaphysics and measurement*. Cambridge, MA: Harvard University Press.
Koyré, A. 1997. *Paracelse*. Paris: Éditions Allia.
Kretzmann, N. (ed.). 1982. *Infinity and continuity in ancient and medieval thought*. Ithaca/London: Cornell University Press.
Kretzman, N., A. Kenny, and J. Pinborg (eds.). 1982. *The Cambridge history of later medieval philosophy*. Cambridge: Cambridge University Press.
Kullmann, W., and S. Föllinger (eds.). 1997. *Aristotelische Biologie: Intentionen, Methoden, Ergebnisse*. Stuttgart: F. Steiner Verlag.
Laing, D. 1857. Notice of a scheme ... for weighing up and recovering ships. *Archaeologia Scotica: Transactions of the Society of Antiquaries of Scotland* 4: 428–439.
Laird, W.R. 1986. The scope of Renaissance mechanics. *Osiris* 2: 43–68.
Laird, W.R., and S. Roux (eds.). 2008. *Mechanics and natural philosophy before the Scientific Revolution*. Dordrecht: Springer.
Lasswitz, K. 1890. *Geschichte der Atomistik vom Mittelalter bis Newton*, 2 vols. Hamburg-Leipzig. Repr. Hildesheim, 1963.
Leijenhorst, C. 2002. *The mechanisation of Aristotelianism. The late Aristotelian setting of Thomas Hobbes' natural philosophy*. Leiden/Boston/Cologne: Brill.
Leijenhorst, C., and C. Lüthy. 2002. The erosion of Aristotelianism. *The dynamics of Aristotelian natural philosophy*, ed. C. Leijenhorst, C. Lüthy, and J.M.M.H. Thijssen, 375–411. Leiden/Boston/Cologne: Brill.
Leijenhorst, C., C. Lüthy, and J.M.M.H. Thijssen (eds.). 2002. *The dynamics of Aristotelian natural philosophy from antiquity to the seventeenth century*. Leiden/Boston/Cologne: Brill.
Lennon, T., J.M. Nicholas, and J.W. Davis (eds.). 1982. *Problems of Cartesianism*. Kingston/Montreal: McGill-Queen's University Press.
Lennox, J. 1985. Theophrastus on the limits of teleology. *Theophrastus of Eresus*, ed. W. Fortenbaugh et al., 143–151. New Brunswick: Transaction Books.
Lenoble, R. 1943. *Mersenne ou la naissance du mécanisme*. Paris: Vrin. Repr. 1971.
Le Soleil à la Renaissance. Sciences et mythes. Colloque international tenu en avril 1963 sous les auspices de la Fédération internationale des instituts et sociétés pour l'étude de la Renaissance, et du Ministère de l'éducation nationale et de la culture de Belgique. Bruxelles/Paris: Presses Universitaires de Bruxelles, 1965.
Lindberg, D. 1976. *Theories of vision from al-Kindi to Kepler*. Chicago: The University of Chicago Press.
Lindberg, D. (ed.). 1978. *Science in the middle ages*. Chicago: Chicago University Press.
Lindberg, D. 1983. *Roger Bacon's philosophy of nature*. Oxford: Clarendon.
Lindberg, D. 1986. The genesis of Kepler's theory of light: Light metaphysics from Plotinus to Kepler. *Osiris* 2: 5–43.
LoLordo, A. 2007. *Pierre Gassendi and the birth of early modern philosophy*. Cambridge: Cambridge University Press.
López Piñero, J.M. 1983a. Gómez Pereira. *Diccionario Histórico*, vol. I, ed. López Piñero et al., 411–414.
López Piñero, J.M. 1983b. Francisco Valles. *Diccionario Histórico*, vol. II, ed. López Piñero et al., 391–394.
López Piñero, J.M. 1983c. Pere Bernat Olesa i Rovira. *Diccionario Histórico*, vol. II, ed. López Piñero et al., 122–126.
López Piñero, J.M. (ed.). 2002. *Historia de la ciencia y de la técnica en la Corona de Castilla*. Junta de Castilla y León: Consejería de Educación y Cultura.

López Piñero, J.M. 2005. The faculty of medicine of Valencia: Its position in Renaissance Europe. *Universites and science in the early modern period*, ed. M. Feingold and V. Navarro, 65–82. Dordrecht: Springer.

López Piñero, J.M., and F. Calero. 1988. *Las Controversias (1556) de Francisco Valles y la medicina renacentista*. Madrid: CSIC.

López Piñero, J.M., and J.A. Garcia Sevilla. 1969. Pere d'Oleza (Petrus d'Olese) y su obra "Summa totius Philosophiae et Medicinae". *Actas del III Congreso Nacional de Historia de la Medicina*, vol. II, 125–129. Valencia: Sociedad Española de Historia de la Medicina.

López Piñero, J.M., T.F. Glick, V. Navarro Brotons, and E. Portela Marco (eds.). 1983. *Diccionario histórico de la ciencia moderna en españa*, 2 vols. Barcelona: Península.

Lüthy, C. 1997. Thoughts and circumstances of Sébastien Basson. Analysis, micro-history, questions. *Early Science and Medicine* 2: 1–73.

Lüthy, C. 2000. What to do with seventeenth-century natural philosophy? *Perspectives on Science* 8: 164–195.

Lüthy, C. 2001a. An Aristotelian watchdog as avant-garde physicist: Julius Caesar Scaliger. *The Monist* 84: 542–561.

Lüthy, C. 2001b. David Gorlaeus's atomism, or: The marriage of Protestant metaphysics with Italian natural philosophy. *Late medieval and early modern corpuscular matter theories*, ed. C. Lüthy, J.E. Murdoch, and W.R. Newman, 245–290. Leiden/Boston: Brill.

Lüthy, C., J.E. Murdoch, and W.R. Newman. 2001a. Introduction. *Late medieval and early modern corpuscular matter theories*, ed. C. Lüthy, J.E. Murdoch, and W.R. Newman, 1–39. Leiden/Boston: Brill.

Lüthy, C., J.E. Murdoch, and W.R. Newman (eds.). 2001b. *Late medieval and early modern corpuscular matter theories*. Leiden: Brill.

Machamer, P. 1998a. Galileo's machines, his mathematics, and his experiments. *The Cambridge companion to Galileo*, ed. P. Machamer, 53–59. Cambridge/New York: Cambridge University Press.

Machamer, P. (ed.). 1998b. *The Cambridge companion to Galileo*. Cambridge: Cambridge University Press.

Macherey, P. 1995. Spinoza, lecteur et critique de Boyle. *Revue du Nord, Lille* 77: 733–774.

Maclean, I. 2002. *Logic, signs and nature in the Renaissance: The case of learned medicine*. Cambridge: Cambridge University Press.

Maecling, E.E. 1977. The doctrine of light in the philosophy of Francesco Patrizi. M. Phil. dissertation, University of London.

Maier, A. 1966. *An der Grenze von Scholastik und Naturwissenschaft, 2*, 2nd ed. Rome: Ed. di Storia e Letteratura (1st ed., 1952).

Maier, A. 1982. *On the threshold of exact science. Selected writings of anneliese Maier on late medieval natural philosophy*, ed. S.D. Sargent. Philadelphia: University of Pennsylvania Press.

Maierù, L. 2007. *John Wallis: una vita per un progetto*. Soveria Mannelli: Rubbettino.

Malet, A. 1989. Studies on James Gregorie (1638–1675). Ph.D. thesis, Princeton University.

Malet, A. 1997. Isaac Barrow on the mathematization of nature: Theological voluntarism and the rise of geometrical optics. *Journal of the History of Ideas* 58: 265–287.

Malherbe, M. 1984. L'induction baconienne. De l'échec métaphysique à l'échec logique. *Francis Bacon. Terminologia e fortuna nel XVII secolo*, ed. M. Fattori, 179–200.

Mandelbaum, M. 1964. *Philosophy, science and sense perception*. Baltimore: Johns Hopkins University Press.

Manning, G. 2008. Naturalism and un-naturalism among the Cartesian physicians. *Inquiry* 51(5): 441–463.

Manning, G. Forthcoming. *Descartes and the Bologna affair*.

Manning, G. Forthcoming. *Descartes, other minds and impossible human bodies*.

Manzo, S. 2001. Francis Bacon and atomism: A reappraisal. *Late scholastic and early modern corpuscular theories*, ed. C. Lüthy, J.E. Murdoch, and W.R. Newman, 209–243. Leiden, Boston, and Cologne: Brill.

Martin, H.-J. 1969. *Livre, pouvoirs et société à Paris au xviie siècle*, 2 vols. Genève: Droz. Repr. 1999.
Martin, C. 2002. Francisco Valles and the Renaissance reinterpretation of Aristotle's *Meteorologica IV* as a medical text. *Early Science and Medicine* 7: 1–31.
Martínez, R. 2001. Il manoscritto ACDF, Index, Protocolli, vol. EE, f. 291r-v. *Acta Philosophica* 10: 215–242.
Mateo-Seco, L.F. 2001. Galileo e l'Eucaristia. La questione teologica dell'ACDF, Index, Protocolli, EE f. 291r-v. *Acta Philosophica* 10: 243–256.
McClaughin, T. 1979. Censorship and defenders of the Cartesian faith in mid-seventeenth century France. *Journal of the History of Ideas* 40(4): 563–581.
McEvoy, J. 1979. The metaphysics of light in the middle ages. *Philosophical Studies* 26: 126–143.
McGrade, A.S. (ed.). 2003. *The Cambridge companion to medieval philosophy*. Cambridge: Cambridge University Press.
McGuire, J.E. 1978. Newton on place, time, and God: An unpublished source. *British Journal for the History of Science* 11: 114–129.
McMahon, W. 2000. The categories in some post-medieval Spanish philosophers. *Medieval and Renaissance logic in Spain*, ed. I. Angelelli and P. Pérez-Ilzarbe, 355–370. Hildesheim and New York: G. Olms.
McMullin, E. 1978. Structural explanations. *American Philosophical Quarterly* 15: 139–147.
McMullin, E. 2009. Hypothesis in early modern science. *The significance of the hypothetical in the natural sciences*, ed. M. Heidelberger and G. Schiemann, 7–37. Berlin/New York: Walter de Gruyter.
Meinel, C. 1988a. Les *Météores* de Froidmont et les *Météores* de Descartes. *Libert Froidmont et les résistances aux révolutions scientifiques*, ed. A.-C. Bernès, 105–129. Haccourt: Association des Vieilles Families de Haccourt.
Meinel, C. 1988b. Early 17th-century atomism: Theory, epistemology and the insufficiency of experiment. *Isis* 79: 68–103.
Menéndez y Pelayo, M. 1953. La Antoniana Margarita de Gómez Pereira. *La ciencia española*, 188–280. Santander: Editora Nacional.
Menn, S. 2003. Metaphysics: God and being. *The Cambridge companion to medieval philosophy*, ed. A.S. McGrade, 147–170. Cambridge/New York: Cambridge University Press.
Middleton, W.E.K. 1994. *The history of the barometer*. Trowbridge: Baros Books (1st ed. 1964).
Millburn, J.R. 1988. *Wheelwright of the heavens: The life and work of James Ferguson, FRS*. London: Vade-Mecum Press.
Montesinos, J., and C. Solís (eds.). 2001. *Largo Campo di filosofare*. La Orotava: Fundación Canaria Orotava de Historia de la Ciencia.
Morrison-Low, A.D. 2002. Feasting my eyes with the view of fine instruments: Scientific instruments in Enlightenment Scotland, 1680–1820. *Science and medicine in the Scottish Enlightenment*, ed. C.W.J. Withers and P. Wood, 17–53. East Linton: Tuckwell Press.
Murdoch, J.E. 1982. Infinity and continuity. *The Cambridge history of later medieval philosophy*, ed. N. Kretzman, A. Kenny, and J. Pinborg, 564–591. Cambridge: Cambridge University Press.
Murdoch, J.E. 1990. From the medieval to the Renaissance Aristotle. *New perspectives on Renaissance thought*, ed. J. Henry and S. Hutton, 163–177. London: Duckworth.
Murdoch, J.E. 2001. The medieval and Renaissance tradition of minima naturalia. *Late medieval and early modern corpuscular matter theories*, ed. C. Lüthy, J.E. Murdoch, and W.R. Newman, 91–133. Leiden/Boston: Brill.
Murr, S. (ed.). 1997. *Gassendi et l'Europe*. Paris: Vrin.
Navarro Brotons, V. 1997. Descartes y la introducción de la ciencia moderna en España. *La filosofía de Descartes y la fundación del pensamiento moderno*, 225–253. Salamanca: Sociedad Castellano-Leonesa de Filosofía.
Navarro Brotons, V. 2002. De la filosofía natural a la física moderna. *Historia de la ciencia y de la técnica en la Corona de Castilla*, ed. J.M. López Piñero, 383–437. León: Junta de Castilla y León, Consejería de Educación y Cultura.

Navarro Brotons, V. 2007a. El movimiento *novator* en la España de finales del siglo XVII y las disciplinas físico-matemáticas. *La ciència en la història dels Països Catalans*, vol. 2. Barcelona: Institut d'Estudis Catalans.
Navarro Brotons, V. 2007b. "La filosofía natural (Sixteenth Century)," and "El movimiment *novator* de les ciències fisicomatemàtiques". *La ciencia en la història dels Països Catalans*, ed. J. Vernet and R. Parés. Barcelona: Institut d'Estudis Catalans. 113–141 and 381–413.
Navarro Brotons, V., and W. Eamon. 2007a. Spain and the Scientific Revolution: Historiographical questions and conjectures. *Beyond the black legend*, ed. Brotons Navarro and E. Eamon, 27–41. Valencia: Instituto de Historia de la Ciencia y Documentación López Piñero.
Navarro Brotons, V., and W. Eamon (eds.). 2007b. *Beyond the black legend: Spain and the Scientific Revolution*. Valencia: Instituto de Historia de la Ciencia y Documentación López Piñero.
Neuburger, M. 1932. *The doctrine of the healing power of nature throughout the course of time*. New York: New York Homeopathic College.
Newman, W.R. 1994. *Gehennical fire: The lives of George Starkey, an American alchemist in the Scientific Revolution*. Cambridge, MA: Harvard University Press.
Newman, W.R. 1999. *The summa perfectionis of Pseudo-Geber: A critical edition, translation and study*. Leiden: Brill.
Newman, W.R. 2001. Experimental corpuscular theory in Aristotelian alchemy: From Geber to Sennert. *Late medieval and early modern corpuscular matter theories*, ed. C. Lüthy, J.E. Murdoch, and W.R. Newman, 291–331. Leiden/Boston: Brill.
Newman, W.R. 2006. *Atoms and alchemy. Chymistry and the experimental origins of the Scientific Revolution*. Chicago/London: The University of Chicago Press.
Newman, W.R., and L.M. Principe. 2002. *Alchemy tried in the fire*. Chicago: The University of Chicago Press.
Nonnoi, G. 2000. Galileo Galilei: quale atomismo? *Atomismo e continuo nel XVII secolo*, ed. E. Festa and R. Gatto, 109–149. Naples: Vivarium.
Owens, J. 1978. *The doctrine of being in the Aristotelian metaphysics*. Toronto: Pontifical Institute of Mediaeval Studies.
Pabst, B. 1994. *Atomtheorien des lateinischen Mittelalters*. Darmstadt: Wissenschaftliche Buchgesellschaft.
Palmerino, C.R. 2000. Una nuova scienza della materia per la *scienza nova* del moto. La discussione dei paradossi dell'infinito nella prima giornata dei *Discorsi* galileiani. *Atomismo e continuo nel XVII secolo*, ed. E. Festa and R. Gatto, 275–319. Naples: Vivarium.
Palmerino, C.R. 2001. Galileo's and Gassendi's solutions to the Rota Aristotelis Paradox. A bridge between matter and motion theories. *Medieval and early modern corpuscular matter theories*, ed. C. Lüthy, J.E. Murdoch, and W.R. Newman, 381–422. Leiden/Boston: Brill.
Palmerino, C.R. 2003. Two Aristotelian responses to Galilei's science of motion: Honoré Fabri and Pierre Le Cazre. *The new science and jesuit science*, ed. M. Feingold, 187–227. Dordrecht/Boston: Springer.
Palmerino, C.R. 2011. The isomorphism of space, time and matter in seventeenth-century natural philosophy. *Early Science and Medicine* 16: 296–330.
Palmerino, C.R., and J.M.M.H. Thijssen (eds.). 2004. *The reception of the Galilean science of motion in seventeenth-century Europe*. Dordrecht/Boston/London: Kluwer Academic Publishers.
Parkinson, G.H.R. (ed.). 1993. *Routledge history of philosophy*, vol. IV. New York: Routledge.
Pastorino, C. 2008. Francis Bacon. *Dictionary of scientific biography*, vol. XIX, 137–142. Detroit: Charles Scribner's Sons.
Penrose, S.B.L. 1934. *The reputation and influence of Francis Bacon in the seventeenth century*. Ph.D. thesis. New York: Columbia University.
Pérez-Ramos, A. 1988. *Francis Bacon's idea of science and the maker's knowledge tradition*. Oxford: Clarendon.
Peterschmitt, L. 2005. Bacon et la chimie. À propos de la réception de la philosophie naturelle de Francis Bacon aux XVIIe et XVIIIe siècles. *Methodos. Savoirs et textes* 5. To be consulted at http://methodos.revues.org/document385.html

Peterschmitt, L. 2007. The Cartesian chemistry: Cordemoy, Rohault, Régis. *Chymists and chymistry: Studies in the history of alchemy and early modern chemistry*, ed. L.M. Principe, 193–202. Sagamore Beach: Watson Publishing.

Pichot, A. 1993. *Histoire de la notion de vie*. Paris: Gallimard.

Pintard, R. 1943. *Le libertinage érudit dans la première moitié du xviie siècle*. Genève/Paris: Boivin. Repr. Genève: Slatkine, 1983.

Porto, P.A. 2002. *Summus atque felicissimus salium*: The medical relevance of the *Liquor alkahest*. *Bulletin of the History of Medicine* 76: 1–29.

Principe, L.M. 1998. *The aspiring adept: Robert Boyle and his alchemical quest*. Princeton: Princeton University Press.

Principe, L.M. 2007. *Chymists and chymistry: Studies in the history of alchemy and early modern chemistry*. Sagamore Beach: Watson Publishing.

Prytanée National Militaire (ed.). 1997. *René Descartes (1596–1650): célébrations nationales du quadricentenaire de sa naissance. Actes du colloque universitaire de La Flèche, 12–13 avril 1996*. La Flèche: Prytanée National Militaire.

Puliafito, A.L. 1987. Per uno studio della *Nova de universis philosophia* di Francesco Patrizi da Cherso. *Atti e Memorie dell'Accademia Toscana di Scienze e Lettere La Colombaria* 38: 159–199.

Purver, M. 1967. *The Royal Society. Concept and creation*. Cambridge, MA: MIT Press.

Pyle, A. 1995. *Atomism and its critics. Problem areas associated with the development of the atomic theory of matter from Democritus to Newton*. Bristol: Thoemmes Press.

Pyle, A. (ed.). 2000. *The dictionary of seventeenth-century British philosophers*, 2 vols. Bristol: Thoemmes Press.

Rabassini, A. 1997. La concezione del Sole secondo Marsilio Ficino. Note sul *Liber de sole*. *Momus. Rivista di Studi umanistici* VII–VIII: 115–134.

Rabassini, A. 2006. 'Amicus lucis'. Considerazioni sul tema della luce in Marsilio Ficino. *Marsilio Ficino: fonti, testi*, ed. S. Gentile and S. Toussaint, 255–294. Rome: Edizioni di storia e letteratura.

Rabassini, A. Forthcoming. *Il vincolo dell'universo. Temi e fonti della filosofia della luce in Marsilio Ficino*. Naples: Istituto Italiano per gli Studi Filosofici.

Ramond, C., and M. Dennehy (eds.). 2009. *La philosophie naturelle de Robert Boyle*. Paris: Vrin.

Rattansi, P., and A. Clericuzio (eds.). 1994. *Alchemy and chemistry in the 16th and 17th centuries*. Dordrecht: Kluwer Academic Publishers.

Recipi, L. 1990. Limits of teleology in Theophrastus? *Archiv für Geschichte der Philosophie* 72: 182–213.

Redondi, P. 1983. *Galileo eretico*. Turino: Einaudi.

Redondi, P. 1985. Atomi, indivisibili e dogma. *Quaderni Storici* 59(2): 529–573.

Redondi, P. 2004. Fede Lincea e teologia tridentina. *Galilaeana* 1: 117–141.

Rees, G. 1975. Matter theory: A unifying factor in Bacon's natural philosophy. *Ambix* 22: 161–173.

Rhigini Bonelli, M.L., and W.R. Shea (eds.). 1975. *Reason, experiment and mysticism in the Scientific Revolution*. New York: Science History Publications.

Ribard, D. 2000. Philosophe ou écrivain? Problèmes de délimitation entre histoire littéraire et histoire de la philosophie en France, 1650–1850. *Annales HSS* 2: 355–388.

Rochot, B. 1944. *Les travaux de Gassendi sur Épicure et sur l'atomisme, 1619–1658*. Paris: Vrin.

Rodis-Lewis, G. 1990. *L'anthropologie cartésienne*. Paris: Presses universitaires de France.

Rodis-Lewis, G. 1997. Un élève du Collège jésuite de La Flèche: René Descartes. *René Descartes (1596–1650): célébrations nationales du quadricentenaire de sa naissance*, ed. Prytanée National Militaire, 25–36. La Flèche: Prytanée National Militaire.

Roger, J. 1993. *Les sciences de la vie dans la pensée française au xviiie siècle*. Paris: Albin Michel.

Rossi, P. 1986. Formiche, ragni, epistemologi. *I ragni e le formiche*, 95–117. Bologna: Il Mulino.

Roux, S. 1996. La philosophie mécanique. Ph.D. thesis, EHESS, Paris.

Roux, S. 1998. Le scepticisme et les hypothèses de la physique. *Revue de synthèse*, 4th ser., 2–3, 211–255.

Roux, S. 2000. Descartes atomiste? *Atomismo e continuo nel XVII secolo*, ed. E. Festa and R. Gatto, 211–273. Naples: Vivarium.

Roux, S. 2002. La philosophie naturelle d'Honoré Fabri (1607–1688). *Les Jésuites à Lyon*, ed. É. Fouilloux and B. Hours, 75–94. Lyon: ENS.

Roux, S. Forthcoming. Pour une conception polémique du cartésianisme. Ignace-Gaston Pardies et Antoine Dilly dans la querelle de l'âme des bêtes. *Qu'est-ce qu'être cartésien ?* ed. D. Kolesnik.

Sabra, A.I. 1981. *Theories of light from Descartes to Newton*. Cambridge: Cambridge University Press.

Sanhueza, G. 1997. *La pensée biologique de Descartes dans ses rapports avec la philosophie scolastique. Le cas Gomez-Pereira*. Paris: L'Harmattan.

Sargent, R.-M. 1995. *The diffident naturalist: Robert Boyle and the philosophy of experiment*. Chicago: University of Chicago Press.

Schabas, M. 2001. David Hume on experimental natural philosophy, money, and fluids. *History of Political Economy* 33: 411–435.

Schmitt, C. 1963. *Aristotle and the Renaissance*. Cambridge, MA: Harvard University Press.

Schmitt, C. 1985. Aristotle among the physicians. *The medical renaissance of the sixteenth-century*, ed. A. Wear, R.K. French, and I.M. Lonie, 1–15. Cambridge: Cambridge University Press.

Scott, J.F. 1938. *The mathematical work of John Wallis*. London: Taylor and Francis.

Shapin, S. 1988. Robert Boyle and mathematics: Reality, representation, and experimental practice. *Science in Context* 2: 23–58.

Shapin, S. 1996. *The Scientific Revolution*. Chicago: University of Chicago Press.

Shapin, S., and S. Schaffer. 1985. *Leviathan and the air-pump. Hobbes, Boyle, and the experimental life*. Princeton: Princeton University Press.

Shapiro, A. 1974. Light, pressure, and rectilinear propagation: Descartes's celestial optics and Newton's hydrostatics. *Studies in History and Philosophy of Science* 5: 239–296.

Shapiro, L. 2003. The health of the body-machine? Or seventeenth century mechanism and the concept of health. *Perspectives on Science* 11(4): 421–442.

Sharp, L.G. 1977. Sir William Petty and some aspects of seventeenth-century natural philosophy. Unpublished Ph.D. dissertation, University of Oxford.

Shea, W. 1970. Galileo's atomic hypothesis. *Ambix* 17: 13–27.

Shea, W. 2001. Galileo e l'atomismo. *Acta Philosophica* 10: 257–272.

Shepherd, C.M. 1974. Philosophy and science in the arts curriculum of the Scottish universities in the 17th century. Ph.D. thesis, University of Edinburgh.

Shepherd, C.M. 1983. University life in the 17th century. In *Four centuries: Edinburgh university life 1583–1983*, ed. G. Donaldson, 1–15. Edinburgh: University of Edinburgh.

Simmons, A. 2001. Sensible ends: Latent teleology in Descartes' account of sensation. *Journal of the History of Philosophy* 39(1): 49–75.

Sirri, R., and M. Torrini (eds.). 1992. *Bernardino Telesio e la cultura napoletana*. Naples: Guida.

Slowik, E. 2009. Newton's metaphysics of space: A "*Tertium Quid*" betwixt substantivalism and relationism, or merely a 'God of the (rational mechanical) Gaps'? *Perspectives on Science* 17: 429–456.

Slowik, E. 2011. Newton, the parts of space, and the holism of spatial ontology. *HOPOS: The Journal of the International Society for the History of Philosophy of Science* 1(2): 249–272.

Smith, A.M. 1976. Galileo's theory of indivisibles: Revolution or compromise? *Journal of the History of Ideas* 37: 571–588.

Solomon, J.R., and C. Gimelli Martin (eds.). 2005. *Francis Bacon and the refiguring of early modern thought. Essays to commemorate the Advancement of Learning (1605–2005)*. Aldershot: Ashgate.

Solís, C. 2001. Los cometas contra Copérnico: Brahe, Galileo y los Jesuitas. *Theoria* 16(2): 353–385.

Spurr, J. 1988. 'Latitudinarianism' and the Restoration church. *The Historical Journal* 31: 61–82.

Stewart, A.G. 1901. *The academic gregories*. Edinburgh: Oliphant/Anderson/Ferrier.

Stewart, D. 1934. Dissertation first exhibiting the progress of metaphysical, ethical and political philosophy since the revival of letters (1817 and 1822). *The reputation and influence of Francis Bacon in the seventeenth century*, ed. S.P.L. Penrose. New York: [s.n.].

Stewart, M.A. (ed.). 1990. *Studies in the philosophy of the Scottish Enlightenment*. Oxford: Oxford University Press.

Stewart, L. 1992. *The rise of public science*. Cambridge: Cambridge University Press.

Stewart, I.G. 2012. Forthoming. *Res, veluti per Machinas, Conficiatur*: Natural history and the 'Mechanical' reform of natural philosophy. *Early Science and Medicine* 17: 87–111.

Stroup, A. 1990. *A company of scientists: Botany, patronage, and community at the seventeenth-century Parisian Royal Academy of Sciences*. Berkeley: University of California Press.

Stroup, A. 2002. Censure ou querelles scientifiques: l'affaire Duclos (1675–1685). *Règlement, usages et science dans la France de l'absolutisme*, ed. C. Demeulenaere-Douyère and É. Brian, 435–452. Paris/London: Éd. Tec & Doc.

Taub, L. 2003. *Ancient meteorology*. London: Routledge.

Thijssen, J.M.M.H. 1991. Some reflections on continuity and transformation of Aristotelianism in medieval (and Renaissance) natural philosophy. *Documenti e studi sulla tradizione filosofica medievale* 2(2): 503–528.

Thijssen, J.M.M.H. 1992. David Hume and John Keill and the structure of the continua. *Journal of the History of Ideas* 53: 271–286.

Thonnard, F.-J. 1962. La notion de lumière en philosophie augustinienne. *Recherches augustiniennes* 2: 125–175.

Todericiu, D. 1974. Sur la vraie biographie de Samuel Duclos (Du Clos) Cotreau. *Revue d'histoire des sciences* 27(1): 64–67.

Turnbull, H.W. (ed.). 1939. *James Gregory tercentenary memorial volume*. London: G. Bell and Sons.

Vallance, J. 1988. Theophrastus and the study of the intractable: Scientific method in *De lapidibus* and *De igne*. *Theophrastean studies,* Vol III, ed. W. Fortenbaugh and R. Sharples, 25–40. New Brunswick: Transaction Books.

van Berkel, K. 1983a. *Isaac Beeckman (1588–1637) en de Mechanisering van het Wereldbeeld.* Amsterdam: Rodopi.

van Berkel, K. 1983b. Beeckman, Descartes et la philosophie physico-mathématique. *Archives de philosophie* 46: 620–626.

van Berkel, K., A. van Helden, and L. Palm. 1999. *A history of science in the Netherlands*. Leiden: Brill.

Van Damme, S. 2002. *Descartes*. Paris: Presses de Sciences Po.

Vasoli, C. 1998. Su alcuni temi della filosofia della luce nel Rinascimento: Ficino (*De sole* e *De lumine*) e Patrizi (L. I della *Panaugia*). *Annali della Facoltà di Lettere e Filosofia dell'Università di Cagliari* IX(46): 63–89.

Verbeek, T. 1988. *La querelle d'Utrecht*. Paris: Les Impressions Nouvelles.

Verbeek, T. 1992. *Descartes and the Dutch: Early reactions to Cartesian philosophy, 1637–1650*. Carbondale: Southern Illinois University Press.

Verbeek, T. (ed.). 1993. *Descartes et Regius: Autour de l'*Explication de l'esprit humain. Amsterdam: Rodopi.

Vernet, J., and R. Parés (eds.). 2007. *La ciencia en la història dels Països Catalans*. Valencia-Barcelona: Universitat de Valencia, Institut d'Estudis Catalans.

Voss, S. 2000. Descartes: Heart and soul. *Psyche and Soma: Physicians and metaphysicians on the mind-body problem from antiquity to enlightenment*, ed. J.P. Wright and P. Potter, 173–196. Oxford: Oxford University Press.

Wagner, P. (ed.). 2002. *Les philosophes et la science*. Paris: Gallimard.

Wallace, W.A. 1981. *Prelude to Galileo: Essays on medieval and sixteenth-century sources of Galileo's thought*. Dordrecht/London: Reidel.

Ward, J. 2008. *Aristotle on homonymy: Dialectic and science*. Cambridge: Cambridge University Press.

Wear, A., R.K. French, and I.M. Lonie (eds.). 1985. *The medical Renaissance of the sixteenth century*. Cambridge: Cambridge University Press.
Webster, C. 1967a. The college of physicians. 'Solomon's House' in Commonwealth England. *Bulletin of the History of Medicine* 41: 393–412.
Webster, C. 1967b. Henry Power's experimental philosophy. *Ambix* 14: 150–178.
Webster, C. 1970. *Samuel Hartlib and the advancement of learning*. Cambridge: Cambridge University Press.
Webster, C. 2002. *The Great Instauration. Science, medicine and reform 1626–1660.* Bern: Lang.
Weeks, S. 2007. Francis Bacon and the art-nature distinction. *Ambix* 54: 101–129.
Weeks, S. 2008. The role of mechanics in Francis Bacon's *Great Instauration*. *Philosophies of technology,* vol. II, ed. C. Zittel, G. Engel, R. Nanni, and N.C. Karafyllis, 133–195. Leiden and Boston: Brill.
Weisheipl, J.A. 1978. The nature, scope, and classification of the sciences. *Science in the Middle Ages*, ed. D. Lindberg, 461–482. Chicago: University of Chicago Press.
Westfall, R.S. 1971. *The construction of modern science: Mechanisms and mechanics*. Cambridge: Cambridge University Press. First published New York: Wiley, 1971.
Westman, R.S., and J.E. McGuire (eds.). 1974. *Hermeticism and the Scientific Revolution. Papers read at a Clark Library Seminar, March 9, 1974*. Los Angeles, William Andrews Clark Memorial Library: University of California at Los Angeles.
Wilson, M. 1978. *Descartes*. New York: Routledge.
Withers, C.W.J., and P. Wood (eds.). 2002. *Science and medicine in the Scottish Enlightenment*. East Linton: Tuckwell Press.
Wolfe, C.T., and O. Gal (eds.). 2010. *The body as object and instrument of knowledge: Embodied empiricism in early modern science*. Dordrecht: Springer.
Wood, P.B. 1988. Science and the Aberdeen enlightenment. *Philosophy and science in the Scottish Enlightenment*, ed. P. Jones, 39–66.
Wood, P.B. 1990. Science and the pursuit of virtue in the Scottish Enlightenment. *Studies in the philosophy of the Scottish Enlightenment*, ed. M.A. Stewart, 127–149. Clarendon Press/Oxford University Press: Oxford/New York.
Wood, P.B. 1994. Science, the universities, and the public sphere in eighteenth-century Scotland. *History of Universities* 13: 99–135.
Wright, J.P., and P. Potter (eds.). 2000. *Psyche and soma: Physicians and metaphysicians on the mind-body problem from antiquity to enlightenment*. Oxford: Oxford University Press.
Yates, F. 1964. *Giordano Bruno and the hermetic tradition*. Chicago: Chicago University Press.
Yeo, R. 2007. Between memory and paperbooks: Baconianism and natural history in seventeenth-century England. *History of Science* 45: 1–46.
Yerushalmi, Y.H. 1991. *Dalla Corte al Ghetto. La vita, le opere, le peregrinación del marrano Cardoso nell'Europa del Seicento*. Milano: Garzanti.
Zittel, C., G. Engel, R. Nanni, and N.C. Karafyllis (eds.). 2008. *Philosophies of technology: Francis Bacon and his contemporaries*, 2 vols. Leiden: Brill.

Author Index

S. Roux and D. Garber (eds.), *The Mechanization of Natural Philosophy*, Boston Studies in the Philosophy of Science 282, DOI 10.1007/978-94-007-4345-8,

Subject Index

S. Roux and D. Garber (eds.), *The Mechanization of Natural Philosophy*, Boston Studies in the Philosophy of Science 282, DOI 10.1007/978-94-007-4345-8,

MIX
Papier aus verantwortungsvollen Quellen
Paper from responsible sources
FSC® C105338

If you have any concerns about our products,
you can contact us on
ProductSafety@springernature.com

In case Publisher is established outside the EU,
the EU authorized representative is:
Springer Nature Customer Service Center GmbH
Europaplatz 3, 69115 Heidelberg, Germany

Printed by Libri Plureos GmbH
in Hamburg, Germany